AF574261

Ion-Containing Polymers

Physical Properties and Structure

POLYMER PHYSICS

Edited by R. S. STEIN

Polymer Research Institute
University of Massachusetts
Amherst, Massachusetts

Volume 1/Introduction to Polymer Physics *R. S. Stein and Joseph Powers*

Volume 2/Ion-Containing Polymers: Physical Properties and Structure *A. Eisenberg and M. King*

Ion-Containing Polymers

PHYSICAL PROPERTIES AND STRUCTURE

Polymer Physics

Volume 2

A. Eisenberg and *M. King*

DEPARTMENT OF CHEMISTRY
MCGILL UNIVERSITY
MONTREAL, CANADA

MEAKINS CHRISTIE LABORATORIES
MCGILL UNIVERSITY
MONTREAL, CANADA

ACADEMIC PRESS New York San Francisco London 1977
A Subsidiary of Harcourt Brace Jovanovich, Publishers

ACADEMIC PRESS, INC.
111 Fifth Avenue, New York, New York 10003

United Kingdom Edition published by
ACADEMIC PRESS, INC. (LONDON) LTD.
24/28 Oval Road, London NW1

Library of Congress Cataloging in Publication Data

Eisenberg, Adi.
Ion-containing polymers.

(Polymer physics treatise ; v. 2)
Includes bibliographical references.
1. Polymers and polymerization. I. King, M., joint author. II. Title. III. Series.
QD381.E37 547'.84 76-19491
ISBN 0-12-235050-2

PRINTED IN THE UNITED STATES OF AMERICA

Contents

Preface

In the past decades, the field of ion-containing polymers has experienced a very rapid growth both in terms of industrial applications and academic interest. Reflecting this high level of activity, a number of very useful reviews covering a variety of topics have appeared in the literature. A single unified treatment of the physical aspects of the subject has not yet appeared. This work attempts to fill this gap.

In a field that is growing as rapidly as this one, the selection of material is obviously as much a matter of the interests and prejudices of the writers as it is the state of the field. In selecting the topics for inclusion, we were guided by the desire to include all the areas we thought would be of interest to polymer physicists, within the limitation of our own interests and availability of publications on the subject.

Polymers containing ions have been referred to both as "ionic polymers" and "ion-containing polymers." Both apellations are obviously correct. However, some confusion may exist, since the phrase "ionic polymers" has also been used to denote polymers synthesized by ionic polymerization mechanisms. For this reason, the phrase "ion-containing polymers" seemed preferable. Also, the latter allows the inclusion of nonionic polymers containing added ionic material of low molecular weight as well as some charge-transfer complexes, all of which can be classified as ion-containing.

Acknowledgments

It is a particular pleasure to acknowledge the benefit of valuable comments and suggestions from friends and colleagues who have read all or part of the manuscript, especially Drs. Z. Alexandrowicz, F. R. Eirich, H. Eisenberg, I. Hodge, M. Mandel, E. P. Otocka, M. Pineri, T. Tanaka, Y. Wada, and K. Wissbrun. This acknowledgment obviously does not imply complete agreement by all these readers with all the contents, and naturally, we assume full responsibility for any remaining errors, omissions, or lack of clarity. It is also a pleasure to thank Drs. D. B. James, R. E. Wetton, C. T. Meyer, and M. Pineri for permission to present material prior to publication. Special thanks are due to Miss O. Gotts and Mrs. C. Brown for their patient typing and retyping (and retyping) of the manuscript and to Mrs. N. King for proofreading. Thanks are also due to the Graphics Department of the Weizmann Institute of Science and to Miss D. Brooks for invaluable assistance with the art work. Parts of this book were written while both of us were visitors in the Polymer Department of the Weizmann Institute of Science, and we would like to express our appreciation for their hospitality. Finally, thanks are due to the American Chemical Society, the American Institute of Physics, the Chemical Society, Hüttig & Wepf Verlag, John Wiley and Sons Inc., Marcel Dekker Inc., North-Holland Publishing Company, Plenum Publishing Corporation, and the Society of Polymer Science, Japan, for permission to use copyrighted material.

List of Symbols

A Absorbance *or* length of link in equivalent chain *or* Helmholtz free energy *or* sample area *or* material dependent constant in relation between maximum relaxation time, temperature and molecular weight *or* material constant relating viscosity and molecular weight

$A_{\perp}, A_{\parallel}$ Absorbance perpendicular or parallel to direction of elongation

a Distance between centers of charge

$a_M{}^J$ Shift factor for compliance with respect to molecular weight

a_T, a_τ Shift factor for time–temperature superposition in viscoelastic measurements

AA Acrylic acid

AN Acrylonitrile

B Butadiene

B Constant relating viscosity or maximum relaxation time with molecular weight *or* constant relating molecular weight and viscosity *or* constant related to second virial coefficient

B_i Formation constant for ion multiplets (i = number of ions in multiplet)

b Ionic sheath radius

b, b_0 Length of segment in polymer model

b_T Shift factor for time–temperature superposition for viscoelastic measurement for a second relaxation mechanism

Bu Butyl

BMA Butyl methacrylate

C Constant defined by Eq. (4), Chapter IV

C_1, C_2 Constants of the WLF equation

C_I, C_M Constants in Eq. (36), Chapter IV

C_p Concentration of polymer

C_s Concentration of added salt

c Concentration

c_f Concentration of backbone charges

c_s Concentration of added salt

CMC Carboxymethylcellulose

CPS Counts per second

D Dichroic ratio *or* elongational compliance *or* diffusion coefficient
d Diameter of cylindrical polymer
d_a Distance between scattering sites
d_{Bragg} Bragg distance
$\Delta Dr - \Delta Do$ Difference in RDF for salt and acid
DEP Diethyl phthalate
DLS Dynamic light scattering
DMF Dimethyl formamide
DMSO Dimethyl sulfoxide
DOP Dioctyl phthalate
DSC Differential scanning calorimetry
DVB Divinyl benzene

E Ethylene
E Young's modulus
E' Real part of complex Young's modulus *or* apparent Young's modulus in gel in equilibrium with swelling medium
E'' Imaginary part of complex Young's modulus
E_0 Young's modulus of unswollen network
E_{act} Activation energy
E_{el} Electrostatic interaction energy
E_{visc} Activation energy for viscous flow
e Electron charge
$E_r(t)$ Time-dependent Young's modulus from relaxation measurements
EA Ethyl acrylate
EHA Ethylhexyl acrylate
EO Ethylene oxide
EPR Electron paramagnetic resonance

F_{el} Electrostatic force
f Fraction of functional chain ends *or* force on chain ends *or* force constant (of Hookean spring) *or* frequency *or* orientation function
f_c Critical frequency
f_e Internal energy component of force of deformation
f^i Orientation at i% elongation
FA Formamide
$f(m)$ Function of mean charge of bound counterions

G' Real part of (or storage) shear modulus
G'' Imaginary part of (or loss) shear modulus
GA Glutamic acid
GL Glycerine
$G_r(t)$ Time-dependent shear modulus from relaxation measurements

H Hookean force constant
ΔH Enthalpy difference *or* linewidth in NMR
ΔH_a Activation enthalpy
H_v Polarization mode with vertically polarized incident light and horizontally polarized analyzer
h End-to-end distance of polymer chain
h^* Most probable end-to-end distance of polymer chain
$\langle h^2 \rangle, \overline{h^2}$ Mean square end-to-end distance of polymer chain
$\overline{h^2_{fr}}$ Mean square end-to-end distance of freely rotating chain
$H(\tau)$ Distribution of relaxation times

$I, I(s)$ Intensity of scattered radiation
$I(0)$ Intensity of incident radiation
I', I'' Real and imaginary components of light scattering in rheooptical tests
I_m Melt index
i $\sqrt{-1}$
IR Infrared

J Shear compliance
J_e Equilibrium shear compliance
$J_r(t)$ Time-dependent shear compliance from relaxation measurements

$J_s(J_s^0)$ Compliance function (zero shear value)

K Molecular-weight-dependent constant relating viscosity and ion concentration *or* dimensionless quantity in X-ray scattering equation *or* proportionality constant *or* constant in Mark–Houwink equation

K' Real part of dynamic strain–optical coefficient *or* internal field correction factor

K_a Equilibrium constant

K_o Mark–Houwink constant in theta solvent

K_s Static strain–optical coefficient

K_w Dissociation constant for water

k Boltzmann constant

k_e Constant of order 1

k_H Huggins constant

L Segment length of polymer coil *or* sample length *or* major axis of prolate ellipsoid or cylindrical polymer

L_1 Principal axis of ellipsoid

l C—C bond length = 1.54Å *or* length of chain link

LL L-Lysine

M Metal (usually as ion)

M Molecular weight

M_c Molecular weight between ionic groups

M_n Number average molecular weight

M_n^0 Number average molecular weight of unassociated polymer

M_0 Monomer molecular weight

M_v Viscosity average molecular weight

M_w Weight average molecular weight

M_w^A Weight average molecular weight of unassociated polymer

m Slope of curve of log E versus T *or* molecular fraction

m_s Molecular weight of monomeric segment

MA Methyl acrylate or methacrylate

MAA Methacrylic acid

MMA Methyl methacrylate

MVP Methyl vinyl pyridine (or pyridinium)

MOSA 3-Methacryloyloxypropane-1 sulfonic acid

MQMVP Methyl-quaternized MVP

N, N_t Number, total number

N_{Av} Avogadro's number

n Molecular fraction *or* number of molecules or units *or* number of ion pairs in cluster

n_c Number of carboxyls per scattering site

n_0 Number of ion pairs in multiplet

NMR Nuclear magnetic resonance

P Polymer (as in PS = polystyrene) *or* degree of polymerization

P_p Degree of polymerization between crosslinks for Pth component

p Probability of intermolecular association

PO or PrO Propylene oxide

q Ionic charge (usually of cation) *or* charge on bead

R Spherulite radius *or* distance between clusters

$\overline{R}$ Average cluster radius

R_0 Distance between multiplets

R_s Hydrodynamic radius

$\overline{R^2}$, $\langle R^2 \rangle$ Mean square radius of gyration

r Distance between centers of charge *or* distance from scattering center *or* fraction of free counterions *or* end-to-end distance of polymer chain in rubber

r_{ij} Distance between ith and jth charge

r_m radius of multiplet

$\overline{r^2}$ Mean square end-to-end distance of chains

$\overline{r^2}_0$ $\overline{r^2}$ in unstrained, unswollen network

$\overline{r^2}_f$ $\overline{r^2}$ in unconstrained state

$\overline{r^2}_{fr}$ $\overline{r^2}$ in freely rotating configuration

$\overline{r^2}_\theta$ $\overline{r^2}$ in θ solvent

$\overline{r_p^2}_0$ $\overline{r^2}_0$ for component p

$\overline{r_p^2}_f$ $\overline{r^2}_f$ for component p

RDF Radial distribution function

RH Relative humidity

S Styrene

S_{ch} Contact surface area of chain

S_m Surface area of multiplet

ΔS Entropy of deformation of the swollen network due to external strain

ΔS_0 Entropy of deformation due to swelling

$\Delta S_0'$ $(= \Delta S_0 + \Delta S)$

$\Delta S_p'$ $\Delta S_0'$ for component p

s Scattering vector *or* number of monomer units in equivalent chain link *or* radius of gyration

SAXS Small-angle X-ray scattering

SR Stress relaxation

SSA Styrene sulfonic acid

T Temperature

T_c Cluster decomposition temperature

T_g Glass transition temperature

T_m Peak temperature, usually in $\tan\delta$ versus T plots *or* temperatures at which the WLF parameters change

T_0, T_{ref} Reference temperature

T_1 Spin lattice relaxation time

$T_{1\rho}$ Spin lattice relaxation time in NMR rotating frame experiment

$T_\alpha, T_\beta, \ldots$ Peak temperatures for $\alpha, \beta, \ldots$ dispersions

t Time

u Counterion mobility

V Volume *or* volume of swollen rubber

V' Volume per carboxyl group

V_m Volume of multiplet

V_p Volume of ion pair

V_r Volume fraction of polymer *or* volume fraction of rubber

V_s Volume of equivalent sphere

V_0 Unstrained, unswollen volume

V_1 Unstrained, swollen volume

v Volume fraction of polymer in the swollen stretched state

v_0 Volume fraction of polymer in the swollen unstretched state

VBTACl Vinyl benzyl trimethyl ammonium chloride

VP Vinylpyridine

VSA Vinylsulfonic acid

W Probability distribution function *or* interaction energy *or* work

W' Electrostatic energy per ion pair in cluster formation

W_{ch} Work of chain stretching

W_{el} Electrostatic work

w Weight fraction *or* frequency (instead of ω) *or* resonance frequency

Δw Band width in resonance experiment

WLF Williams, Landel, and Ferry

y Salt concentration/polymer concentration

Z Number of links in equivalent chain

z Valence *or* total concentration of free ions

z_{el} Excluded volume parameter due to electrostatic interaction

α Degree of neutralization *or* highest-temperature peak or relaxation process in dynamic studies *or* viscoelastic relaxation mechanism involving entire chain without rupture (as opposed to χ mechanism) *or* extension ratio or expansion ratio

α_b Strain at break

α_g Linear expansion coefficient in glassy state

α_N Expansion ratio determined by computer simulation

α_η Expansion ratio determined viscometrically

$\alpha_{\eta e}$ Electrostatic part of α_η

$(\alpha\lambda)_r$ Ultrasound energy dissipated per wavelength

β Second highest temperature peak or relaxation process in dynamic studies *or* breadth of distribution of relaxation times *or* excluded volume per segment *or* thermal expansion coefficient of swollen stretched gel

β' Peak or relaxation process near β (see above)

β_0 Thermal expansion coefficient of swelling medium

γ Third highest peak or relaxation process in dynamic studies *or* degree of counterion association *or* constant in Eq. 49, Chapter V

$\dot{\gamma}$ Rate of shear

γ_i Extension ratio of *i*th coordinate due to both swelling and external strain

Δ Damping constant or decrement in mechanical loss measurements ($= \pi \tan \delta$) *or* increment or difference

δ Loss or phase angle (as in $\tan \delta =$ loss tangent)

ε Dielectric constant

ε' Real part of complex dielectric constant

ε'' Imaginary part of complex dielectric constant

ε^* Complex dielectric constant

$\varepsilon_0, \varepsilon_0'$ Low-frequency limit of dielectric constant

$\varepsilon_\infty, \varepsilon_\infty'$ High-frequency limit of dielectric constant

$\Delta\varepsilon_1, \Delta\varepsilon_1', \Delta\varepsilon$ Relaxation strength $\varepsilon_0 - \varepsilon_\infty$

$\Delta\varepsilon_2, \Delta\varepsilon'_2$ Relaxation strength of second mechanism

$\Delta\varepsilon_1''$ Excess loss due to first mechanism

$\Delta\varepsilon_2''$ Excess loss of solution over solvent (i.e., $\varepsilon'' - \varepsilon''_{H_2O}$)

ζ Linear charge density

η Viscosity

$[\eta]$ Intrinsic velocity

η' Real part of complex viscosity

η'' Imaginary part of complex viscosity

η^* Complex viscosity

$\check{\eta}$ Shear-dependent steady flow viscosity

η_0 Steady flow viscosity

η_s Solvent viscosity

η_{sp} Specific viscosity

θ Scattering angle *or* bond angle *or* "ideal" solvent

$\theta(\theta^0)$ Primary normal stress function (zero shear value)

θ_m Angle of maximum scattering

κ Rate of shear *or* Debye–Hückel parameter

Λ_n Distribution function of *n*th power of κ

λ Wavelength of radiation *or* degree of ionization *or* fraction of energy of ion pair formation released per ion pair upon cluster formation *or* relaxation time *or* extension ratio

λ_i Extension ratio of *i*th coordinate due to external strain only

λ_0 Relaxation time of uncharged bead–spring system

λ_1, λ_2 Relaxation times of 3 bead–2 spring system

μ Dipole moment *or* Azimuthal angle in light scattering *or* Poisson's Ratio

μ' Poisson's Ratio for gel in equilibrium with swelling medium

ξ Charge density parameter

ρ Density *or* charge density

ρ_o Density at reference temperature

σ Chain expansion factor $(\overline{h^2}_0/\overline{h^2}_{fr})^{1/2}$

σ_b Stress at break

σ_{H_2O}, σ_s Constants in Eq. 107, Chapter V, (subscript s refers to added salt)

σ'_j $= \lambda^{1/2} V_0 \sigma_j$ (j = H_2O or s)

σ_{21} Shear stress

$\sigma_{11} - \sigma_{22}$ Normal stress difference

τ, τ' Relaxation time

τ_c, τ_{conf} Conformational relaxation time

τ_m Maximum (viscoelastic) relation time

$\tau_r, \tau_R, \tau_{rot}$ Rotational relaxation time

τ_Q Relaxation time for axial diffusion process

τ_1, τ_2 Relaxation time of first, second dispersion regions

$\Delta \log \Delta\tau$ Half-width at half-height of peak of distribution of relaxation times

ν Frequency ($= \omega/2\pi$) *or* total number of charges on polymer chain *or* exponent in Mark–Houwink equation

ν_{max} Frequency of maximum energy dissipation, usually in dielectric or dynamic mechanical studies

Φ Universal parameter for viscosity relationship

ϕ Mole fraction *or* osmotic coefficient *or* phenyl ring (in chemical equations instead of C_6H_5) *or* angle between transition moment of absorbing group and segment axis *or* dihedral angle in polymer chain

χ Chemical (as in χ) mechanism or process *or* effective radius of exclusion

χ_c Degree of crystallinity

ψ Distribution function of coordinates of beads *or* electrostatic potential

ψ_0 Zero shear distribution function of coordinates of beads

ω Frequency

ω_0 Resonance frequency

Chapter I

Introduction

In spite of the fact that the common inorganic glasses, which comprise one class of ion-containing polymers, were the first macromolecular synthetics, it is only very recently that the large family of ionic polymers has received systematic attention. The primary reason for this recent upswing of interest, which is reflected in the appearance of several thousand articles and patents in the field in less than two decades, is undoubtedly the enormous range of applications of these materials. These applications, in turn, are due to the dramatic changes in properties of polymeric materials which result from the introduction of strong intra- and intermolecular Coulombic interactions resulting from the presence of ions. This book concerns itself with an in-depth discussion and analysis of several physicochemical aspects of the properties of such materials, based on both organic and inorganic systems.

The field of ion-containing polymers includes both some of the most ancient and the most modern work in macromolecular science. Until relatively recently, only two classes of materials were receiving extensive scientific attention. These comprise the network glasses and the polyelectrolytes.

Because of the commercial importance of the first group of materials, an extensive literature is available which deals with both the scientific and commercial aspects of the field. Only a few representative references can be mentioned here [1]. The field of polyelectrolytes has attracted considerable attention both from theorists and experimentalists, and two monographs have appeared [2, 3], one of which dates back to 1961. The primary areas of interest here were the thermodynamic and configurational properties in solution, although some commercial applications are of quite early origin, i.e., the ion-exchange resins [4].

Due to the enormous range of applications, it seems appropriate, at this point, to present an outline of the types of uses to which ion-containing polymers have been put in the past. This outline is due to M. F. Hoover [5] and is presented here with only very slight modification.

A. APPLICATIONS OF ION-CONTAINING POLYMERS

1. Processing aids
    a. Solids–liquids separation aids (coagulants–flocculants)
    b. Dewatering aids (sewage and mineral sludge)
    c. Flotation aids
    d. Scale and deposit inhibitors
    e. Biocides
    f. Catalysts
    g. Pigment retention aids in paper
    h. Drainage aids in paper
    i. Sheet formation aids in paper
    j. Demulsifiers
2. Process modifiers
    a. Rheological modification of fluids
        (1) Friction reducers (turbulence suppressors)
        (2) Mobility control in oil field flooding for secondary recovery
        (3) Gelling
    b. Dispersants for pigments, clays, sludge, etc., in both water- and oil-based systems.
3. Product additives
    a. Dry-strength resins in paper
    b. Functional coatings for paper (electroconductive, adhesive, photosensitive)
    c. Consumer products (shampoos, antistats, cosmetics, etc.)
    d. Textile finishing (soil release, antistatic)
    e. Detergents (PO_4 replacement)
    f. Lube-oil additives for detergency
4. Ionics as products
    a. Permeable membranes (electrodialysis)
    b. Biomedical applications
    c. Cosmetics (gels, films)
    d. Electrodeposition coatings
    e. Ion-exchange or metal-chelation resins
    f. Polysalt complexes
    g. Modified plastics ("ionomers")
    h. Modified fibers
    i. Inorganic glasses
    j. Ceramics

B. RECENT SYMPOSIA AND REVIEWS IN THE FIELD

It is not surprising that the wide range of applications mentioned above has led to a flood of publications and symposia. Between 1972 and 1975 five major national or international symposia have dealt with this area (outside the framework of aqueous solutions of polyelectrolytes), the first on the topic having taken place only in 1968 [6]. In 1972, a symposium on water-soluble polymers was held in New York under the auspices of the American Chemical Society (ACS) [7]. In 1973, two symposia took place, one in Pasadena in May [8], and the other in Chicago in August, the latter again within the framework of the ACS [9]. Two meetings were also held in 1975, one in Uxbridge, England, in May, and the other in Jerusalem, in July, the latter as part of the IUPAC Symposium on Macromolecules.

The range of ionic materials that have been synthesized and studied to a greater or lesser extent is truly enormous. The synthetic aspects of some classes have been reviewed in two extensive recent papers [10, 11], which cover over 500 references. The mechanical aspects, on the other hand, have received far less attention. The first review on that topic appeared in 1967 [12], and concentrated primarily on the polyphosphates, since that system was the only one that had been explored extensively up to that point. The second review appeared in 1971 [13], and indicated a growing interest in the mechanical properties of organics. One additional short review dealt only with the glass transition [14]. Very recently, a collection of review articles has appeared [15]. Another aspect that has not received extensive attention is the study of the structure of ion-containing polymers in bulk and in non-aqueous solutions. Clearly the supermolecular structure must influence the properties profoundly, and a few recent investigations have concentrated on this effect.

C. SCOPE OF THIS WORK

At this time, despite the fact that the number of papers dealing with the structural and physical aspects is relatively small, it is felt that the level of understanding of the interrelation of these two is sufficiently advanced to warrant the preparation of a book in this still developing field. In addition, a range of complementary topics is also discussed. Thus, Chapter II reviews the structure and glass transition of ion-containing polymers. Chapter III deals with viscoelastic and dielectric properties of completely ionizable homopolymers, and Chapter IV with those of copolymers. Finally, in

Chapter V, configuration-dependent properties in dilute solution are treated in an introductory manner for the sake of completeness. These topics form a unit which, it is hoped, will serve not only to introduce and review the field, but also to underline the fact that many of the areas covered here have received insufficient experimental and theoretical attention.

It should be pointed out that some topics had to be omitted from this work. Foremost among these are the biological polymers, the structure and properties of which have been receiving considerable attention. Electrical properties are also omitted. This treatment therefore concentrates primarily on the structure, elasticity, and viscoelasticity of synthetic ion-containing polymers.

D. GENERAL CLASSIFICATION

Because of the enormous range of polymeric materials which contain ions or to which ions can be attached, the problem of a comprehensive classification is very complex. Only for a few selected classes of materials do comprehensive schemes exist, such as for the materials based on quaternary ions (ammonium, sulfonium, or phosphonium) by Hoover [10]. A more general approach has been taken by Holliday [15], who divides the materials into networks, long covalent chains, and short covalent chains, and within each category into organics, inorganics, hybrids (organic–inorganic), and polyelectrolytes. The scheme adopted in this book does not differentiate between organics or inorganics, but rather is based on the degree of covalent cross-linking and the degree of ionic character, which are treated as two more or less independent variables. Figure 1.1 outlines this approach schematically, using the two variables mentioned above as the two dimensions.

1. Nonionics

Starting with the nonionic or patent polymers, we can differentiate two main types—the networks on the one hand and the linear or branched systems on the other. It is noteworthy that most of the networks are inorganic, i.e., SiO_2, P_2O_5, As_2S_3, but that organic materials also belong to this category, i.e., the styrene–divinylbenzene networks which can be regarded formally as the parent polymer of the ion-exchange resin based on poly(styrene sulfonic acid) in the various salt forms. Vulcanized rubber can serve as another example of an organic material in this group, and many more can be imagined. As examples of the second group of nonionics, one can mention the commercially important organics, i.e., polystyrene, polyisoprene, poly-

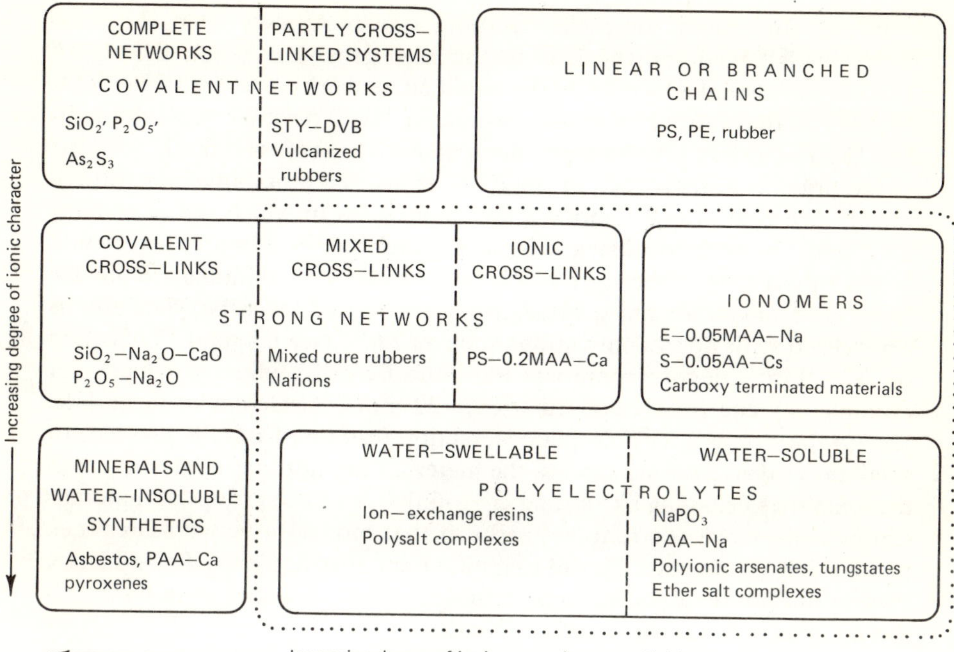

FIG. 1.1. Classification by structure/property similarities. See p. 9 for an explanation of the nomenclature of ionic copolymers and polyelectrolytes.

ethylene, etc., as well as some inorganics such as poly(phosphoric acid) $(HPO_3)_x$. A cautionary comment is in order at this point. The nonionics have been represented as two distinct families. It would be more correct, although less convenient, to describe these materials in terms of a more gradual progression from the truly linear, through the slightly branched, strongly branched, lightly cross-linked, strongly cross-linked, all the way to the complete networks like SiO_2. These differences in classification, however, are quite minor. For all of the above-mentioned materials, a vast literature exists. Since they do not contain ions, they will not be discussed here further.

2. Partly Ionic Materials

Both the networks and the linear or branched polymers can be thought of as parent materials for the partly ionic compositions, i.e., those in which the number of ionic groups per repeat unit is appreciably less than one. Those materials in which that number approaches one or exceeds it will be grouped together under polyelectrolytes.

Within this partly ionic classification, one progresses from the most highly cross-linked (on the left) to the least cross-linked (on the right). The most cross-linked would be the modified networks, prepared from SiO_2 or P_2O_5 by the addition of small amounts of Na_2O or other network modifiers [1]. The ions in these compositions do not act as cross-links, in contrast to the organic systems; i.e., in these materials, the cross-links are still covalent. The next group comprises those networks in which the cross-links are mixed, i.e., both covalent and ionic. Perhaps the best example here would be the mixed-cure rubbers, in which both covalent C–C links are present along with ionic cross-links, which usually are based on carboxylate anions neutralized with polyvalent cations such as Zn^{2+} (see Chapter IV, Section B). The third category comprises the ionically cross-linked polymers, for example, styrene copolymerized with ~20 mole % calcium methacrylate. Due to the strong intermolecular interaction resulting from the presence of small polyvalent cationic species, the materials are not thermoplastics, with the ionic sites behaving like strong cross-links. All three groups of materials can be considered as strong networks in that moderate temperatures (i.e., up to ~200°C) are not sufficient to render them thermoplastic. By and large, they would not be expected to be soluble in either organics or in water, although they might swell in some solvents.

On the same level with respect to the degree of ionic character one can place the weak networks. These are best exemplified by copolymers of normal organic monomers with ionizable or ionic monomers. The materials are thermoplastic at relatively low temperatures, in contrast to the strong networks. In the recent literature they have been referred to as ionomers. These materials are discussed in Chapter IV.

Finally, the bottom row of the table is occupied by the highly ionic materials, i.e., those that are either water soluble or swellable by virtue of their high ion content or that contain one or more ions per repeat unit but that remain insoluble by virtue of their high lattice energy, molecular weight, or other factors. The latter materials, placed in the lower left-hand corner of the table, comprise both minerals and synthetics. The minerals are exemplified by such materials as asbestos or the pyroxenes, while calcium silicate $(CaSiO_3)_x$ or poly(calcium acrylate) may serve as examples of the synthetics.

The second group of highly ionic materials can be described as polyelectrolytes, and these can be divided into two categories—those that swell in water, like the polysalt complexes or the ion-exchange resins, and those that dissolve in water, like the ionic homopolymers, for example, poly(sodium acrylate) or poly(sodium phosphate).

Clearly, the preceding classification scheme is primarily one of convenience, and thus may be expected to lead to some difficulties for some specific materials. It does show, however, the large degree of horizontal, vertical, and diagonal connectivity between the various families of materials. For example, an organic network (Fig. 1.1) can be converted by the inclusion of a small amount of ionic comonomer into a mixed cross-link material, and by a further increase in the degree of ionic substitution into an ion-exchange resin. To pick another example, a soluble linear polyelectrolyte can be converted, by the incorporation of a small amount of covalent cross-links, into an ion-exchange resin, or, by neutralization with a "strong" cation such as Ca^{2+} into a water-insoluble synthetic. To illustrate diagonal relationships, a linear polymer can be converted by copolymerization with calcium acrylate (as in the previous examples, this is not meant to suggest a synthetic route, but a conceptual one) to an ionically cross-linked strong network, which with increasing calcium acrylate concentration, eventually becomes the completely ionic but water-insoluble poly(calcium acrylate) in the lower left corner. This volume concerns itself primarily with the materials located in the lower right side of Fig. 1.1 in the region surrounded by the dotted line. In the figure itself, and in this brief explanation, only very few examples could be introduced. Furthermore, since a vast array of materials has been synthesized to date in the area of ionomers and water-soluble polymers, further subdivision is clearly needed. This will be attempted below.

E. DETAILED CLASSIFICATION AND EXAMPLES OF LINEAR AND BRANCHED IONICS

As was mentioned before, the range of materials which could be listed here as examples is enormous. No attempt will be made to provide a comprehensive list of examples: only representative types within each category will be chosen. As suggested in Fig. 1.1, the materials are divided into two main categories, namely, those with a low ion content and those in which a major fraction (or all) of the repeat units carry charges.

1. Polymers of Low Ion Content

(*a*) *Strong Networks*

The materials in this category are copolymers, which behave like cross-linked systems either by virtue of the presence of covalent cross-links which are reinforced by weak or strong ionic interactions, or because of the presence of strong ionic interactions alone, or by a combination of effects whereby the backbone is insoluble in those solvents that plasticize the ionic groups or vice versa. Two main subgroups can be distinguished:

i. Materials possessing mixed cross-links

Mixed-cure rubbers. These materials are based on normal commercial rubbers containing some carboxyl groups. They are cross-linked by any of the customary vulcanizing agents in addition to, for example, zinc oxide. Some of the Zn^{2+} ions act as strong cross-links. These materials are discussed in Chapter IV, Section B.

The Nafion-type systems. The Nafions themselves are ionics based on polytetrafluoroethylene with sulfonic acid or salt groups placed at the end of side chains. In those solvents that might have dissolved the nonionic prepolymer, the ionic aggregates (see Chapter II) are very stable and act as cross-links. Conversely, highly polar solvents, which interact with the sulfonate groups, do not solvate the polymer. Thus, either the ionic regions or the nonionic areas act as cross-links that make the material insoluble although it does absorb a range of solvents to varying extents. A number of materials of this type can be imagined; for example, it can be anticipated that polystyrene containing ~10–15 mole % of metal carboxylate groups as a copolymer might have generally similar solution properties, even if the cations are monovalent and large. No systematic study of these systems has been published, although the Nafions are receiving considerable attention and are described in Chapter IV, Section A2.

ii. Materials possessing strong ionic cross-links only

In the case of organic copolymers containing even a small concentration of acidic groups (carboxylic, sulfonic, or phosphonic), neutralization with polyvalent cations of relatively small size (e.g., Ca^{2+}) makes the materials behave like cross-linked systems. Also the materials are not thermoplastic, since the cross-links acquire sufficient mobility only at temperatures at which the polymer decomposes. The solubility behavior might resemble that of some of the Nafion-type materials, but in this case primarily because of the very strong ionic interaction, not because of the large number of them as in the previous case.

(*b*) *Ionomers*

Again, the number of materials that can be included here is enormous. Practically all the common organic monomers can be copolymerized with a range of ionizable monomers, so that the number of combinations far exceeds the number of known polymers. In Chapter IV, which is devoted to the properties of ion-containing copolymers, the materials are divided into two groups, i.e., those based on noncrystalline and on crystalline polymers, the former being further divided into materials of high glass transition and into rubbers. For the sake of convenience, the strong networks are also included in Chapter IV as are the diionically terminated short chains.

In general, the ionomers are thermoplastic materials containing up to ~10% of ionic comonomers and are thus insoluble in water in most cases.

For both noncrystalline and crystalline polymers, a special case can be envisaged whereby the ionic groups are present only at the chain end. Several studies on materials of this type have been carried out, some of which are reported in Chapter II.

As has been mentioned, a very wide range of backbone materials and ionic monomers can be copolymerized to yield ionomers. Furthermore, the ionomers can be completely or partially neutralized and can also contain plasticizer. Sample designations would thus become extremely long. For this reason, a shorthand way for the designation of ionomers is proposed below. An ionizable copolymer (or terpolymer) is denoted by: (i) its composition, with the mole fraction(s) of the minor component(s) specified, (ii) the counterion and the fraction of the total ionizable material neutralized (where applicable), and (iii) the plasticizer (where applicable), expressed as the weight fraction of the total plasticized weight. The method is illustrated by the following example:

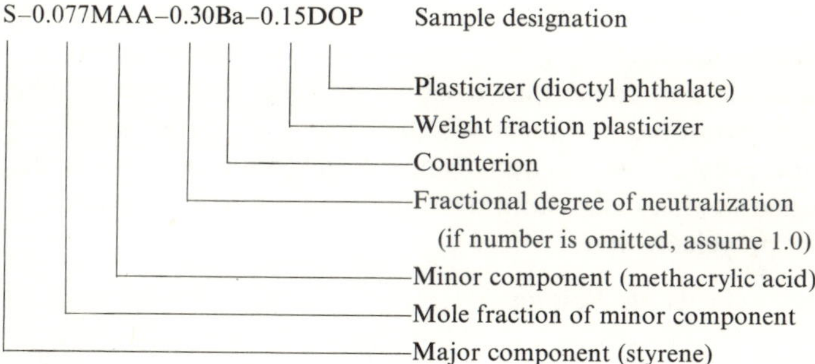

The following symbols for common monomeric components are employed: S—styrene; B—butadiene; E—ethylene; AA—acrylic acid; MAA—methacrylic acid. PS, PAA, etc., indicate the corresponding homopolymers. Partially neutralized homopolymers are indicated as PMAA–0.25Na, for example. Other abbreviations are introduced where first used. The above scheme represents a modification of the scheme proposed by Navratil and Eisenberg [16].

i. Noncrystalline copolymers

Materials of high glass transition. An excellent example here would be the copolymers of styrene and sodium methacrylate or of poly(methyl methacrylate) with methacrylic acid. These materials are discussed in Chapter IV, Section A1.

Rubbery materials. Any of the common rubbers, such as polybutadiene, copolymerized with acrylic or methacrylic acid, can serve as an example of macromolecules in this category. They are discussed in Chapter IV, Section B.

ii. Crystalline copolymers

The best-known examples in this category are the copolymers of ethylene and acrylic acid or methacrylic acid. These have been investigated extensively and are discussed in Chapter IV, Section C1. A wide range of other materials is, of course, also possible.

2. Polyelectrolytes

The materials in this category are usually homopolymers of ionizable or ionic monomers, although in some cases water solubility is retained even if only 20 mole % of an ionic comonomer is incorporated into the material. Since most of the studies to date are devoted to the homopolymers, the examples below will be limited to that class. Two main groups can be distinguished, those that swell in water but do not dissolve completely due to the presence of either covalent cross-links or strong ionic groups which act as cross-links, and those that dissolve in water completely. There exist a number of nonionic polymers which dissolve in water, such as polyacrylamide and poly(ethylene oxide). Since these do not carry ions, they will not be included in this discussion, although some schemes do refer to them as nonionic polyelectrolytes.

(*a*) *Water-Swellable Systems*

i. Homopolyelectrolyte-based networks

Any of the polymers to be discussed under Section B2 can be copolymerized with a small amount of a bifunctional cross-linking agent to yield a network system which would be expected to swell extensively in water in the ionized form. These types of materials are used as ion-exchange resins, although copolymers of a lower ion content may also be employed.

ii. Polysalt complexes

A wide range of polyanionic homopolymers in aqueous solution can be mixed with dissolved polycationic materials to yield stoichiometric polysalt complexes which, generally, are insoluble in water although they can swell considerably depending on the specific polyions used. These materials are described in Chapter IV, Section D. They are treated as copolymers since they are made up of more than one homopolymer and are thus analogous to the nonionic blends.

(*b*) *Water-soluble systems*

In this section, again, only representative examples of the different types of materials can be cited. The materials are discussed in Chapters III and V.

i. Inorganics

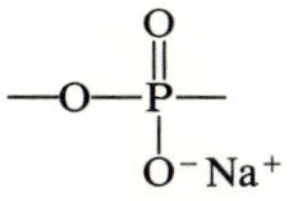

$$\text{—O—}\overset{\overset{\displaystyle O}{\|}}{\underset{\underset{\displaystyle O^- Na^+}{|}}{P}}\text{—}$$

$NaPO_3$

$$\text{—O—}\overset{\overset{\displaystyle O^- Na^+}{|}}{\underset{\underset{\displaystyle O^- Na^+}{|}}{Si}}\text{—}$$

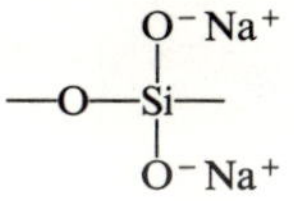

Na_2SiO_3

ii. Organics

Anionics

Carboxylic acid salts

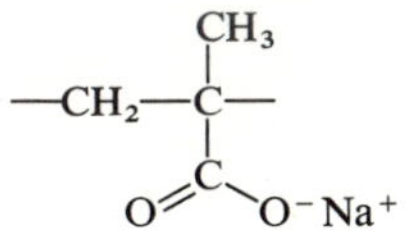

$$\text{—CH}_2\text{—}\overset{\overset{\displaystyle CH_3}{|}}{\underset{\underset{\displaystyle O{=}C{-}O^- Na^+}{|}}{C}}\text{—}$$

poly(sodium methacrylate)

Sulfonic acid salts

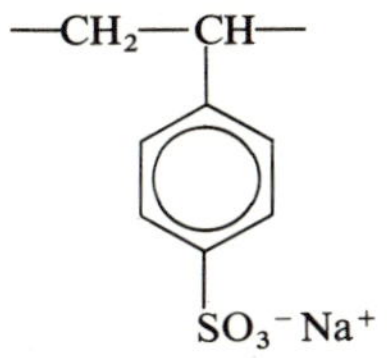

$$\text{—CH}_2\text{—}\underset{\underset{\displaystyle C_6H_4SO_3^- Na^+}{|}}{CH}\text{—}$$

poly(sodium styrene sulfonate)

Phosphonic acid salts

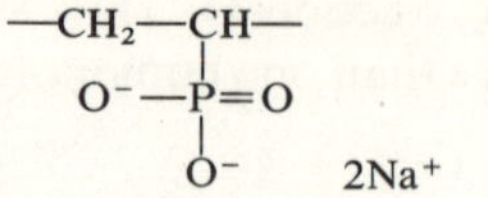

$$-CH_2-\underset{\underset{O^-}{|}}{\overset{|}{CH}}- \quad O^- - P = O \quad 2Na^+$$

poly(vinyl sodium phosphonate)

Cationics The organic chemistry and applications of these materials were reviewed by Hoover [10]; the interested reader is referred to this article for details. Only examples of the most important structural types can be given here.

Polymers containing ammonium groups

(*i*) *Integral quaternaries* These can be either on a linear chain or a cyclic (aliphatic or aromatic) species; in any case they form part of the backbone. Three examples follow:

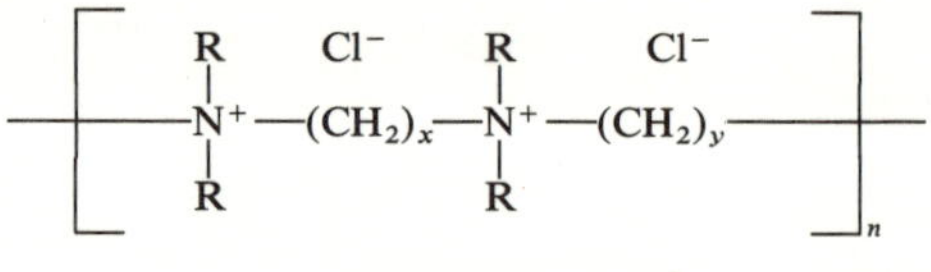

$$\left[-\overset{R}{\underset{R}{N^+}}-(CH_2)_x-\overset{R}{\underset{R}{N^+}}-(CH_2)_y- \right]_n \quad Cl^- \; Cl^-$$

X, Y, ionene

$$\left[-{}^+N\langle\text{ring}\rangle-(CH_2)_x-\langle\text{ring}\rangle N^+-(CH_2)_y- \right]_n \quad Cl^- \; Cl^-$$

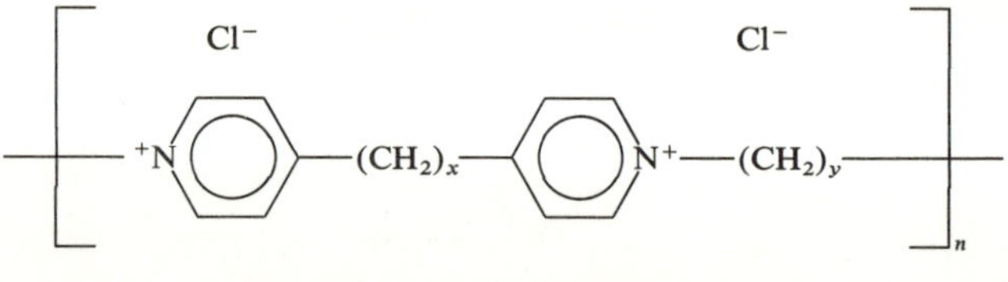

An aromatic ionene (a polypyrazine)

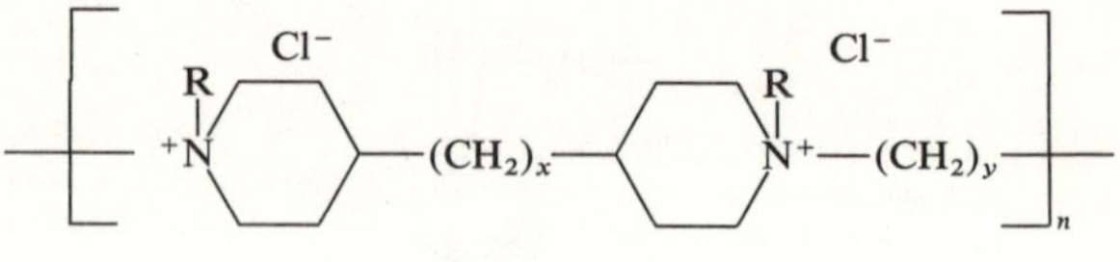

$$\left[-{}^+\overset{R}{N}\langle\text{ring}\rangle-(CH_2)_x-\langle\text{ring}\rangle\overset{R}{N^+}-(CH_2)_y- \right]_n \quad Cl^- \; Cl^-$$

A polypiperazine

(*ii*) *Polymers containing pendant quarternaries* The ionic groups may be attached to an aliphatic chain or to a cyclic backbone directly or through various intermediates. The quarternary ion itself may be aliphatic or part of a heterocycle. Several examples follow:

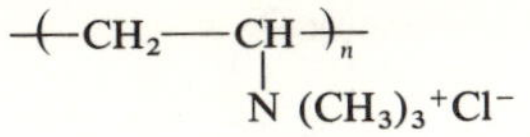

$$-\!\!\left(CH_2-\underset{\displaystyle N(CH_3)_3^{+}Cl^{-}}{\underset{|}{CH}}\right)_n\!\!-$$

Poly(vinyl trimethyl ammonium chloride)

$$-\!\!\left(CH_2-CH\right)_n\!\!- \quad (\text{CH attached to } C_6H_4-CH_2-N(CH_3)_3^{+}Cl^{-})$$

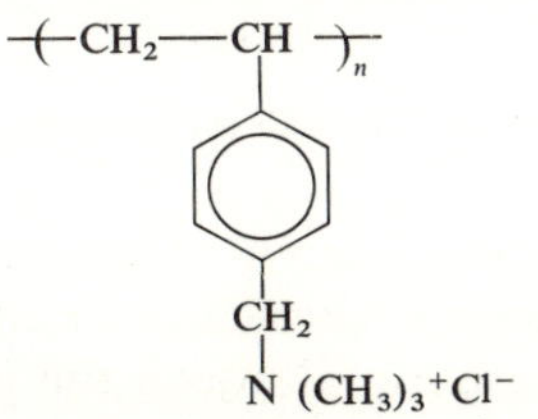

Poly(vinyl benzyl trimethyl ammonium chloride)

$$-\!\!\left(CH_2-CH\right)_n\!\!- \quad (\text{CH attached to pyridinium ring, } N^{+}-CH_3\ Cl^{-})$$

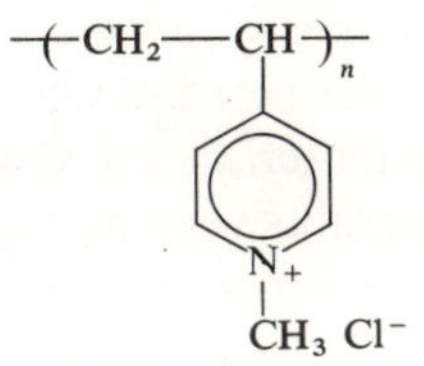

Poly(*N*-methylvinylpyridinium chloride)

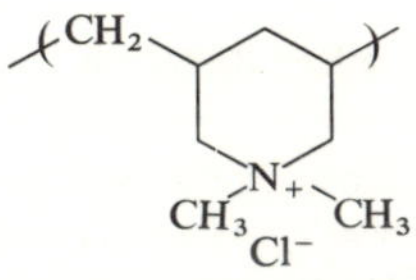

$$\left(CH_2 \ldots\right) \quad \text{(piperidinium ring, } N^{+}(CH_3)(CH_3)\ Cl^{-})$$

Poly(diallyl dimethyl ammonium chloride)

Polymers containing sulfonium groups

The selection of materials in this category is much smaller than in the preceding one. Structurally, similarities do exist (except for sulfonium ions as part of aromatic ring systems). One example:

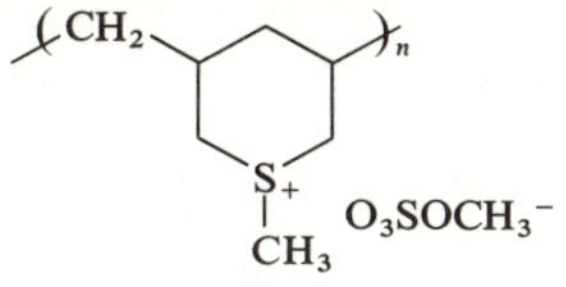

$$\left(CH_2 \ldots\right)_n \quad \text{(thianium ring, } S^{+}-CH_3\ \ O_3SOCH_3^{-})$$

Poly(methyldiallyl sulfonium methylsulfate)

Polymers containing phosphonium groups

Also in this case, the total number of examples that have actually been prepared is quite small. One example of a possible compound is:

$$-\!\!(\mathrm{CH_2-CH})_n\!\!-$$
$$\mathrm{C_6H_4}$$
$$\mathrm{CH_2}$$
$$\mathrm{P(CH_3)_3^+\ Cl^-}$$

Poly(vinyl benzyl trimethyl phosphonium chloride)

Mixed anionic–cationic copolymers Although the synthetic techniques may not be simple, it is possible to synthesize a range of materials containing both anionic and cationic groups, although the pH of the solution would determine which predominates.

Inorganic salts in polar polymers Several examples of this type of material have been investigated, the best known being solutions of $LiClO_4$ in poly(propylene oxide). These materials are discussed in Chapter II and Chapter III. The small cation interacts very strongly with the ether oxygens giving the polymers many of the properties of either ionomers or polyelectrolytes.

REFERENCES

1. W. Eitel, "Silicate Science." Academic Press, New York, 1965; W. A. Weyl and E. C. Marboe, "The Constitution of Glasses." Wiley (Interscience), New York, 1964.
2. S. A. Rice and M. Nagasawa, "Polyelectrolyte Solutions." Academic Press, New York, 1961.
3. F. Oosawa, "Polyelectrolytes." Dekker, New York, 1971.
4. J. A. Marinsky (ed.), "Ion Exchange: A Series of Advances," Vols. I and II; J. A. Marinsky and Y. Marcus (eds.), "Ion Exchange and Solvent Extraction," Vols. III–VI. Dekker, New York, 1973, 1974.
5. A. Eisenberg and M. F. Hoover, "Ion Containing Polymers." Amer. Chem. Soc. Short Course Handout, Amer. Chem. Soc., New York, 1972.
6. San Francisco Amer. Chem. Soc. Meeting, *Polym. Preprints* **9** (1968).
7. N. M. Bikales (ed.), "Water-Soluble Polymers." Plenum Press, New York, 1973.
8. A. Rembaum and E. Sélégny (eds.), "Polyelectrolytes and Their Applications." Reidel, Dordrecht, 1975.
9. A. Eisenberg (ed.), Ion-containing polymers, *J. Polym. Sci. Polym. Symp.* **45** (1974).
10. M. F. Hoover, *J. Macromol. Sci.* **A4,** 1327 (1970).
11. M. F. Hoover, *J. Polym. Sci. Polym. Symp.* **45,** 1 (1974).
12. A. Eisenberg, *Adv. Polym. Sci.* **5,** 59 (1967).
13. E. P. Otocka, *J. Macromol. Sci.* **C5,** 275 (1971).
14. A. Eisenberg, *Macromolecules* **4,** 125 (1971).
15. L. Holliday (ed.), "Ionic Polymers." chapter I. Applied Science Publishers, London, 1975.
16. M. Navratil and A. Eisenberg, *Macromolecules* **7,** 84 (1974).

Chapter II

Supermolecular Structure and Glass Transitions

Although the two topics to be discussed here are related only indirectly, they are combined because both exert a profound influence on the mechanical properties, which comprise the bulk of Chapters III and IV. First the structural aspects will be described, and subsequently the glass transition. Since the latter topic is the subject of an existing review [1], only those aspects that have received insufficient attention in the review will be covered extensively here.

A. SUPERMOLECULAR STRUCTURE

The supermolecular structure of ion-containing polymers, particularly those aspects dealing with microphase separation, present a particularly challenging problem in that none of the usual tools of structure determination is readily applicable. Furthermore, the results obtained by some of those techniques that have been applied are subject to different interpretations by the various groups working in the field.

In this discussion, the presentation is divided into three parts, arranged in order of increasing complexity and uncertainty. The first part deals with the thermodynamics and electrostatics of ion association in macromolecular organic systems. The energy of formation of pairs and higher multiplets is computed to illustrate the fact that ion association of this type is to be expected in polymers on theoretical grounds. The limits of multiplet formation are also discussed. Subsequently, relevant experimental evidence is cited, the evidence coming primarily from the viscous and viscoelastic properties of a wide range of ion-containing polymeric materials.

The next problem to be discussed is far more involved; it concerns itself with the possibility of cluster formation, i.e., of the aggregation of the multiplets into larger clusters, probably incorporating both ionic and nonionic material and exhibiting some of the properties of a new phase. Here experimental evidence is cited first, including results of X-ray diffraction, electron microscopy, and NMR. A number of rheological studies are also mentioned (anticipating the more detailed discussion of Chapters III and IV), which can be interpreted only in terms of a multiphase structure.

The last section on this topic is devoted to a discussion of some theoretical aspects of cluster formation. It will be shown that by balancing the electrostatic forces involved against the elastic effects and assuming several different cluster geometries, reasonable cluster sizes and intercluster distances can be computed which are in qualitative agreement with the experimental results.

1. Multiplet Formation

(*a*) *Theoretical Considerations*

Water is one of the very few solvents that have a high enough dielectric constant to solvate ions to the point where predominantly single ions are encountered. If, however, a calculation is performed in regard to the energetics of ion association in solvents of low dielectric constant (for polyethylene $\varepsilon = 2.5$ [2]), it is seen that the interaction energy increases to the point where the probability of encountering a single ion is exceedingly rare. As an illustration, one can utilize the equation for opposite point charges of equal magnitude e, the electron charge, i.e.,

$$W = -e^2/\varepsilon r \tag{1}$$

(where W is the interaction energy and r the distance between the charges) and compare W with the value of kT at room temperature. If $r = 1.5$ Å (a reasonable value for the charge separation between COO^- and Na^+ if contact of the ions is assumed), then $W = 7 \times 10^{-12}$ erg, while kT at room temperature is approximately 4×10^{-14} erg. A calculation employing Boltzmann's relation (neglecting entropy effects) reveals that under those conditions not even one ion pair per mole is dissociated. Clearly this treatment is highly oversimplified; the trend, however, is unmistakable.

At this point, it is advisable to inquire into the possibility of further ion association. This was done by a number of workers for ions in both low-molecular-weight organic solvents and in polymers. One of the studies of the first type, which are most relevant to the present discussion, was performed by Pettit and Bruckenstein [3], who considered the energetics of association for ionic aggregates such as ion pairs, triplets, quartets, up to sextets. (The nomenclature of this book differs somewhat from that of Pettit and Bruckenstein.) Their calculation was an improvement over preceding work in that orientation entropy effects were included. The salts considered in the study were Bu_4NCl, Bu_4NClO_4, Bu_4NBr, $NaClO_4$, $HClO_4$, and KCl (where Bu is the butyl group); solvents of dielectric constant between 2.27 and 7.38 were included. If polarizabilities are neglected and internuclear distances are taken to be 4 Å (a reasonable average value for the large ions considered in the study), the formation constants B_i for ion pairs, triplets, quartets, and sextets are given by

$$\begin{aligned} \log B_2 &= 52.1/\varepsilon - 3.7 \\ \log B_3 &= 75.1/\varepsilon - 6.9 \\ \log B_4 &= 121.3/\varepsilon - 7.5 \\ \log B_6 &= 193.2/\varepsilon - 12 \end{aligned} \tag{2}$$

The effect of varying the dielectric constant at two different salt concentrations on the distribution of the various species is shown in Fig. 2.1. It indicates that in solvents of low dielectric constant, higher aggregates are possible [e.g., $(AB)_3$], although their concentrations may be low at low overall salt concentrations.

In the case of a polymer, the situation is further complicated by the fact that all of the ions of one sign are fixed to the chain. This leads to an ex-

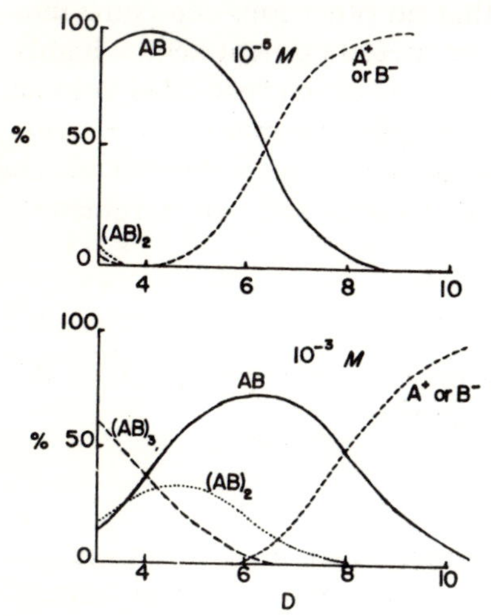

FIG. 2.1. The effect of dielectric constant on the association of a simple salt A^+B^- at two concentrations at 25°C. AB is an ion pair. $(AB)_2$ a quartet, $(AB)_3$ a sextet. [3].

cluded volume effect which puts an upper limit on the multiplet size in the absence of special geometrical arrangements such as sheets or rods. A theoretical analysis performed recently [4] considers a spherical multiplet which contains only ionic material, with all the organic chain material relegated to the surface of this multiplet. It is assumed in the derivation that neighboring ionic groups separated by several nonionic units are not incorporated into the same multiplet, because of the unfavorable entropic effect of ring closure. If the radius of the multiplet is r_m, then the surface area of the multiplet S_m is $4\pi r_m{}^2$, while the volume of the multiplet V_m is $4\pi r_m{}^3/3$. The total number of ion pairs in the multiplet n_0 can be expressed as the ratio of V_m to V_p, the volume of an ion pair, or as the ratio of S_m to S_{ch}, the contact surface area of the polymer segment to which the ion is attached. Each ion-containing segment coats the surface area of the sphere in such a manner that no ion can come closer to this sphere than the thickness of a polymer chain. By setting the two ratios equal, i.e.,

$$V_m/V_p = S_m/S_{ch} \qquad (=n_0) \tag{3}$$

and inserting the appropriate relations, one obtains

$$r_m = 3V_p/S_{ch} \tag{4}$$

Specifically for the case of ethylene-(sodium methacrylate) copolymers, approximate values for V_p and S_{ch} are 12 Å^3 and 12 Å^2, respectively, yielding a value of r_m of $\sim$3 Å and V_m of $\sim$100 Å^3. This implies that a maximum value of n_0 for the spherical multiplet is eight ion pairs ($\simeq$100 Å^3/12 Å^3), but probably less since perfect volume occupancy is highly unlikely. It is worth recalling that once such a spherical multiplet is formed, it is completely coated with nonionic material, so that no other ions can come into contact with it. It should be stressed that the preceding treatment is highly approximate and serves primarily to illustrate the drastic excluded volume problem to be expected in spherical or nearly spherical multiplets. Finally, from simple stoichiometric considerations, an equation is derived which correlates the multiplet separation R_0 with various relevant parameters. The equation is

$$R_0 = (n_0 M_c/\rho N_{av})^{1/3} \tag{5}$$

where M_c is the average molecular weight between ionic groups along the chain, ρ the density, and N_{av} Avogadro's number.

(b) *Experimental Confirmation*

The most direct evidence for multiplet formation comes from an analysis of viscoelastic and rheological studies on ion-containing polymers. Investigations on two distinct classes of materials are relevant here—carboxy terminated "low" polymers and copolymers containing a small percentage of ions. The data on the latter will be presented in much greater detail in Chapter IV; here only those features that reflect the presence of ion multiplets are mentioned.

Several groups of workers have demonstrated that dicarboxylic acids of low molecular weight can associate under some conditions to yield materials that resemble high polymers in their physical properties, i.e., in their increased viscosity, fiber-forming ability, etc. [5–8]. In one study [8], a wide range of aliphatic and aromatic dicarboxylic acid salts with a number of divalent cations was investigated. The materials were found to be crystalline, with melting points above 175°C, but showed stability in air until well above 300°C. The authors concluded that the materials must be polymeric because of the high viscosity of the melt, and attempts were made to determine the degree of association from empirical equations. The addition of monovalent ions was found to decrease the viscosity; this was taken as evidence that sodium ions act as chain terminators. A parallel study involving interchange of various diacids (azaleic and dodecanedioic) from different regions of the sample (as followed by IR spectroscopy) revealed that ion interchange was rapid. In view of our present understanding, we can interpret the findings in the following way: Chain extension (i.e., an increase in the molecular weight, and possibly even some degree of cross-linking) is encountered when dicarboxylic acids are neutralized with divalent cations. In the melt, the rate of ionic bond interchange is rapid and is accelerated by the presence of sodium. An excellent review on the synthesis, structure, properties, and uses of metal dicarboxylates has been published recently [9].

As the subsequent study will show, monovalent cations also act as chain extenders at low temperatures. Otocka, Hellman, and Blyler [10] investigated the viscosity of carboxy-terminated butadiene polymers of low molecular weight neutralized with mono- and divalent cations and compared the results with those obtained with methyl-ester-terminated polymers. They showed that multiplet formation occurs even with monovalent cations; in addition, they determined the average enthalpy of association of the multiplets formed with monovalent cations. The most important part of the study involved the determination of the viscosity of the salts as a function of the degree of neutralization and type of cation. The weight-average molecular weights of the associated polymer were determined from a modified form of the equation of Kraus and Gruver [11], i.e.,

$$M_w^{A} = M_w^{0} + fp\alpha M_n^{0}/(1 - fp\alpha) \tag{6}$$

where M_w and M_n refer to the weight- and number-average molecular weights, the superscripts "A" and "0" denote the associated polymer and the polymer prior to association, f is the fraction of functional chain ends, p is the probability of intermolecular association, taken as 1 at 25°C, and α is the degree of neutralization. Viscosities were measured on a Brookfield viscometer and on a capillary rheometer; a correlation of the viscosities of the monovalent salts with M_w^A is shown in Fig. 2.2 for 25 and 75°C.

Several important facts emerge from a study of this figure. The first of these is that the slope of the curve at 25°C is very close to the value of 3.4 expected for entangled polymers, confirming the validity of the concept of chain extension via multiplet (quartet in this case) formation and of the applicability of the equation used to calculate M_w^A. At 75°C, the data show a lower slope and curvature, but the data can still be fitted to a 3.4 slope if p is decreased to 0.9. At still higher temperatures, even lower values of p are needed to obtain a slope of 3.4. The p values can be related to an equilibrium constant for the association reaction, and from a plot of the equilibrium constant as function of temperature (ln K_a versus $1/T$), the enthalpy of association can be computed. A value of $\Delta H \simeq 25$ kcal was obtained, very close to the value obtained in a calculation (based on simple electrostatics) of energy required to separate an ion quartet, either in a pyramidal form [12] or in a square planar conformation [13] into two ion pairs. It should be added that the authors [10] also found that the values of p, K_a, and ΔH varied from one ion to the next in the monovalent series, and that the viscosities were non-Newtonian, possibly because of a stress-induced ion-interchange reaction. The concept of ion interchange will be discussed more extensively in subsequent chapters. The divalent salts yield a slope somewhat higher than 3.4, implying, most probably, that associations beyond the quartet stage, i.e., cross-links, are formed.

EPR of butadiene–(methacrylic acid) copolymers neutralized by Cu(II) also supports the suggestion of ion aggregation. Pineri *et al* [14] found that their results on a copolymer containing 9 mole % acid neutralized with Cu(II) could be explained only on the basis of the presence of two Cu(II) ions in close proximity. This material will be discussed further in Chapter IV.

Results of studies of mechanical or viscoelastic properties also support the existence of ion multiplets. These studies will be discussed at length in Chapters III and IV; only those aspects that bear directly on the problem of ion association are presented here.

Otocka and Eirich [15] investigated the modulus as a function of temperature of a series of butadiene copolymers containing methacrylic acid or 2-methyl-5-vinylpyridine and the corresponding salts. The lithium salts of the acid copolymer revealed a pronounced inflection point in the modulus resembling a short rubbery plateau (Fig. 2.3), the height of which was related

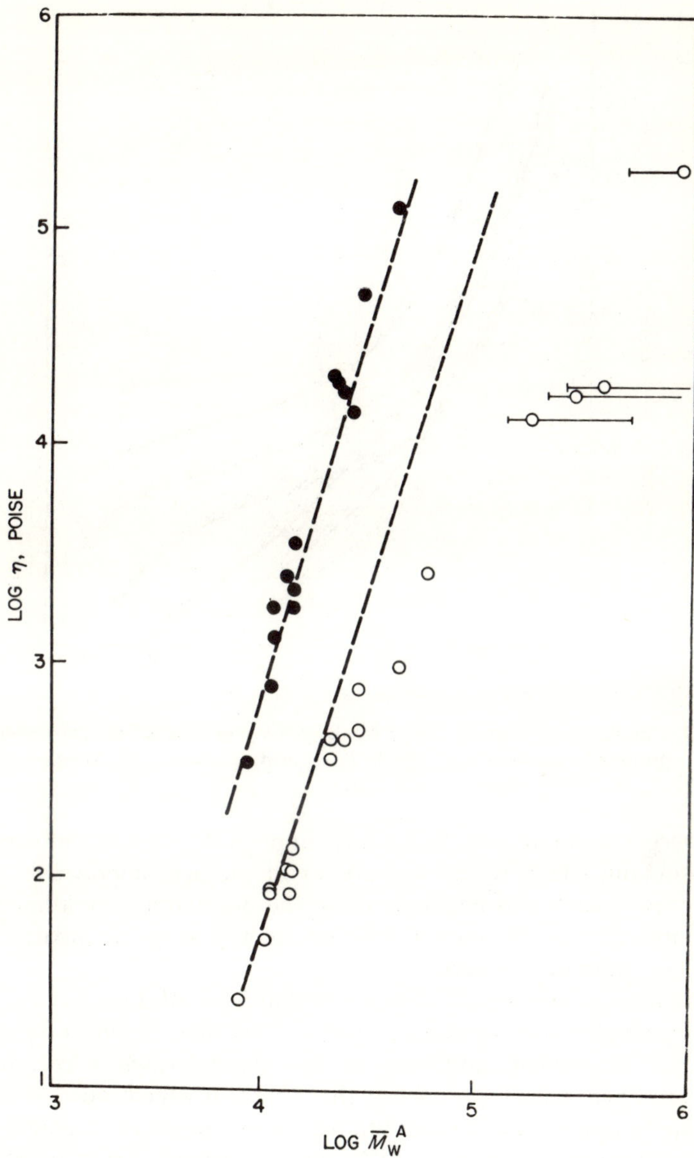

FIG. 2.2 Viscosity–molecular weight relations at 25°C (●) and 75°C (○) for carboxy-terminated polymers neutralized with monovalent cation. The dashed line represents the 3.4 power law. [10].

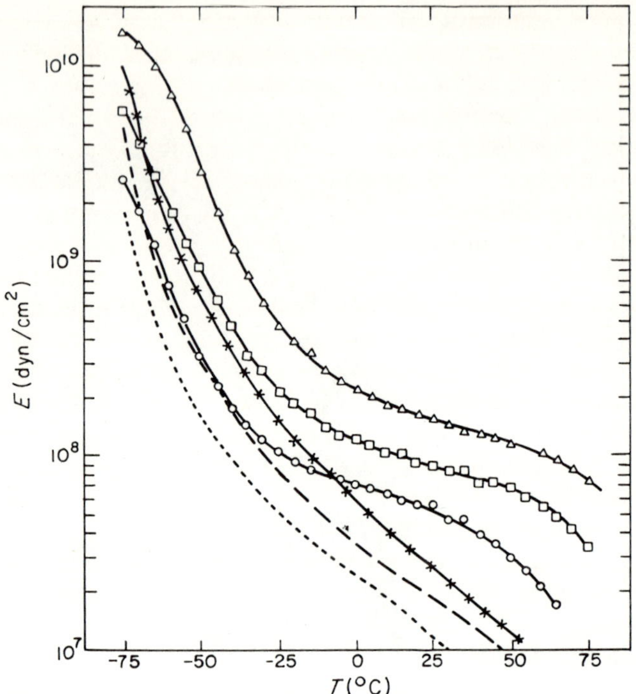

FIG. 2.3. Modulus-temperature curves for B–MAA copolymers and corresponding Li salts, 90% neutralized. Symbols: 4.7% MAA: (---) acid, (○) salt; 7.7% MAA: (——) acid, (□) salt; 11.6% MAA: (×) acid, (△) salt. [15].

to the concentration of ions. This clearly implied the existence of some type of cross-linking effect, which was absent in the acid copolymer. A very similar phenomenon was encountered in the quarternary pyridinium salts when compared with the parent polymer, as well as in the mixture of the anionic and cationic polymers.

Apparent cross-link densities were calculated for all three types of materials using simple rubber elasticity theory. They were found to be strongly temperature dependent, and only at low temperatures were cross-link densities reached which approached the region in which most of the ions participated in cross-link formation. Very low temperatures could not be utilized because of the nearness of the glass transition. From a plot of the logarithm of the cross-link density against the reciprocal temperature, apparent enthalpies of network dissociation were calculated. These ranged from several hundred calories to ~ 10 kcal, depending on the system, the temperature region, and the ion concentration. While the absolute values of ΔH may be seriously in error because of the method used in its determination, it is most significant that strong evidence for network formation is present, and that the network is most probably formed through association of ion pairs, i.e., through multiplet formation.

In a later study, Eisenberg and Navratil [16] investigated the stress relaxation behavior of styrene–(sodium methacrylate) copolymers as a function of ion concentration. Two inflection points in the modulus–time curve were found (Fig. 2.4), one at $\log E = 6.5$, corresponding to the normal entanglement spacing in polystyrene, with the second at a higher level, its position dependent on the ion content. Up to 6 mole % of the ionic component, the position of the higher inflection point could be predicted (within a factor of two) from a simple calculation based on the theory of rubber elasticity, assuming that each ion pair acts as a cross-link, i.e., that it is incorporated into a multiplet of at least two ion pairs. Above 6 mole % evidence of ion clustering was found, which will be discussed in Section A2.

The determination of multiplet sizes is not straightforward, and some contradictions in the quantitative interpretation of experimental results exist. As a consequence, the data, which some workers find as evidence for higher aggregate (cluster) formation, are interpreted by others as evidence for simple multiplets. This will be discussed further in Section A2.

2. Cluster Formation

It was pointed out in the previous section that an ion multiplet of at most eight ion pairs would be completely coated with nonionic material, which would prevent intimate contact with any additional ionic material. The term "ion cluster" is taken to mean an ionic aggregate of greater size than a

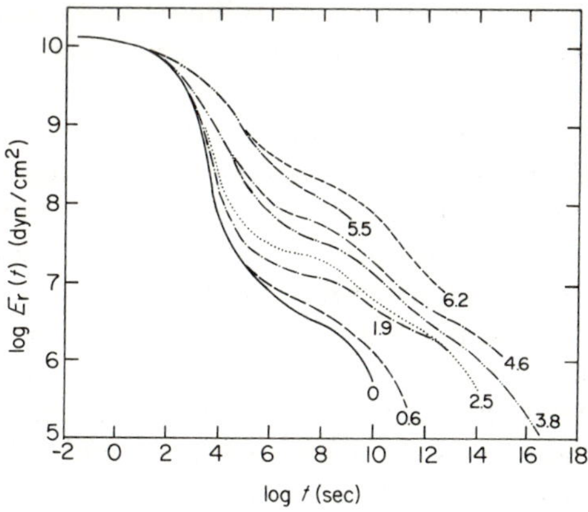

Fig. 2.4. Stress relaxation master curves for completely neutralized styrene ionomers of different ion content. Mole % NaMa in copolymers indicated on each curve. [16].

single multiplet; it requires, therefore, that a considerable amount of nonionic macromolecular material be incorporated in it. The cluster microstructure can be conceptualized in several ways: One example would be a central multiplet, coated with its nonionic skin and surrounded by other multiplets or ion pairs; another would be a collection of multiplets separated by nonionic material. Still others are, of course, possible. In any case, it implies the existence of an overall structure in which ion-rich regions alternate in some fashion with ion-poor regions.

(*a*) *Experimental Evidence*

At this time, the experimental evidence concerning cluster formation is contradictory, and no unanimity has been achieved even in regard to the existence of clusters. On the whole, the bulk of the evidence favors the existence of clusters, at least in some temperature and concentration regions. However, some negative evidence also exists, and both lines will be discussed in this section.

The most convincing evidence for the existence of clusters of some type comes from the analysis of small-angle X-ray scattering (SAXS) data. The scattering of X-rays is due to electron density fluctuations within a material. In addition to independent scattering, consisting of coherent scattering and Compton scattering, X-ray scattering also includes an interference component which is related to the radial distribution function (RDF) of electron density through the Fourier transform of the angular dependence of the scattered intensity. Thus maxima in the RDF can be correlated with preferred separations of electron-dense regions within a material.

The molecular interpretation of RDF peaks depends on the choice of a suitable scattering model; for example, the dimensions obtained from scattering data may be related to either the size of individual scattering sites (intraparticle interference) or to the preferred separation of scattering sites (interparticle interference). The Fourier-transform analysis for the RDF requires accurate intensity data over a wide range of scattering angles (normally both wide angle and small angle). If scattering data are not available over a wide range, dimensional information can still be obtained from the analysis of X-ray intensity peaks. However, the results must be interpreted with caution, and often several interpretations are possible.

For example, if the scattering is due to intraparticle interference, the average size of the particles can be obtained from the Guinier approximation, which relates the scattered intensity $I(s)$ to the scattering vector $s = (4\pi \sin \theta)/\lambda$, where 2θ is the scattering angle and λ is the wavelength of the radiation. Assuming a dilute two-phase system of uniform particles, the relationship is [17]

$$I(s) = I(0) \exp(-\tfrac{4}{3}\pi s^2 \overline{R^2}) \tag{7}$$

where $\overline{R^2}$ is the mean-square radius of gyration of the scattering particles. Other methods of computation, which give different weighted means, also exist.

For interparticle interference scattering, one can relate the average separation between sites to the angular position of the scattering maximum. This separation is related by a constant factor to the primary Bragg spacing [17], i.e.,

$$d_a = K \cdot d_{\text{Bragg}} = K\lambda/(2 \sin \theta) \tag{8}$$

where K is a dimensionless quantity of the order of 1. The value of $K = 1.22$ has been computed for a random distribution of volumeless scattering sites on a quasi-crystalline lattice [17]. In real scattering situations, this factor may not apply.

Wilson *et al.* [18] and Longworth and Vaughan [19] have investigated copolymers of ethylene and methacrylic acid as a function of copolymer composition and degree of neutralization. In the case of a polymer containing 6 mole % MAA, no peaks were found below $2\theta = 10°$, while in the same material neutralized to the extent of 90%, a new peak at $2\theta \simeq 4°$ was found. This is shown in Fig. 2.5 for (a) pure polyethylene, (b) a sample containing 6% MAA, and (c) the acid copolymer neutralized to the extent of 90%. Further studies have shown that the peak height increases with acid content even if the ion content is kept constant, i.e., by using samples of progressively higher acid content neutralized only to such an extent as to give a total of

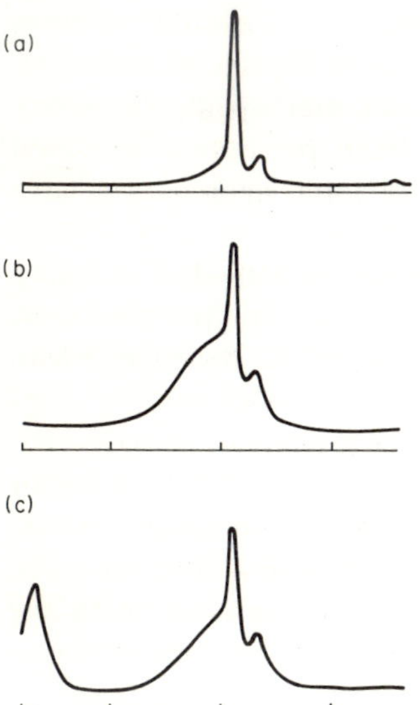

FIG. 2.5. X-ray diffraction scans from: (a) branched polyethylene; (b) copolymer, E–0.06MAA; (c) ionomer, E–0.06MAA–0.90Na [18].

5 mole % salt in the polymer. At the highest degree of neutralization (90%) for samples containing a total of 1.7 and 2.8 mole % acid, the Bragg spacing corresponding to the peak position was found to be 24 and 27 Å, respectively, with a decrease in the spacing with increasing acid content at a constant salt content. For instance, in a sample containing 6.3 mole % acid and neutralized to the extent of 43% (2.7 mole % salt), the spacing went down to 20 Å (cf. 27 Å above).

A similar low-angle peak was observed for all the alkali metals from Li through Cs and also for Zn, Ba, and NH_4^+. The peak persists till over 300°C and does not exhibit any orientation effects even if the polyethylene matrix is highly stretched and the PE crystallites manifest the effect of stretching quite strongly. Water, however, does show a pronounced effect, in that low levels of humidity enhance the peak, while saturating thè sample with water destroys it.

All the above observations are consistent with the hypothesis that a new phase is present in these ionomers, which incorporates the ionic units and much of the acid; such a structure was first proposed by Bonotto and Bonner [20]. If water were present in small quantities, it would be absorbed into the ionic phase, while larger quantities of water would solvate the ions and weaken their attraction for each other to the point where the cluster would fall apart.

In a subsequent investigation, Delf and MacKnight [21] obtained SAXS data for films of neutralized and unneutralized E–0.04MAA copolymers. For the sample E–0.04MAA–0.59Cs, in addition to a 20-Å peak, they observed a peak at an angle corresponding to $d_{\mathrm{Bragg}} = 83$ Å, i.e., considerably higher than in the previous study. Subsequently [22] this spacing has been ascribed to the lamellar thickness, not the intercluster distances.

Various interpretations of the origin of the "ionomer" peak, occurring in the range $2° < 2\theta < 4°$, have subsequently been presented, and as a result different conclusions regarding the state of ionic aggregation have been drawn.

Binsbergen and Kroon [23] suggested that the ionomer peak is due to a favored separation between ionic domains, consistent with the view that the ionic aggregates are randomly distributed through the material. From the data of Wilson *et al.* [18], they calculated the number of carboxyl groups per domain as a function of the copolymer content, assuming complete incorporation of carboxyl groups and also assuming that the constant factor K in Eq. (8) equals unity. Their calculated values were in the range of 5 to 11 carboxyl groups per domain. This led to the conclusion that the ionic domains responsible for the scattering maximum are consistent with the ionic multiplets described in Section A1, whose predicted maximum size is eight ion pairs.

Marx *et al.* [24] take essentially the same view regarding the origin of the ionomer peak as Binsbergen and Kroon. However, their analysis of the ionic domain size differs quantitatively from the former study. The observations by Marx *et al.* regarding the ionomer peak at $2\theta \simeq 4^\circ$ in neutralized E–MAA copolymers confirm the eariler findings of Wilson *et al.* [18] with respect to the effect of temperature and water adsorption. They conclude, as before, that the scattering is ionic in origin.

The Bragg spacing corresponding to the ionomer peak was related to an average spacing between scattering sites by Eq. (8). Marx *et al.* [24] considered several values for the parameter K. Assuming that volumeless scattering sites are arranged on a paracrystalline lattice, the value of K has been estimated [17] to be 1.22. Marx *et al.* estimated the value $K \simeq 0.9$, by comparison with a light-scattering analog—Fraunhofer diffraction. They provided a second estimate for K in the following way: From the behavior of methacrylic acid and acetic acid upon neutralization, they concluded that the carboxyl groups of the partially neutralized acids aggregate as trimers. They then related the computed average spacings between carboxyl trimers with the observed Bragg spacings, and determined that $K \simeq 0.77$ gave the best fit. They used this value to relate the Bragg spacing of the ionomers to the average separation between scattering sites, and from this obtained estimates for the number of carboxyls per scattering site, assuming complete incorporation.

Marx *et al.* concluded that the state of aggregation of carboxyl groups in ethylene ionomers varies from dimer to tetramer over a composition range of 5 to 18 wt % MAA, as shown in Table I. They made similar conclusions regarding the state of aggregation of butadiene ionomers. In the homopolymer PMAA–Na they determined an aggregation state of 7, which they found to be consistent with the predicted limit of eight for ionic clusters, assuming perfect packing. They found no evidence for aggregates of more than seven carboxyls, and indeed they suggest that the onset of clustering at 6 mole % ions seen in the styrene ionomers [16] corresponds to the transi-

TABLE I

Range of Carboxyl Group Aggregation in Ethylene Ionomers

MAA content					No. of carboxyls/scattering site $n_c = (Kd_{Bragg})^3/V'$	
(wt %)	(mole %)	Ion content	d_{Bragg} (Å)	Vol/carboxyl group $V' \times 10^{-2}$ (Å^3)	with $K = 0.77$	with $K = 1.22$
5	1.7	0.90Na	24	31.3	2	8
10	3.6	0.50Na	22	15.0	3	13
17	6.3	0.43Na	20	9.0	4	16

tion from dimer to trimer aggregates. In regard to the calculated spacings obtained by Marx *et al.*, if the value $K = 1.22$ had been chosen to estimate the mean separation between scattering sites, then considerably different values for the number of carboxyl groups per site would have been obtained, since the volume per scattering site is dependent on K^3. These alternate values of carboxyl aggregation state are also given in Table I.

Marx *et al.* also observed the effect of water absorption on the aggregate spacing. They found that the number of acid groups per aggregate increased linearly with the water content, expressed as the ratio $[H_2O]/[Na^+]$.

In other investigations, analysis of the SAXS data from cesium-neutralized ethylene ionomers utilized a different approach. Roe [25] obtained X-ray scattering data for an E–AA copolymer containing 5.35 mole % acid and its 78% neutralized cesium salt. He determined the RDFs of the acid copolymer and its salt by the Fourier-transform method. The difference in the RDFs of the salt and acid forms was taken to represent the contribution

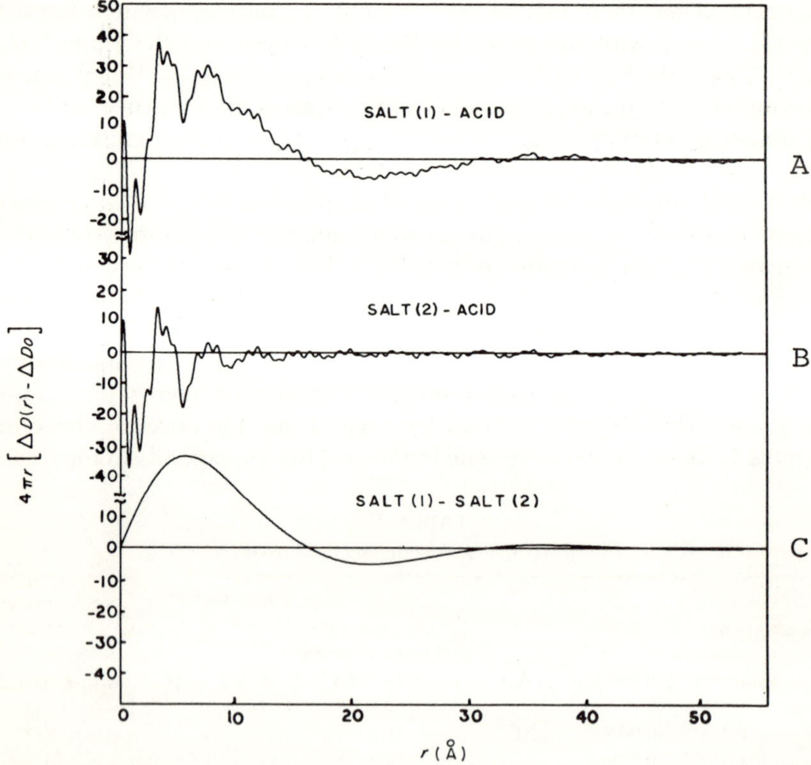

FIG. 2.6. $4\pi R[\Delta D(r) - \Delta Do]$ versus r, for the annealed cesium salt of an E–0.038MAA copolymer. Curve A includes the ionomer peak, while curve B omits this datum. Curve C represents the difference between the top curves. [26].

of the cesium atoms to the interference scattering. The difference function was similar to that shown in Fig. 2.6(b) [26], and indicated a Cs–O spacing (3.1 Å) and probable Cs–Cs spacings (4.5 and 5.0 Å). However, spacings between 6 and 16 Å were absent. From these observations and from other arguments, Roe concluded that there was no evidence for the existence of ionic clusters of ~15 Å or smaller, but that there was evidence for dimer formation.

The data employed by Roe in his RDF calculations did not include the small-angle "ionomer" peak at $2\theta \simeq 2^\circ$, as shown in Fig. 2.7. Kao *et al.* [26] followed the same method of analysis as Roe on a similar polymer: an E–MAA copolymer containing 3.8 mole % acid and its 64% neutralized cesium salt. They showed that by ignoring the ionomer peak, one obtains essentially the same difference RDF as Roe did. However, with the contribution of the ionomer peak accounted for, the difference RDF shown in Fig. 2.6(a) is obtained. This latter RDF indicates a preferred separation of cesium atoms in the 4- to 10-Å range and was taken as evidence for ionic cluster formation, since this separation is inconsistent with a uniform distribution.

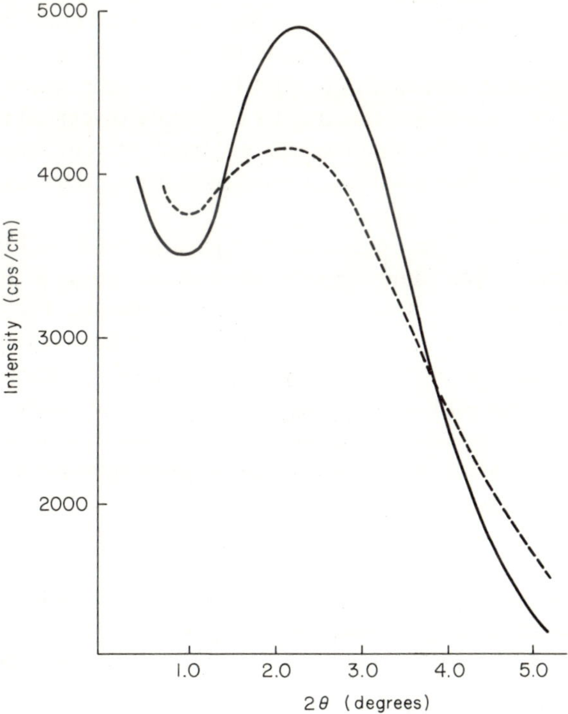

FIG. 2.7. Ionomer peak for E–0.038MAA–0.69Cs in the solid state (25°C) (solid line) and in the melt (120°C) (dashed line). [27].

Kao *et al.* tried to fit the data to a two-phase hard-sphere model corresponding to clusters of 7- to 9-Å radius. They found the best fit for clusters of 9-Å radius, containing 68 cesium atoms per cluster, although none of the models was found to reproduce adequately the detailed shape of the experimental RDF. The difference RDF due to the ionomer peak alone, shown in Fig. 2.6(c), becomes positive again at distances larger than 30 Å and is suggestive of a shallow, broad maximum at about 35 Å. This second peak may be expected to arise from extracluster contributions to the cesium RDF and can probably be identified with the ionomer peak maximum at $2\theta \simeq 2^\circ$.

In a subsequent study, MacKnight *et al.* [27] analyzed the SAXS data from a number of ethylene ionomers. In line with earlier findings [18, 19, 24], they observed a small-angle intensity maximum in each of the dry copolymer salts, while essentially no scattering occurred in the small-angle region for the unionized copolymers. The position of the peak corresponds to a Bragg spacing of either 25 or 35 Å, depending on the counterion and copolymer composition. The ionomer peak was found to persist virtually unchanged in the melt; it was unaffected by orientation, but was completely obliterated by water saturation.

The scattering profiles of the cesium ionomers were analyzed by various SAXS techniques to obtain an estimate of the size of the scattering sites. They found values of the order of 8–10 Å for the Guinier radius of gyration, assuming that the scattering was due to randomly dispersed ionic clusters. The peak maximum was assumed to arise not from intercluster contributions, but from the interference resulting from short-range ordering of matrix ions around a central cluster.

MacKnight *et al.* [27] proposed the model illustrated in Fig. 2.8 for the ionomer structure. The ionic clusters, of average radius 8–10 Å, are surrounded primarily by a shell of hydrocarbon chains. Presumably nonclustered ions would be attracted from the surrounding matrix, and a preferred distance, corresponding to $20\ \text{Å} < d_{\text{Bragg}} < 35\ \text{Å}$, would be established.

No direct estimate of the intercluster separation was made; however, the data were found to be consistent with a volume fraction of clusters of 0.02 to 0.05. The corresponding range of intercluster separations depends, of course, on the fraction of ions incorporated in clusters. For 50% incorporation and a cluster radius of 10 Å, intercluster separations of 70–90 Å are obtained.

Since the interpretations of ethylene ionomer X-ray data are at variance, both qualitatively and quantitatively, the conflict remains to be resolved. On the whole, the evidence presented by MacKnight and coworkers, involving the computation of RDFs, appears to be based on a firmer foundation than alternate explanations. It is interesting to note that if the shallow

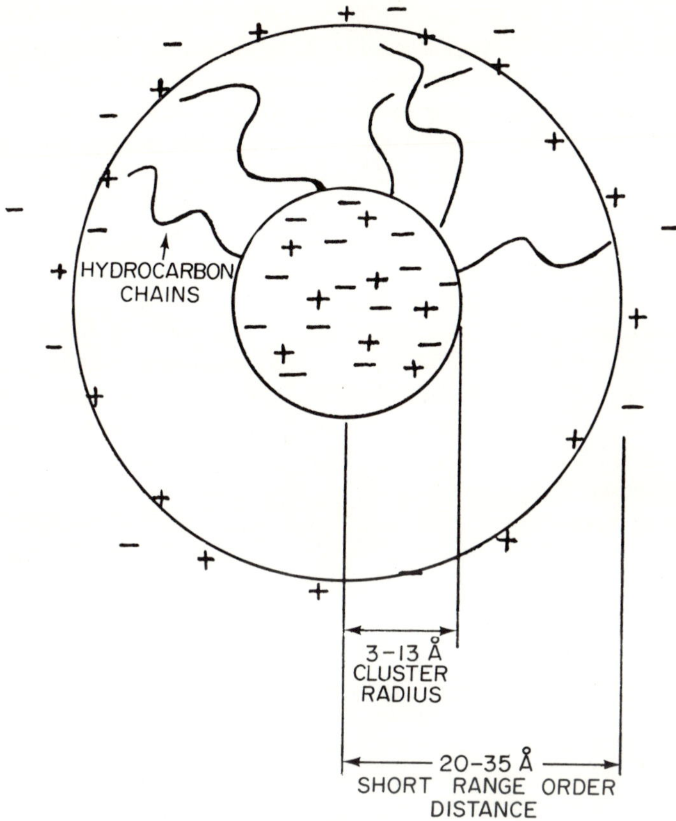

FIG. 2.8. Ionic cluster model proposed by MacKnight *et al.* [27].

maximum at about 35 Å in Fig. 2.6C corresponds to the ionomer peak maximum, which for the particular ionomer studied corresponds to a Bragg distance of 35 Å, this would suggest that the value of K in Eq. (8) is very close to 1.0. If a more accurate RDF could be obtained for this region, it would provide the most direct determination of K.

X-ray evidence for ion clustering has been seen in other ionomer series as well. Eisenberg and Navratil [16,28] found SAXS peaks in styrene–(cesium methacrylate) copolymers containing more than 6 mole % of the ionic comonomer. For instance, in a sample containing 7.9 mole % CsMA, a peak at $2\theta \simeq 1.5°$, corresponding to a Bragg repeat distance of $\sim$60 Å, was found. Similar spacings were observed in other S–MAA–Cs copolymers above 6 mole % (Fig. 2.9) and were totally absent in the unionized material or pure polystyrene. In view of the high electron density of cesium, these

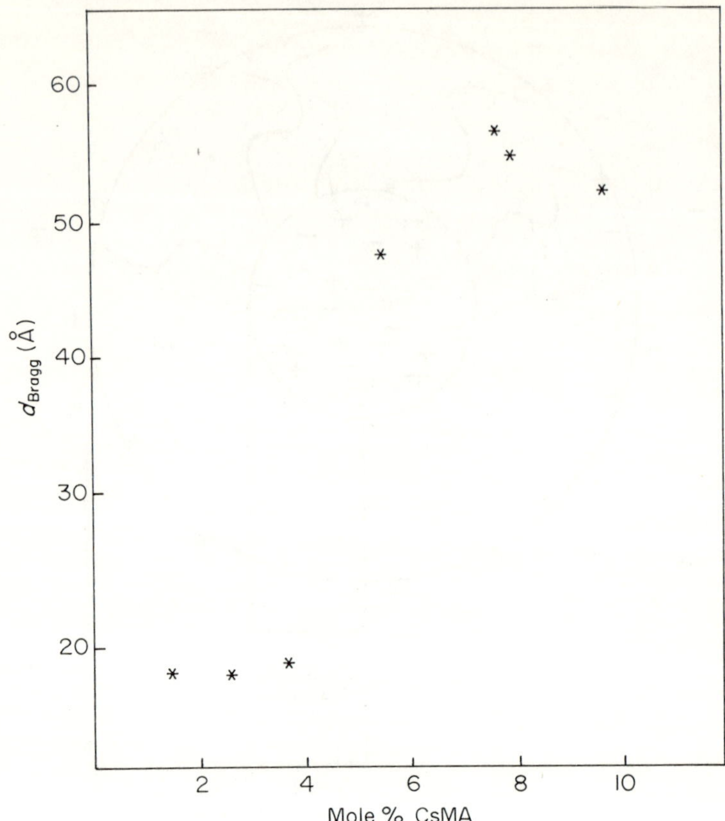

FIG. 2.9. X-ray scattering maxima for S–MAA–Cs ionomers. [28].

low-angle peaks in a material devoid of crystallinity can only be rationalized on the basis of ionic cluster formation.

Below 6 mole % CsMA, Bragg spacings of ~19 Å were seen in the S–MAA–Cs copolymers. These results can possibly be interpreted in terms of simple multiplet formation, as opposed to clustering. The only other peaks in the scattering intensity for these materials were those corresponding to $d_{Bragg} = 4.8$ and 9.3 Å ($2\theta \simeq 18.5°$ and $9.5°$, respectively), which were seen in both the ionized and unionized copolymers; these smaller spacings were assumed to be characteristic of polystyrene and not of any internal arrangement within the ionic domains.

Pineri *et al.* [14] found evidence for ionic clustering in butadiene-(methacrylic acid) ionomers, as well as in carboxy-terminated butadienes. In a low-molecular-weight butadiene polymer ($\overline{M}_n = 4.4 \times 10^3$) containing

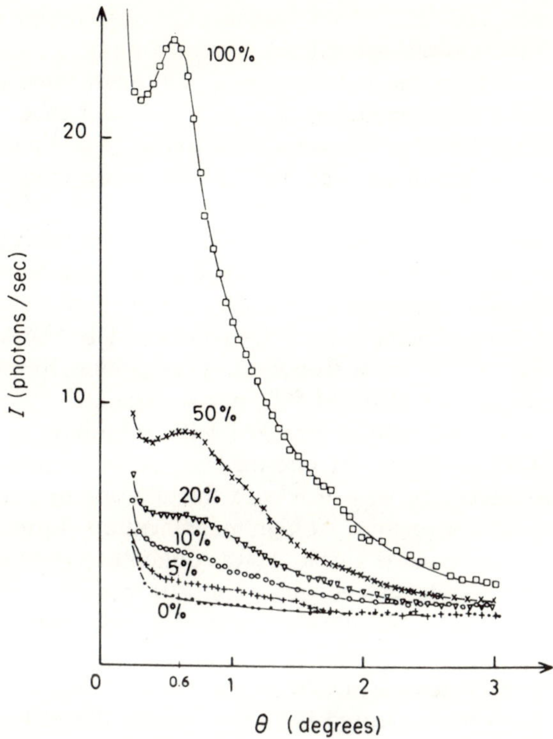

FIG. 2.10 Plots of I (photons/sec) versus θ for carboxy-terminated butadiene neutralized with manganese acetate. [14].

terminal acid groups (2.2 mole %), it was observed (Fig. 2.10) that the SAXS peak at $2\theta = 1.2^\circ$, corresponding to $d_{\text{Bragg}} \simeq 70$ Å, increased in intensity as the degree of neutralization rose, but did not change position. From the shapes of the scattering profiles above $2\theta = 1.2^\circ$, they determined an average cluster size at each degree of neutralization, as well as the total cluster volume. They found that the total volume of clusters was proportional to the degree of neutralization, and that above 15% ionization, the cluster size was approximately constant ($\bar{R} = 5.7$ Å). Since the cluster separation was also constant, they concluded that each sample was a mixture of two distinct macrophases: the unneutralized fraction of the polymer on the one hand and very large domains of the neutralized polymer containing smaller, nearly equidistant ionic clusters on the other.

In the previously mentioned study by Marx *et al.* [24], the X-ray scattering patterns of B–MAA–Na ionomers containing 5–12 mole % MAA were analyzed. In each case, Bragg spacings of the order of 20–25 Å were observed, while the scattered intensity at low angles (down to $2\theta \simeq 0.4^\circ$) was devoid of maxima. The 20- to 25-Å spacings are very similar to those seen in the lithium- and sodium-neutralized styrene ionomers containing more than 6 mole % ions. In the latter material, although ionic clusters were clearly indicated by other evidence, no spacings larger than ~20 Å were seen in the X-ray scattering. Thus, no conclusions regarding the existence of ionic clusters in butadiene ionomers can be made from the study of Marx *et al.*

Electron microscopy is another source of information concerning clustering, although the interpretation of results is even more difficult than in the case of SAXS. Several studies on ionomers have been performed. In an early example of this work, Davis *et al.* [29] investigated E–MAA copolymers over a wide range of variables. Copolymers containing 10 mole % of the acid were embedded in a blend of 50% polystyrene and 50% poly(methyl methacrylate), sectioned, and neutralized with rubidium by floating the sections on RbOH solutions. At intermediate magnification (1 μm = 12 mm), a granular structure was seen with fluctuations in electron density extending over ~300-Å regions. At high magnification (1 μm = 50 mm), a fine-grained structure became visible. It was difficult to interpret, however, because it was only visible with adequate contrast when the image itself was out of focus. It is conceivable that the grains may have been the individual clusters.

Very recently, an extensive review of the electron microscopy work of the Du Pont group has appeared [30]. This review illustrates very clearly the profound effect of ions on the properties of polymers. It is shown that acid samples yield photomicrographs which are very similar to those of low-density polyethylene. By contrast, the neutralized copolymers show the presence of grains of 100 to 1000 Å in size, accompanied by the destruction of the spherulitic structure of the acid copolymer. The granular structure seems to be well developed already at ~3% of salt. The structure appears both in sectioned bulk ionomers and in samples prepared by casting dilute acid copolymer from solution onto aqueous CsOH or other bases.

It should be noted at this time that MacKnight [31] has found that the occurrence of the graining structure depends on the detailed method of preparation of the samples. On the one hand, this find may indicate that the grains do not represent the individual clusters; on the other hand, it may mean that the degree of development of the clusters depends on sample history. The latter possibility seems much more likely.

In a subsequent investigation of B–MAA copolymers, Marx *et al.* [32] prepared samples (containing 7 to 18 wt % acid) by neutralizing the copolymer with sodium methoxide, depositing a few drops of the polymer solution on mercury, and, after evaporation, placing the sample on an electron microscope grid. After thorough drying, the samples were stained

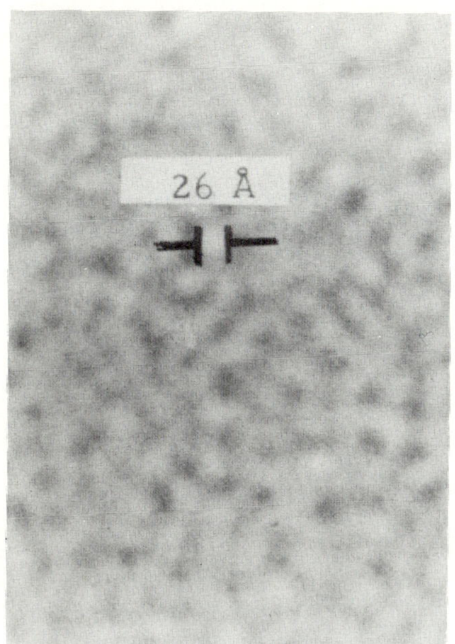

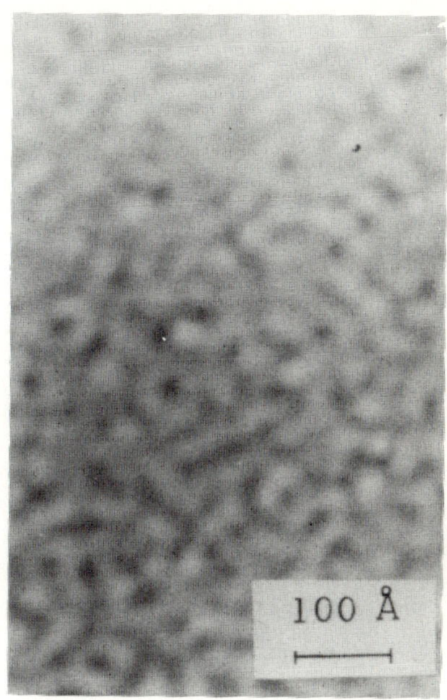

FIG. 2.11. Electron micrographs of two B–MAA copolymers completely neutralized with sodium: (*left*) 7 wt % acid, (*right*) 18 wt % acid. [32].

with osmium tetroxide, which becomes incorporated into the diene. A typical result is shown in Fig. 2.11 for samples containing 7 and 18% of the acid comonomer. Heterogeneities of the order of 25 Å are clearly visible, although in the same sample, as little as 1 μm apart, much smaller periodicities were encountered. Two periodicities were found in all the samples; in the 18% sample they were of the order of 25 and 13 Å, the other samples giving similar results. Although the clustering mechanism may be somewhat different in the butadiene system than in the ethylene- or styrene-based ionomers, the work illustrates clearly that heterogeneities in ion-containing polymers can be observed by electron microscopy.

In a recent study, Pineri *et al.* [33a] investigated by electron microscopy the structure of a complexed terpolymer of butadiene, styrene, and 4-vinylpyridine (VP). The starting terpolymer, B–0.10S–0.048VP, was complexed in benzene solution by mixing with methanolic $FeCl_3$, giving samples of different iron concentrations. Electron microscopy showed the existence of heterogeneously distributed iron-rich regions, which varied in size over a wide range from less than 100 Å to more than 1000 Å. This can be seen in Fig. 2.12, for the sample containing 1.4 pyridinium groups per iron atom. The iron-containing domains were found to be noncrystalline.

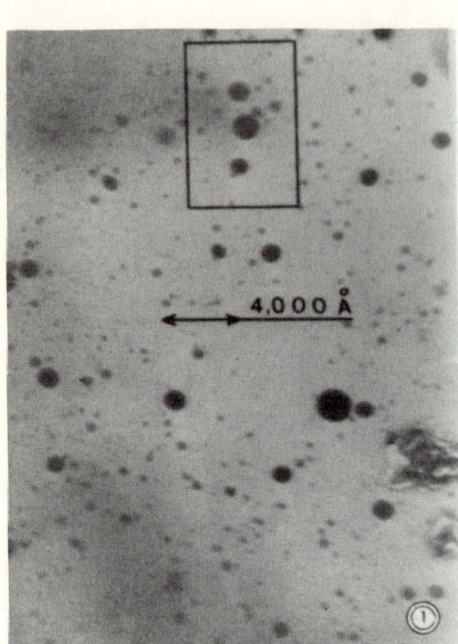

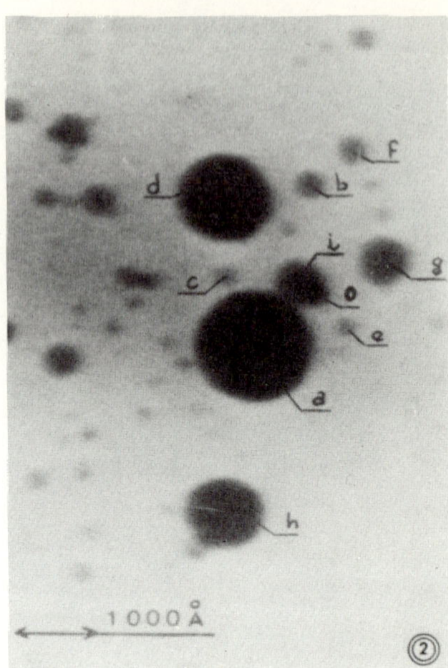

FIG. 2.12. Electron micrographs of terpolymer B–0.10S–0.048VP containing 1.4 pyridinium groups per iron atom. [33].

A very recent Mössbauer study of the terpolymer discussed above, coupled with magnetic susceptibility measurements [33b], gave some very detailed information on the types of aggregates encountered in these materials. Iron atoms in three distinct environments were detected. The first corresponds to magnetically ordered iron in fine grains. At room temperature, the Mössbauer spectrum shows a doublet that is very large and asymmetric, suggesting that the environments of the iron atoms in these clusters differ appreciably. When the temperature is lowered, a six-line spectrum appears, suggesting that the grains are paramagnetic with very small distances between iron atoms. The paramagnetic character was confirmed by thermoremanence and by obedience to the Langevin law in magnetic susceptibility measurements. From the behavior of the Mössbauer spectrum upon application of an external magnetic field, it was shown that the structure of these clusters is speromagnetic, i.e., that it involves an ordered spin structure with a statistical distribution of spin directions. This is consistent with the amorphous structure of the clusters as seen in electron diffraction. For the fully complexed polymer, $\sim 60\%$ of the iron atoms are in clusters of this type, and half of these clusters have radii of ~ 8 Å, involving about thirty iron atoms. From these data, intercluster distances of ~ 60 Å were calculated. Neutron scattering yields a value of 80 Å.

The second type of aggregate is deduced from the appearance of a doublet at $T < T_g$. The very large value of the quadrupole splitting suggests the existence of different types of ligands (including possibly Cl and H_2O) on the Fe. From the behavior of the spectrum upon application of an external magnetic field, it was shown that the structure is a dimer, involving a pair of octahedra sharing one edge, the total spin of which is zero. About 20–30% of all the iron is in structures of this type.

The final type of species is suggested by the appearance of a doublet at very low temperatures. The iron atoms here may be in isolated complexes with no coupling between them. About 10–20% of the metal is incorporated in these structures.

The remaining evidence concerning cluster formation comes from an analysis of rheological properties. As mentioned before, these will be discussed in detail in Chapters III and IV, so the presentation of the material here will be very brief.

A comparison of the dynamic mechanical behavior of E–MAA copolymers and their salts is most convincingly interpreted in terms of microphase separation or clustering. A polymer containing 4.1 mole % carboxylic acid groups was studied extensively by MacKnight *et al.* [34], who found that upon neutralization a tan δ peak at $\sim$25°C (β') disappears and is replaced by two others, one at $-10°$ (β) and the other between 50° and 100°C (α). While the β relaxation is not dependent on the nature or concentration of the cations (it should be mentioned, however, that Otocka and Kwei [35] found a slight dependence), the α relaxation shows a strong dependence both on their type and quantity, suggesting that it is due to changes occurring in regions containing the ionic material. By contrast, the relative insensitivity of the β relaxation to changes in ion content or type suggests that it is due to motions occurring in regions containing predominantly nonionic material. The coexistence of these two relaxations, in turn, reinforces the idea that the regions in which these two types of materials are encountered are separate, suggesting the validity of the concept of microphase separation. An alternate interpretation based on other findings for similar materials [35] will be given later.

Melt rheology studies were performed by Sakamoto *et al.* [36] on E–MAA copolymers containing 4.1 mole % of the acid above the crystalline melting point of polyethylene. The results indicate that for the acid there exists a good correlation between the frequency dependence of the complex viscosity and the shear-rate dependence of the apparent viscosity; also, time–temperature superposition is valid for G' values as a function of frequency for different temperatures. By contrast, this is not the case for either the sodium or the calcium copolymers. These results are again consistent with the occurrence of phase separation involving the ionic component, since simple cross-links, even labile ones, would not be expected to lead to the observed results.

In a recent study of S–MAA–Na copolymers as a function of ion concentration [16], several lines of evidence were found which strongly support the idea of clustering or microphase separation. The behavior of these materials below 6 mole % ions was already discussed. Above 6%, it was found that time–temperature superposition is no longer applicable. Failure of time–temperature superposition in noncrystalline, thermally stable polymers can arise from only very few causes, such as side-chain relaxation [37], a chemical bond-interchange reaction along the backbone [38], or microphase separation [39] as it is found, for instance, in block copolymers of styrene and butadiene. In the latter case, failure of superposition is due to the fact that the polystyrene, which is the dispersed phase, provides an added compliance which changes the shape of the compliance or modulus curve for the rubbery phase. Due to the different temperature dependence of the two contributions, time–temperature superposition fails.

Exactly the same interpretation can be given to the S–MAA–Na copolymers above 6 mole % ions. Clustering is further reinforced by small-angle X-ray diffraction data which yield a peak corresponding to a Bragg distance of ~60 Å, in qualitative agreement with a theory to be described below. Also, water uptake experiments indicate that below the clustering region each sodium ion is capable of accepting only one water molecule, whereas above that region more than one molecule can be absorbed; for these samples equilibrium is not achieved even after 5000 hr for prisms of ~1 mm thickness, and by that time 3 to 6 water molecules per sodium atom have been absorbed. Finally, dynamic mechanical studies (~1 Hz) show two tan δ peaks, one at ~130°C and another at ~200°C. Both are quite high in intensity and are probably due to glass transitions. The lower T_g is undoubtedly caused by a polystyrene phase containing some ionic material, while the higher one is due to the clustered regions, containing most of the ions but also some of the hydrocarbon material. All these results point strongly toward clustering in these systems.

In still another system, Tobolsky *et al.* [40] found that the viscoelastic behavior of carboxyl-containing rubbers could best be explained by postulating the existence of hard ionic clusters. These phenomena will be discussed more fully in Chapter IV.

Finally, the trend in T_g versus ion content can also be interpreted as favoring the cluster hypothesis, although less directly. Matsuura and Eisenberg [41], in a study of the salts of (ethyl acrylate)–(acrylic acid) copolymers, which will be discussed in more detail below, showed that while the initial rise in T_g with ion content is linear, above ~10% the rate of increase accelerates. This accelerated increase has been interpreted as reflecting the onset of clustering.

The evidence for clustering is by no means unanimous, however. Three lines of evidence can be cited in favor of a homogeneous distribution of ions, perhaps in the form of quartets but not of much higher aggregates.

Otocka and Kwei [35], in contrast to the findings of MacKnight *et al.* [34], found that both the position and the height of the noncrystalline β peak in ethylene ionomers are dependent on the nature and concentration of the cations. They ascribed the β peak to the glass transition of amorphous PE containing unclustered ionic groups which act as cross-links; then, by estimating the T_g of poly(sodium acrylate) as 230°C and by using an empirical relation for the glass transition as a function of copolymer composition, they obtained good agrement between the experimental and calculated values of T_β over a 12-degree range. From this they conclude that the ions are dispersed on a molecular level.

In a broad-line NMR study of the same copolymers, Otocka and Davis [42] investigated the line narrowing as a function of temperature of both the ^{7}Li and ^{1}H nuclei. They found that the decrease in the proton line width occurred at only a slightly lower temperature than that of the ^{7}Li, with only a slight residual Li linewidth remaining after the disappearance of the proton line as shown in Fig. 2.13. From this they again concluded that most of the ions are dispersed on a molecular level.

Finally, the X-ray studies of Roe [25] and of Marx *et al.* [24], which were discussed in detail earlier, also provide evidence against clustering, although the former study has been superseded by the works of Kao *et al.* [26] and MacKnight *et al.* [27].

As can be seen from the preceding summary of experimental results, the problem of clustering is most complex. It is clear that at low ion concentration, only multiplet formation occurs, the multiplets acting simply as cross-links. However, beyond a certain ion concentration (which depends on the material), the multiplets may very well aggregate to form clusters, the detailed structure of which is most probably dependent on the sample history.

On balance, the evidence in favor of clusters seems stronger than the contrary evidence. On the positive side, X-ray diffraction peaks indicating Bragg distances between 20 and 80 Å have been found for various materials. This is perhaps the strongest bit of evidence. Results of electron microscopy also point to microphase separation, although here the total evidence is somewhat weaker. Rheological studies support clustering most strongly, so that in conjunction with the preceding results, the existence of some type of clustering or microphase separation is highly probable.

With regard to the negative evidence, the glass transition results of Otocka and Kwei [35] do not exclude the possibility of clustering. The fact that the copolymerization equation was found applicable only proves that *some*

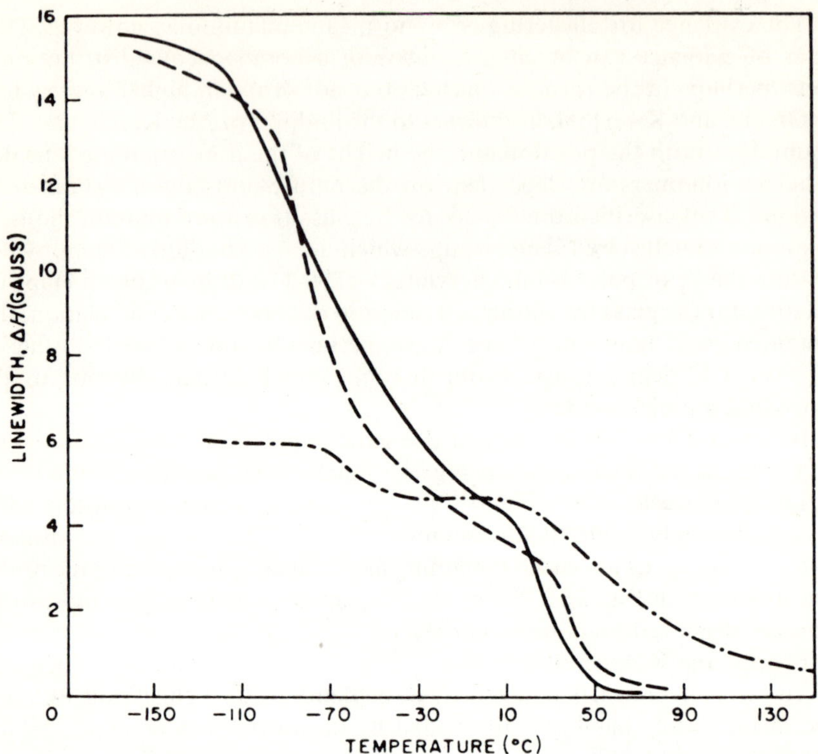

FIG. 2.13. Linewidth in wide-line NMR: (——) ^1H in E–0.49 AA, unneutralized (60 MHz); (– – –) ^1H in E–0.49AA–Li (56.5 MHz); (– · – ·) ^7Li in E–0.49AA–Li. [42].

ionic species are present in the hydrocarbons; a wide range of copolymerization equations could have been used, some of which would undoubtedly have given much worse agreement, and it is by no means clear that the particular equation that was used was actually the best for the system from independent considerations.

The interpretation of the NMR results [42] is also open to question, since the residual linewidth for the ^7Li nuclei may be much more significant than it was assumed to be. The data are very qualitative, and if one is sympathetic to the evidence for clusters, one might suggest that the difference between the ^1H and ^7Li widths confirms that a good fraction of the metal is, indeed, in cluster form since otherwise the relative linewidth due to ^7Li would parallel exactly that for ^1H. This interpretation was also suggested by Longworth [30].

Finally, the Fourier-transform approach [25] does prove, very convincingly, that no clusters exist with repeat distances below ~16 Å. However, all the other X-ray evidence suggests that the Bragg spacings are anywhere between 20 and 80 Å. Thus, on balance, the evidence favoring clustering seems to be stronger than the evidence against it. In the next section, a theoretical approach to clustering will be presented.

For the sake of completeness, the list below gives the approximate ion concentrations above which clustering is apparent. The silicates are included, although it should be noted that in that system the phenomenon has been referred to as spinodal decomposition. In this inorganic system, bond interchange is possible, which facilitates more extensive phase separation.

Material	Critical ion concentration (mole %)
PE	1
PS	4–6
PEA	10–15
PAA with 20 wt% formamide	10–20
Na_2O–$xSiO_2$	≈20
Na_2O–xP_2O_5	Unclustered

(*b*) *A Theory of Clustering*

In view of the strong evidence supporting the existence of ionic clusters, an attempt was made to explore cluster formation from a semiquantitative theoretical point of view [4]. In this theory, it was taken that multiplets, primarily for electrostatic reasons, can aggregate to form clusters. The further assumptions of the theory are as follows:

1. In forming a cluster, work is done to stretch the polymer chains (i.e., segments between ionic groups) from the distance corresponding to the existence of only primary multiplets R_0 to that corresponding to a cluster of size n (ion pairs). It is assumed that the chain segments connect only nearest-neighbor multiplets. The equation for distances between multiplets is also applicable to distances between clusters, and the work of chain stretching can be calculated from the theory of rubber elasticity.
2. Electrostatic energy is released when multiplets aggregate to form a cluster, the amount depending on the geometry of the cluster and the dielectric constant of the medium.
3. The cluster is not infinitely stable. At some temperature T_c the cluster decomposes, and at that temperature the elastic forces and the electrostatic forces just balance each other.
4. In contrast to the multiplet case, it is assumed here that half of all sequential ion pairs incorporated into the same cluster do yield "rings." This assumption is based on the reasoning that ring formation is expected

to occur for only one of the two chains emanating from an ion pair on the surface of a cluster, and that for the most probable cluster sizes, the majority of ion pairs are in contact with the surface. The factor of $\frac{1}{2}$ is arbitrary; it may well be somewhat larger or smaller. Chainging this factor from $\frac{1}{4}$ to $\frac{3}{4}$ does not affect the results appreciably. Those chains that do form rings contract, while those that do not experience an extension.

First, an attempt is made to calculate the elastic forces involved in cluster formation. If the fundamental multiplets consist of two or more ion pairs, it is evident that they will act as cross-links. If these multiplets now form clusters, then, as was pointed out above, some chains will be extended while others will contract, unless one postulates a very rapid reorganization within the multiplet and subsequent "ring" formation, i.e., the incorporation of sequential ion pairs within the same multiplet. Otocka *et al.* [10] found no evidence for rapid quartet reorganization around room temperature, which means that elastic forces must play a significant part in cluster formation, at least for multiplets of two or more ion pairs.

The elastic force f acting on the two ends of a polymer chain is given by

$$f = 3kTh/\overline{h^2} \tag{9}$$

where $\overline{h^2}$ is the mean-square end-to-end distance for the free chain and h the actual separation of the ends. The work of stretching a polymer chain from a distance R_0 to R, the distance corresponding to a cluster size of n, is

$$W = \int_{R_0}^{R} f\, dh = \int_{R_0}^{R} (3kT/\overline{h^2})h\, dh = (3kT/2\overline{h^2})(R^2 - R_0{}^2) \tag{10}$$

where R as a function of n is given by Eq. (5). The work of contracting a chain from a distance of R_0 to 0 is

$$W = \int_{R_0}^{0} f\, dh = \int_{R_0}^{0} (3kT/\overline{h^2})h\, dh = -(3kT/2\overline{h^2})R_0{}^2 \tag{11}$$

Since, as was mentioned before, it is taken that half the chains expand from R_0 to R while the other half contract from R_0 to 0, the net work per chain done upon cluster formation, W_{ch}, is simply the average of the two expressions for expansion and contraction, i.e.,

$$W_{\mathrm{ch}} = (3kT/4\overline{h^2})(R^2 - 2R_0{}^2) \tag{12}$$

It is convenient to express $\overline{h^2}$ in terms of the mean-square end-to-end distance of the equivalent freely rotating chain, $\overline{h^2_{\mathrm{fr}}}$, i.e.,

$$\overline{h^2} = \sigma^2 \overline{h^2_{\mathrm{fr}}} \tag{13}$$

σ is the chain expansion factor, and for vinyl polymers, $\overline{h_{\text{fr}}^2} = 4l^2 M_c/M_0$, where M_c/M_0 is the degree of polymerization between ionic groups and l is the C—C bond length. Substitution of this relationship into Eq. (12) yields

$$W_{\text{ch}} = (3kTM_0/16\sigma^2 l^2 M_c)(R^2 - 2R_0{}^2) \tag{14}$$

Finally, inserting the values for R from Eq. (5), one obtains

$$W_{\text{ch}} = (3kTM_0/16\sigma^2 l^2 M_c)[(nM_c/\rho N_{\text{av}})^{2/3} - 2(n_0 M_c/\rho N_{\text{av}})^{2/3}] \tag{15}$$

The electrostatic energy per ion pair released upon cluster collapse is, as was mentioned before, a function of the geometry of the cluster. Several geometries were considered, and the results are described below. However, for the present, the energy is expressed in the most general terms as the fraction λ of the energy released upon formation of an ion pair from isolated ions for the particular system under consideration. Thus, by comparison with Eq. (1), the electrostatic energy per ion pair upon cluster formation is

$$W' = -\lambda e^2/\varepsilon r \tag{16}$$

At some temperature T_c the cluster becomes thermodynamically unstable, and at that temperature the elastic force and the electrostatic force just balance each other. Since the elastic force was calculated per chain and the electrostatic force per ion pair W_{ch} can be set equal to W' at T_c, thus

$$\lambda e^2/\varepsilon r = (3kT_c M_0/16\sigma^2 l^2 M_c)[(nM_c/\rho N_{\text{av}})^{2/3} - 2(n_0 M_c/\rho N_{\text{av}})^{2/3}] \tag{17}$$

This equation can now be solved for n. The result is

$$n = (\rho N_{\text{av}}/M_c)[(16\sigma^2 l^2 M_c \lambda e^2/3kT_c M_0 \varepsilon r) + 2(n_0 M_c/\rho N_{\text{av}})^{2/3}]^{3/2} \tag{18}$$

Since T_c can, in principle, be determined experimentally, and λ calculated for any particular cluster geometry, n and R can be calculated. The results of such a procedure will be described below. Also, several predictions can be made about the relationship of n and R to various experimental parameters such as the spacing of ions along the chain or the dielectric constant.

In an attempt to apply the theory to real systems, three rather arbitrary cluster geometries were assumed and the corresponding values of n and R calculated. The first model, perhaps the most artificial, assumes that all the ions exist as pairs, not as higher multiplets. It assumes further that the ion pairs are distributed on a cubic lattice, head to tail in any one row, and antiparallel between neighboring rows and that they are separated from each other by the thickness of a polymer chain. The cluster consists of a region of space, which is occupied by this cubic lattice of ion pairs. The second assumes that some of the ions are present in the form of octets, while others exist as pairs, the pairs acting as electrostatic ties between octets. The third postulates that the cluster is made up of spherical multiplets, most of which possess an extra positive or negative charge. The multiplets are therefore attracted to each other by electrostatic forces, and the cluster consists of several of these multiplets, the surfaces of which are separated from each other by at least two chain thicknesses.

The results for these three examples as applied to ethylene ionomers containing 4.5 mole % sodium methacrylate are: For first case, $\lambda = 0.020$, $n = 160$, and $R = 55$ Å; for the second case, $\lambda = 0.063$, $n = 800$, and $R = 95$ Å, while for the third case, $\lambda = 0.008$, $n = 82$, and $R = 44$ Å. While these results are encouraging, it should be stressed that this theory is highly approximate and should be regarded merely as demonstrating feasibility rather than predicting true cluster geometries. Many factors are neglected (such as crystallinity); others are oversimplified (such as the assumption of a unique T_c rather than a temperature range or the inclusion of all ionic groups into the clusters), but the theory does demonstrate that for three widely divergent cluster geometries, reasonable values of R are obtained. It should be pointed out that if this theory is applied to the data for the polystyrene ionomers [28], the result for λ is ~ 0.05, entirely within the range calculated for the ethylene ionomers.

The theory points out one additional factor: At low ion concentration, as the distance between multiplets increases, the interaction between multiplets is expected to decrease to the point where the elastic forces become too large to overcome; below that concentration clustering would not be expected. While this point has not, as yet, been observed in ethylene ionomers, it seems to be encountered in styrene ionomers, at ~ 6 mole % acid, as the rheological data indicate. The concentration at which clustering sets in is dependent in part on the dielectric constant of the parent polymer. With a low dielectric constant, as in PE, clustering is evident at the lowest ion concentrations studied (~ 1 mole %). In PS, clustering begins at 6 mole %, with poly(ethyl acrylate) at $\sim 12\%$. In the polyphosphates, clustering is not seen at all.

Much more recently, Ponomarev and Ionova [43] developed a mathematically highly complex theory of clustering; in its present form, the theory does not lend itself to an easy interpretation as to how the various molecular parameters of the system influence the clustering phenomenon, i.e., how cluster size, separation, volume, etc., are related to the ionic size and charge. It is to be hoped that future developments of the theory will supply this, so that the predictions of the theory can be tested experimentally for a range of materials.

B. GLASS TRANSITIONS

This portion of the book will deal with the effect of the incorporation of ions on the glass-transition temperature. The presentation will be divided into three parts. The first will deal with the effect of incorporating a small concentration of ionic groups into the polymer, either by a process of copolymerization or by dissolving low-molecular-weight salts in polar polymers. The second part will review the phenomena encountered in salts based on completely ionizable polymers, specifically the polyphosphate, silicate, and acrylate systems. Finally, the glass-transition relations found in the aliphatic ionenes will be described.

1. Effect of Ions at Low Concentrations

(*a*) *Inorganic Salts with Polar Polymers*

The effect of polymer–salt interactions on the thermomechanical properties of nonionic polymers has been explored in several studies. Moacanin and Cuddihy [44] measured the internal friction in poly(propylene oxide)—PPO—containing added lithium perchlorate and took the maximum of the damping peak as the glass transition. $LiClO_4$, because of its low lattice energy, dissolves in the polyether and changes a number of properties, among them the glass transition. They observed that at low salt concentrations (up to 16 wt %), the T_g of the solution rises linearly with increasing ion concentration; beyond ~20 wt % $LiClO_4$, the effect of the salt is stronger; i.e., T_g increases even more sharply. This study, which will be discussed in greater detail in Section B1 of Chapter III, indicates that as a result of the strong ion–dipole interactions present in the system, the average segmental mobility is lowered and the glass transition raised over what it would be in the absence of salt.

Hannon and Wissbrun [45] studied another system in which polymer–salt interactions modify the glass transition and the mechanical properties. Calcium thiocyanate, which has appreciable solubility in "Phenoxy," a linear condensation product of 2,2-bis(4-hydroxyl)-propane and 1-chloro-2,3-epoxypropane, was also found to raise the glass-transition temperature of polymer–salt mixtures beyond that of the pure polymer. As before, they found that T_g of the solution increases linearly with added salt concentration up to 10 wt % $Ca(SCN)_2$. At higher concentrations, the rate of increase falls (in contrast to the findings of Moacanin and Cuddihy), and the measured values by DSC and thermal expansion methods diverge considerably. Hannon and Wissbrun, again in contrast to the observations by Moacanin and Cuddihy on the PPO–$LiClO_4$ system, found no negative deviations from volume additivity in Phenoxy–$Ca(SCN)_2$ mixtures. This indicates that the strength of the polymer–salt interaction in the latter case is considerably less and accounts for the smaller effect on T_g (1.8°/mole %, as opposed to 5.5°/mole %) observed by Hannon and Wissbrun.

These latter authors also investigated polymer–salt interactions in a number of other systems [46]. In poly(vinyl acetate), poly(methyl acrylate), and poly(methyl methacrylate) systems containing bivalent metal nitrates (Ca, Cd, Cu, and Zn), appreciable differences were observed in the T_g behavior for different ions. In most cases, the presence of salt increased T_g, but in some instances, most notably with PMMA–$Zn(NO_3)_2$, T_g was significantly decreased upon addition of salt. The authors followed the structural changes in the polymer–salt systems by infrared spectroscopy and proposed a model for the structure of the polymer–salt complex, the specific nature of which is dependent on the particular salt–polymer system involved. Common to all of the structures they observed were the metal ion, with its associated ligands (in this case NO_3^-), the polar macromolecular moeity (polyester), and a tightly bound solvent molecule (usually H_2O). An example of the proposed structure is shown in Fig. 2.14.

Very recently, Wetton *et al.* [47] investigated the glass-transition behavior of poly(propylene oxide) of high molecular weight and of poly(propylene glycol) of low molecular weight with added $ZnCl_2$ or $CoCl_2$. $ZnCl_2$ shows the same effect on both sets of samples, i.e., an initially slow nonlinear increase, with a subsequently more rapid linear rise, at the rate of ~4.0°/mole % between ~10 and ~30 mole % $ZnCl_2$. Zn(II) favors a tetrahedral coordination involving the sp^3 hybrid; since two of the coordination sites

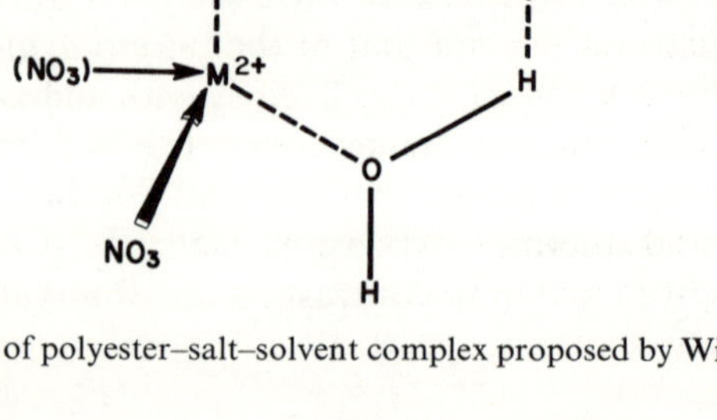

FIG. 2.14. Structure of polyester–salt–solvent complex proposed by Wissbrun and Hannon. [46].

are already occupied by chlorines, the other two are taken up by vicinal oxygens on the polyether. This arrangement, incidentally, leads to an unstrained ring involving part of the chain and the Zn(II) ion. Clearly, the T_g of this species will be higher than that of the uncoordinated polymer. The T_g behavior over the entire concentration range studied could be expressed by an equation for a copolymer, i.e.,

$$T_g = T_g(2) + \phi_1[T_g(1) - T_g(2)] \tag{19}$$

where ϕ_1 is the mole fraction of coordinated units, and $T_g(1)$ and $T_g(2)$ are the glass transitions of the coordinated and uncoordinated polymer, respectively. The best fit is obtained with $T_g(1) = 44°C$ and $T_g(2) \simeq -65°C$.

The situation with $CoCl_2$ differs appreciably from the preceding. The polymer of high molecular weight is unaffected by the presence of the salt, presumably because of the impossibility of finding a mutual solvent. For the polymer of low molecular weight, initially the T_g is found to increase more slowly with added salt, but at ~18 mole % of the metal chloride, a very rapid rise is observed. The more rapid increase may possibly be associated with sp^3d^2 hybridization of the cobalt leading to both intra- and intermolecular complex formation. The latter would be equivalent to cross-link formation.

Two other polymers were investigated in this work, although less extensively. It was found that poly(ethylene oxide) behaves similarly to PPO, but the high degree of crystallinity of the PEO complicates the behavior somewhat. By contrast, poly(tetramethylene glycol) shows only a very slight increase in T_g with added salt content, probably because of the inability of that material to form complexes with the salt involving vicinal oxygens. Such a complex would involve a seven-membered ring, which apparently is much less stable than those formed from poly(propylene glycol) or poly(ethylene glycol).

In the poly(acrylonitrile)–$Fe(ClO_4)_2 \cdot 6H_2O$ system [48], and in a number of similar mixtures of PAN and hydrated metal perchlorates [49], Reich and Michaeli observed a decreased T_g with increasing concentration of added salt. Their T_g data are illustrated in Fig. 2.15. In this case, the salt–water complex appears to act merely as a plasticizer; the salt mobility assumes values typical of those seen in normal liquids [50]. These studies illustrate the fact that an inorganic salt can dissolve in a polar polymer without strong interaction and without causing decreased polymer mobility.

The fact that ionic species can either raise or lower the glass transition of a polymer (depending on the size of the ions) can be explained on the following basis: In general, the introduction of a low-molecular-weight species could be expected to lower T_g as a result of the well-explored plasticization effect, which is due among other things to an increase in the free volume.

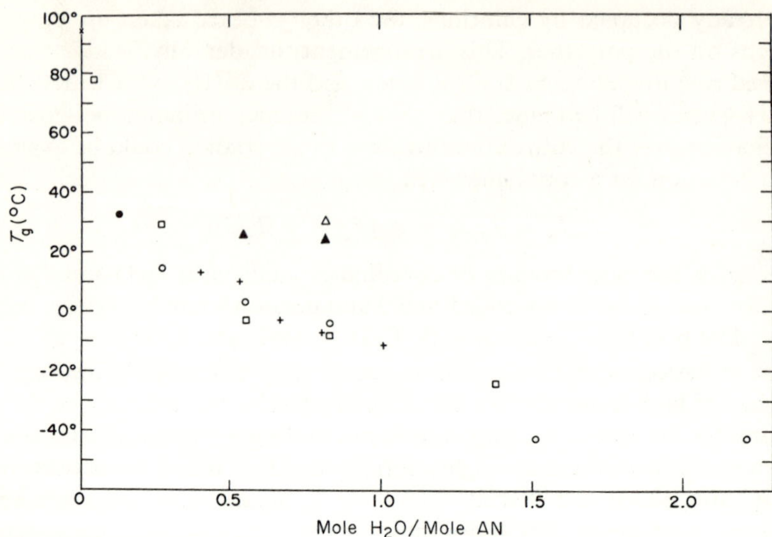

FIG. 2.15. T_g values for PAN–$M(ClO_4)_2 \cdot 6H_2O$ complexes as a function of hydrated salt content, plotted as T_g versus the molar ratio of H_2O to monomeric acrylonitrile. × = PAN, ● = Li^+, ○ = Mn^{2+}, □ = Fe^{2+}, ▲ = Co^{2+}, △ = Ni^{2+}, + = Zn^{2+}. [49].

If, however, there is a strong attractive interaction between the low-molecular-weight species and the polymer, then the additional free volume is more than compensated for by the strong interaction, which reduces the free volume, thus increasing T_g. Hence, a small cation like Li raises the T_g of PPO appreciably; $Ca(SCN)_2$ raises the T_g less strongly in phenoxy polymers, while $M(ClO_4)_2 \cdot 6H_2O$ complexes, as a result of their very large volumes, lower the T_g of PAN.

(b) Ionic Copolymers

Fitzgerald and Nielsen [51] studied the dynamic mechanical properties of sodium salts of styrene-(methacrylic acid) copolymers at three concentrations of NaMA, i.e., 2, 10, and 40 mole %. They observed that the maximum in the logarithmic decrement increased from ~120°C for the 2% sample to 145° for 10% and 185° for the 40% sample. Although this relationship is apparently nonlinear, the trend is quite clear and parallels the one found in the Moacanin and Cuddihy study [44].

Eisenberg and Navratil [16, 28] also investigated the effect of ion content on T_g in the S–MAA system. T_g values were measured by mechanical loss and DSC techniques; both methods gave similar results. It was found that the glass-transition temperature increased linearly with increasing ion content (up to 10 mole %). No discontinuity in the values T_g was seen at 6 mole %, the point of onset of cluster formation. This is illustrated in Fig. 2.16.

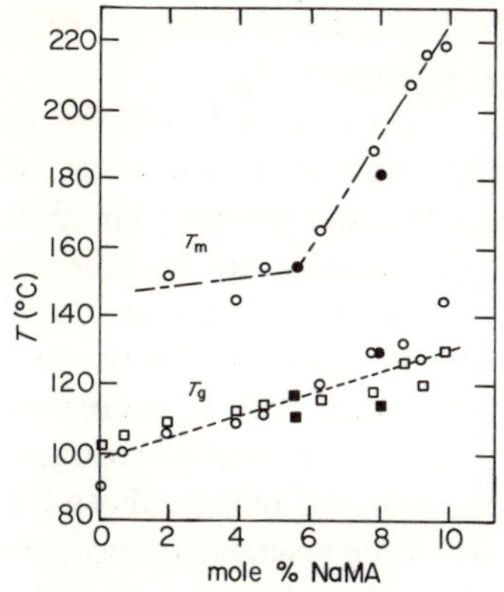

FIG. 2.16. T_g and T_m values in S–MAA copolymers as a function of sodium content. The circles represent transitions observed dynamic mechanically (~1 Hz); the squares by DSC. Filled symbols represent low-molecular-weight polymers; the open symbols represent high-molecular-weight polymers. Above 6 mole % NaMA. T_m corresponds to the point at which tan $\delta = 0.5$. [28].

The dynamic mechanical results of the same study, which will be presented later, reveal the existence of a large energy dissipation region above the glass transition for ion contents greater than 6 mole %. It seems reasonable that this large dispersion is due to the T_g of the clustered ionic regions in these materials. The peak positions of the high-temperature dispersions are also shown in Fig. 2.16.

In a recent study, Ogura *et al.* [52] investigated the glass transition in the sodium salts of an S–0.20MAA copolymer by means of IR spectroscopy and DSC. The rise in T_g with increasing degree of ionization (in the range 0 to 90%) was found to be linear. The values of T_g, as determined from the temperature dependence of the ratio of the peak absorbance of the 1700 and 1745 cm^{-1} bands, were found to agree with T_g values obtained by DSC. It was also found that the enthalpy ΔH of the residual intermolecular hydrogen bonding increased with increasing ion content.

Ogura *et al.* attempted to describe the relationship between T_g and ion content in terms of the simple copolymer equation

$$T_g = n_1 T_{g_1} + n_2 T_{g_2} + n_3 T_{g_3} \tag{20}$$

where n_i and T_{g_i} represent the mole fraction and T_g of each homopolymer component (PS, PMAA, PMAA–Na). The calculated values of T_g were significantly lower than the observed values, indicating that composition effects alone are insufficient to explain the effect of ions on the glass transition. The authors were, however, able to describe the dependence of T_g on the degree of neutralization in terms of the volume shrinkage produced by the formation of cross-links, using the theory of Fox and Loshaek [53]. They concluded that the introduction of ions results in the formation of ionic domains which not only function directly as cross-links but also tend to reinforce the residual hydrogen bonds; in either case the increase in T_g can be attributed to an increase in the cross-linking effect. The possibility that the transition observed at these temperatures could be due to the onset of mobility in ionic domains was ruled out since the absorbance at 1560 cm^{-1} (asymmetric stretching of COO^-) did not change at T_g.

Otocka and Eirich [15], in their extensive studies of ion-containing butadiene copolymers, investigated the glass-transition behavior of three polymer series containing various concentrations of ionic groups. The first system was butadiene–(lithium methacrylate), the second was butadiene–[methyl(2-methyl-5-vinyl)pyridinium iodide], and the third was an equimolar mixture of the first two, with the lithium iodide formed in the process of mixing remaining in the polymer. The authors found that T_g increased linearly with the concentration of ionic groups, the effect being strongest for the pyridinium salts, weakest for the lithium salts, and intermediate for the mixtures. It is of interest to note that in a study of the viscoelastic properties of the quaternary pyridinium salts (cf. Section B in Chapter IV), the authors found a temperature (in all cases far above T_g) below which the ions appeared to form relatively stable cross-links. This indicates that T_g in these materials is due not to the onset of mobility of the ionic groups, but of the other segments, suggesting that the ions act as cross-links and increase T_g by that mechanism, as was also found by Ogura *et al.* [52].

The dynamic mechanical properties of ethylene–(metal methacrylate) copolymers have been investigated in several laboratories [34, 35], the metal being most frequently sodium although others have also been studied. These materials will be discussed much more extensively in Section C of Chapter IV. Otocka and Kwei [35], who studied the sodium and magnesium salts, ascribed the loss maximum in dynamic mechanical tests which occurs between 0° and -20°C (the β peak) to the glass transition for these materials. They found that the β peak temperature shifted to higher temperatures with increasing salt content, and that the variation in T_g with ion content obeyed the simple copolymer relationship [Eq. (20)]. They took the T_g of PE as ~ -20°C, and estimated the T_g values of PAA–Na and PAA–Mg as 230° and 400°, respectively. If the assignment of the β peak as the glass transition is correct, then one can calculate dT_g/dc for both materials in the low salt content regime. The values obtained are 5.7°/mole % and 9.7°/mole %, respectively.

MacKnight *et al.* [34] assigned the high-temperature tan δ peak ($\sim$50°C) to the glass transition in the ion-containing regions of the polymer; they observed that the peak position moves to higher temperatures with increasing ion content. These findings are qualitatively consistent with those of Otocka and Kwei if it is considered that clustering occurs in these materials, and also that Otocka and Kwei investigated materials containing a somewhat lower concentration of ions than those studied by MacKnight *et al.* It is entirely feasible that most of the ions, particularly in the higher concentration ranges, are present in separated regions (clusters), the T_g of which is $\sim$50°C, while some of the ions are present outside of these regions, and thus lead to an increase in the glass transition of branched PE as observed by Otocka and Kwei.

Recently, attention has been given to a fourth ionomer series—the polyester-based ionic copolymers. Williams [54] studied a series of *n*-butyl methacrylate ionomers, with respect to the effect of ion content and moisture on T_g. The ionomers were actually terpolymers of styrene, *n*-butyl methacrylate, and potassium methacrylate, prepared by partial alkaline hydrolysis of the copolymers; styrene was in each case the minority component of the starting material (0 to 39 mole %). For any given styrene content and molecular weight, T_g, as measured by DSC, was seen to rise sharply at very low ion contents and then level off at about 1.5 mole % KMA. The T_g determinations were made at ambient humidity, generally 40–50% RH. Since the moisture content at constant RH increases with increasing ion content, Williams concluded that the leveling off of the T_g curves was due to a balance between the chain-stiffening effect of ionization and the plasticization by atmospheric water.

Tsutsui *et al.* [55] measured T_g values by dilatometry in (methyl acrylate)–(acrylic acid) copolymers as a function of the degree of neutralization. Their results for the sodium salts at three different copolymer compositions are shown in Fig. 2.17. In each case, the initial rapid rise in T_g at low degrees of neutralization was followed by a gradual decline in T_g at higher sodium contents.

Matsuura and Eisenberg [41] investigated the glass-transition behavior of ethyl acrylate ionomers, with regard to ion concentration, type of ion, and water content. The most striking observation in this study was that for every type of ion studied, a plot of T_g as a function of ion content for the dry polymers gave an unusual sigmoidal curve, which could be correlated with

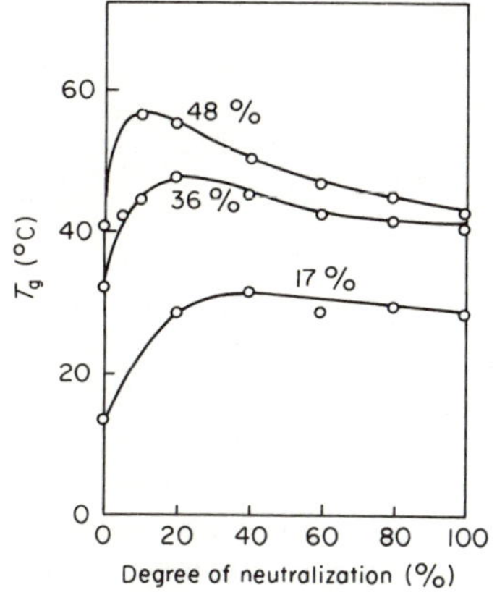

FIG. 2.17. T_g as a function of the degree of neutralization for MA–AA copolymers. The total AA content (mole %) is indicated for each curve. [55].

the onset of the failure of time–temperature superposition in viscoelastic studies (see Section D of Chapter IV). Furthermore, all of the T_g versus ion concentration curves for the various counterions can be superimposed if the plots are made against cq/a, where c is the metal acrylate content, q the cation charge, and a the distance between centers of charge. This can be seen in Fig. 2.18. The fact that T_g can be correlated with q/a is not surprising since it has been observed in other ionizable polymers, such as the polyphosphates to be discussed in Section B2. However, it is significant that the accelerated rise of T_g at $q/a \simeq 0.05$ coincides with the onset of nonsuperposability in viscoelastic behavior. Since the latter phenomenon presumably reflects ion clustering [16], it is tempting to assume that the onset of clustering is also a function of q/a, rather than simply the ion concentration.

Plasticization with water depressed the T_g, as expected. However, as shown in Fig. 2.19, beyond ~20 wt % H_2O, T_g varies linearly with water content and is independent of ion concentration over a wide range (20 to 50 mole % Na). This finding is qualitatively similar to the T_g leveling seen by Williams [54] and by Tsutsui *et al.* [55]. Finally, at Na concentrations greater than 12 mole %, the rate of change in T_g per water molecule per ion pair at constant ion content, $(\partial T_g/\partial n_{H_2O})_c$ was found to be linear, but with different slopes above and below two water molecules per pair. This reflects either a change in the nature of the association between the water molecules and the ions at that concentration, or some type of clustering transition. Since all the samples that exhibited this change in slope had ion concentrations above 12% and were thus presumably clustered, it is quite possible that at two H_2O molecules per ion pair the nature of the cluster changes or the cluster even falls apart. The first possibility, however, seems more probable.

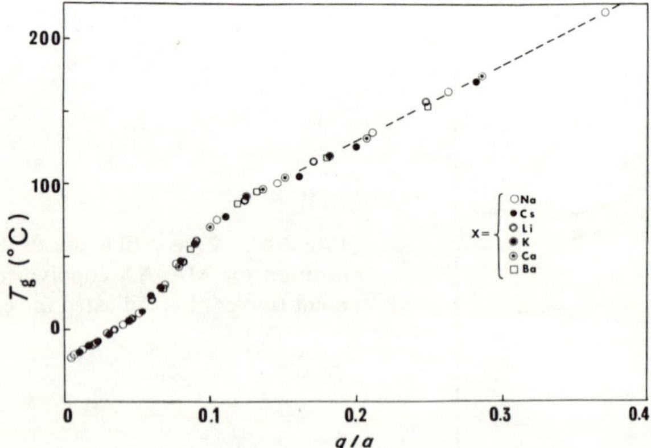

FIG. 2.18. T_g of EA–AA copolymers as a function of cq/a, where c is the carboxylate content, q the cation charge, and a the distance between centers of charge in angstroms. [41].

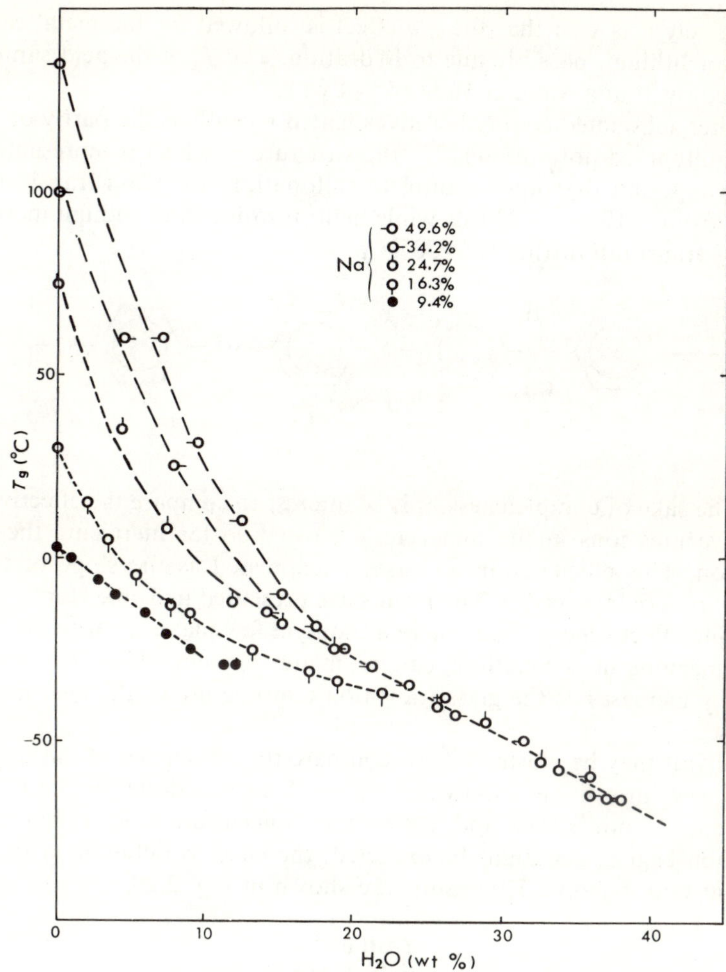

FIG. 2.19. T_g of EA–AA–Na as a function of water content. [41].

The above arguments are highly speculative and remain to be confirmed in further experimentation. The apparent discrepancy in the behavior of the three types of acrylate ionomers studied in different laboratories in regard to T_g as a function of ion content has yet to be explained, although incomplete drying is a possibility, as suggested by Williams.

Very recently, the glass transitions of a sulfonic acid derivative of polytetrafluoroethylene (Nafion) were investigated [56]. For an equivalent weight of 1365, it was found that the acid material had a T_g of 103°C, while in Li, Na, K, and Cs salts, T_g values were 216°, 235°, 225°, and 210°, respectively.

Qualitatively it is seen that the q/a effect is followed for the metal cations except for lithium, possibly due to hydration. The T_g of the acid sample is decreased by 2° at a water content of ~4 wt %.

Another sulfonated copolymer investigated recently is the partly or completely sulfonated polysulfone [57], the structure of which is represented by (I). It was found that upon complete sulfonation, the glass transition increased from ~175 to ~230°C, while neutralization with sodium increased the glass transition further to ~305°C.

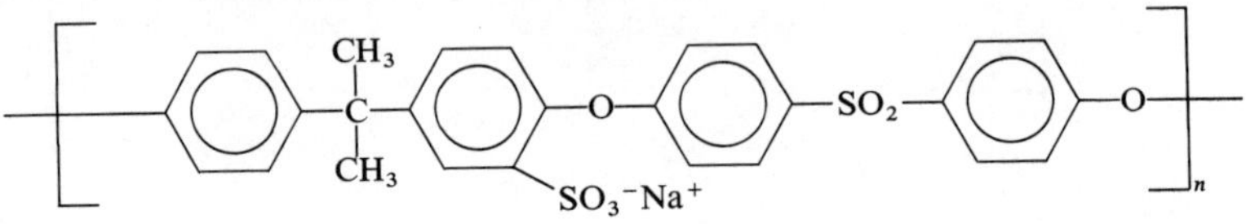

(I)

For the sake of completeness, it is of interest to compare the effectiveness of the various ions in the materials discussed so far in raising the glass transition. This efficiency in all cases is expressed as the slope of the T_g versus c plot (c in mole %). The results are presented in Table II in order of increasing effectiveness. Since there are only a few materials listed in Table II, no meaningful correlations can be made. It seems, however, that the efficiency increases as the glass-transition temperature of the host material decreases.

Finally, it may be worthwhile to compare the efficiencies of various ions in raising T_g in a single copolymer system. This was done for the EA–AA copolymers, both above and below the concentration at which cluster formation begins. As might be expected, the effect is definitely a function of q/a in both regions. The results are shown in Fig. 2.20.

TABLE II

dT_g/dc for Various Ion-Containing Polymers

Nonionic component	Ionic component	Conc. range (mole %)	$(T_g)_0$ (°C)	dT_g/dc (°C/mole %)	Ref.
"Phenoxy"	$Ca(SCN)_2$	0–20	~100	1.8	45
PPO	$ZnCl_2$	12–32	−65	4.0	47
PPO	$LiClO_4$	0–10	−70	5.5	44
Polysulfone	Sulfone–$So_3^-Na^+$	0–100	≃175	1.3	57
EA	NaA	0–12	−20	2.7	41
S	NaMA	0–10	100	3.2	28
B	LiMA	0–8	−90	5.4	15
E	NaMA	0–3	−20	5.7	35
B	MVPI	0–8	−90	8.9	15
E	$Mg_{1/2}MA$	0–2	−20	9.7	35

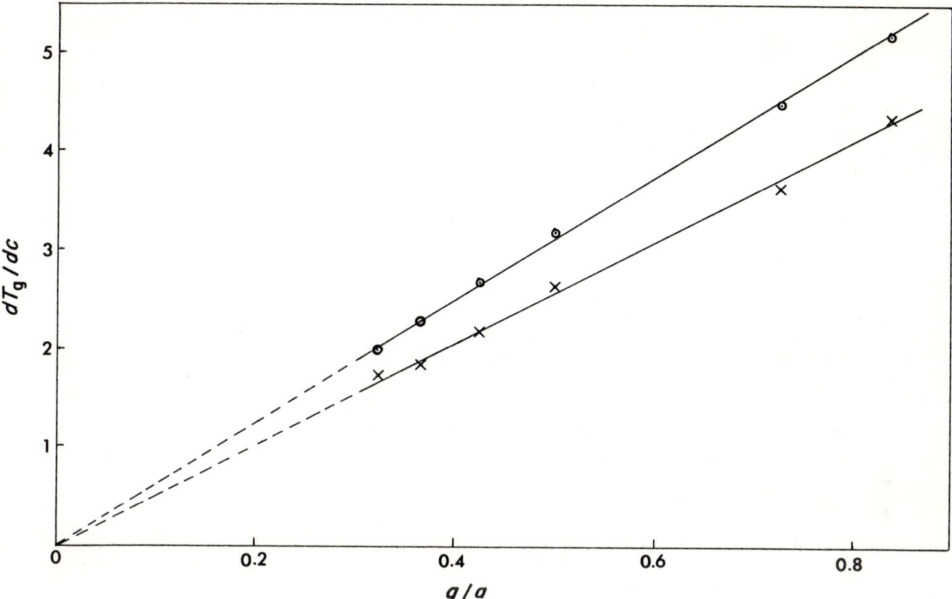

FIG. 2.20. dT_g/dc versus q/a for E–AA ionomers in the low (○) and high (×) ion content regions. In the low ion content region, concentration ranges of 0–5 to 0–10 mole % were used; in the high ion content region (above the clustering point) concentration ranges were from 17–35 to 25–85 mole %. [41].

In view of the work of Ogura *et al.* [52], there is no doubt that ions function as cross-links at low concentrations. However, it seems unlikely that simple cross-linking alone is sufficient to account for the increase in T_g with increasing ion content. Since Otocka and Kwei [35] were able to interpret their results for ethylene ionomers in terms of a simple copolymer effect, it is quite possible that both copolymerization and cross-linking play a role in fixing the T_g of ionomers. This has been observed in organic materials [53]. Furthermore, if ions functioned solely as thermally stable cross-links in the glass-transition range, then one would not expect the counterion (q/a) effect observed in the EA–AA system. Finally, at higher ion concentrations, clustering undoubtedly changes the chain conformation sufficiently to influence the glass transition of the organic regions also.

2. Completely Ionizable Homopolymers

(a) *Counterion Effect*

The effect of the nature of the counterion on the glass transitions of completely ionizable homopolymers has so far been investigated for three systems: the polyphosphates [58, 59], the silicates [60], and polyacrylates [61].

In spite of the drastic differences in the structures of these polymers, the results are surprisingly similar.

The polyphosphate study was the most extensive, with several homopolymers and counterion copolymers included in it [58]. The homopolymers included the nonionic $(HPO_3)_x$, the glass transition of which is $-10°C$, and, among others, the lithium, sodium, and calcium polyphosphates, with glass transitions of 335°C, 285°C, and 520°C, respectively. Altogether, about 50 materials were examined.

An attempt to rationalize the data started with the premise that the glass transition in these systems is determined by the strength of the anion–cation interaction; i.e., this is the factor that limits segmental mobility. It was thought that at T_g, kT is just large enough to allow the anion to leave the coordination sphere of the cation (or vice versa) at a rate which allows volume equilibration to be achieved with reasonable relaxation times. Thus, kT_g must be proportional to the electrostatic work W_{el} of removing an anion from the coordination sphere of a cation, i.e.,

$$T_g \propto W_{el} \propto \int F_{el}\, da \tag{21}$$

where F_{el} is the electrostatic interaction, a the internuclear distance between anion and cation, and the limits of integration are the distance of closest approach and infinity. Since F equals the product of the charges divided by the square of the distance between them, upon integration it is found that $T_g \propto q_+ q_- / a$, where q is the charge. Since the anion charge is constant for any series of materials, the simplest relationship is

$$T_g \propto q/a \tag{22}$$

where q is the cation charge, and a is now the distance between centers of charge for the anion and cation at closest approach. Thus, a plot of T_g versus the ratio of q (in units of 1 electron) over a (in angstroms) yields the straight line given by

$$T_g = 625(q/a) - 12 \tag{23a}$$

For counterion copolymers, the number average q/a value was chosen. The results for the polyphosphate series are shown in Fig. 2.21 along with those for the silicates, acrylates, and ionenes, which will be discussed below.

The silicates at a composition of $M_2O{:}SiO_2 \approx 1$ represent another predominantly linear polymer system which is quite easy to investigate. Only one counterion copolymer system was studied [60], i.e., Na_2SiO_3–$CaSiO_3$;

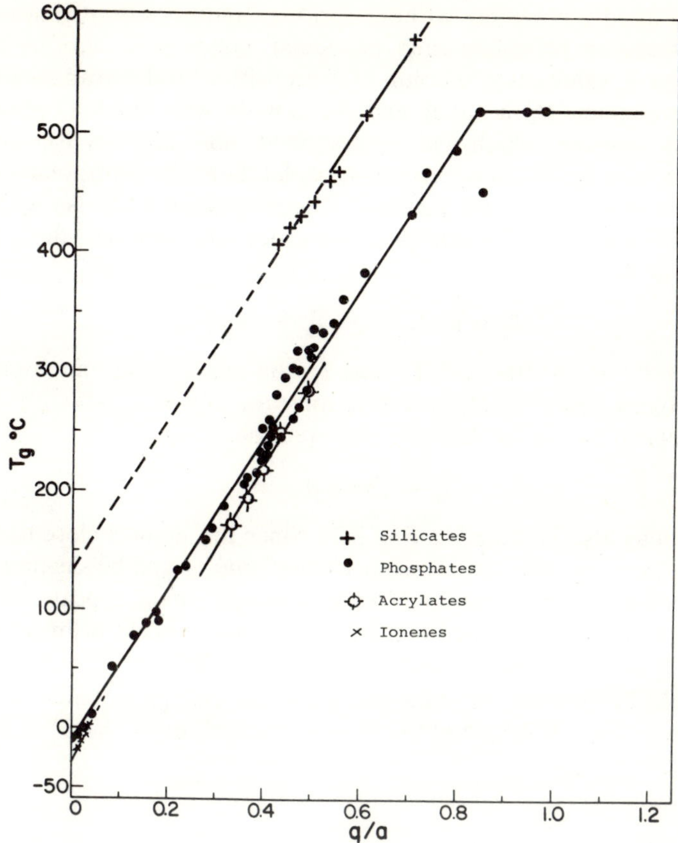

FIG. 2.21. T_g versus q/a for various ion-containing systems. [58, 60, 61, 69].

the results, again, were very similar to those obtained for the polyphosphates. The expression for T_g as a function of q/a is

$$T_g = 635(q/a) + 132 \tag{23b}$$

indicating that the slope is identical to that for the polyphosphates within experimental error. These results are also shown in Fig. 2.21.

More recently, several polyacrylate salts were studied [60], including sodium, potassium, and the sodium–calcium counterion copolymer system. Here, however, the study is much more difficult, because the salts cannot be dried completely without serious decomposition. It was therefore necessary to plasticize each particular polymer to various extents, measure the T_g values as a function of composition, and extrapolate to zero plasticizer content. Both water and formamide were used as plasticizers to avoid a problem, which will be described more fully in Section B3 in connection with the ionenes; with the polyacrylates the extrapolated results for the two plasticizers were identical. The results for PAA–Na with water are shown in Fig. 2.22. The extrapolation procedure employed the following relationship [62]

$$T_g = w_1 T_{g_1} + w_2 T_{g_2} - K w_1 w_2 \tag{24}$$

where w is the weight fraction, T_{g_1} and T_{g_2} the glass-transition temperatures of the polymer and diluent, and K a constant. Again T_g was found to be linearly related to q/a, in this case the equation being

$$T_g = 730(q/a) - 67 \tag{23c}$$

This plot may also be found in Fig. 2.21. Since the slope is close to that for the phosphates or silicates, it suggests that one might be dealing with a general constant for polyanions with relatively small repeat units. The results of a q/a study of the ionenes, which also indicate a linear relation between T_g and q/a, will be described in Section B3.

It should be pointed out that while for the polyphosphate system the partly neutralized HPO_3–$NaPO_3$ system does fit onto the T_g versus q/a

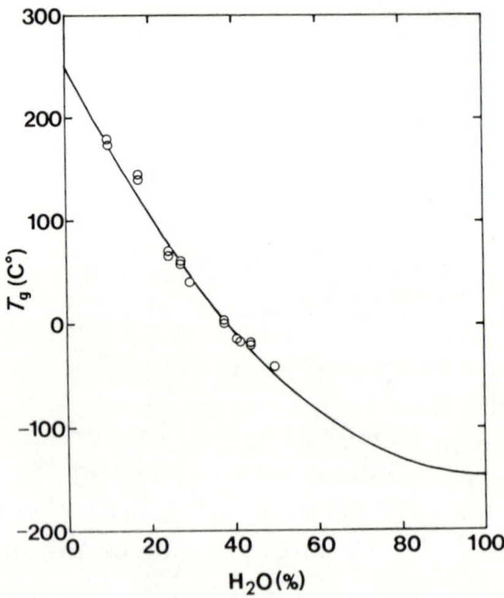

FIG. 2.22. T_g of PAA–Na as a function of water content. [61].

plot, this is not the case for the acrylate system. The extrapolated value for PAA from the T_g versus q/a plot of the metal acrylates would be $-67°C$, while the actual value is $\sim +105°$. This substantial difference is probably due to the fact that in PAA the acid groups are capable of forming a high concentration of highly structured, rigid hydrogen-bonded dimers, or even larger groups, which lower the segmental mobility of the polymer appreciably. Hydrogen bonding exists in phosphoric acid also, but there the acid hydrogen can interact with more than just one oxygen per repeat unit. Highly structured dimers are probably not formed, and segmental mobility is not decreased as strongly as it is in the acrylates.

The applicability of the q/a approach to the ethyl acrylate ionomer series has already been discussed. As seen in Fig. 2.18, above $cq/a \simeq 0.1$, T_g rises approximately linearly with ionic interaction and extrapolates reasonably well to the line for the completely ionized polyacrylates.

(b) *Molecular-Weight Effect*

The effect of molecular weight on the glass transition of an ionic polymer was studied in only one system—sodium polyphosphate [63]. Phosphate polymers with two different types of terminal groups were prepared: OH terminated and O^-Na^+ terminated. With each type, the plot of T_g versus $1/P$ gave a linear relationship, although the two slopes were radically different. A number of theoretical approaches with adjustable parameters were found to fit the data, including the simple copolymer equation [Eq. (20)], taking the chain ends as one comonomer and the middles as the other. Only one theory without adjustable parameters was tested: the Gibbs–DiMarzio theory [64]. It was found to fit only the OH-terminated series. The O^-Na^+ series shows a very unusual type of behavior in that T_g varies only slightly with chain length, probably due to a cross-linking effect involving the end groups.

(c) *Effect of Cross-Link Density*

The change of glass transition with cross-link density was investigated for multicomponent phosphate glasses containing residual OH groups by Ray [65]. It was found that the relationship obeys the equation developed by DiMarzio [66] and is thus quite analogous to the behavior of the organics. The results are shown in Fig. 2.23. For binary phosphate glasses, with increasing P_2O_5 content the results all tend toward 270°C, the T_g of P_2O_5 [67]. Thus, for P_2O_5–BaO mixtures, for P_2O_5 concentrations of 50, 60, and 70 mole %, the T_g values are 470, 400, and 343°C, respectively, while with K_2O as the network modifier the values at the same cross-link densities were 243, 252, and 260°C.

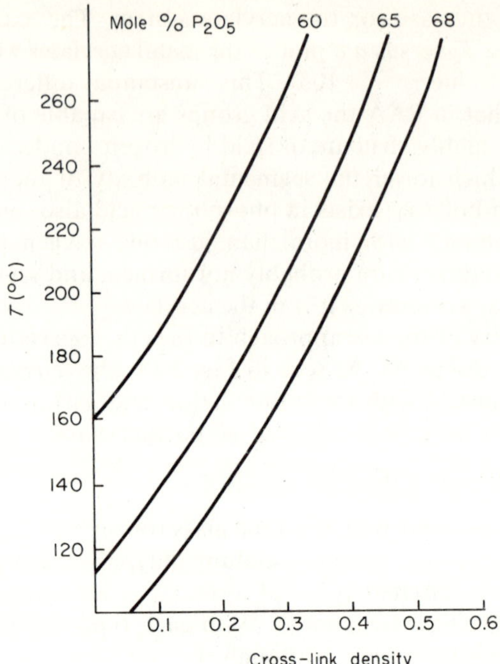

FIG. 2.23 Glass-transition temperature of phosphate glasses as a function of the cross-link density as expressed by mole % P_2O_5. [65].

3. THE IONENES

In only one other series of ionic polymers, the aliphatic ionenes [68], has the glass transition been investigated extensively. These polymers consist of short methylene sequences separated by dimethylammonium ions along the backbone. The structure of a typical ionene is (II), where *m* and *n* can vary from 2 to 16. The glass-transition relations found for these materials [69] do not conform in some respects with those found for the polymers described in the previous sections, so they are described separately.

$$\left[-(CH_2)_m-\overset{CH_3}{\underset{CH_3}{N^+}}-(CH_2)_n-\overset{CH_3}{\underset{CH_3}{N^+}}- \right]_x$$

(II)

Because of their regular structures, ionenes are highly crystalline, and a direct determination of the glass transition is impossible. It is therefore necessary to determine the T_g values of any one polymer with several different concentrations of diluent and to extrapolate to zero diluent concentration using some reasonable equation. The one chosen here was the same one used for the acrylates, i.e., Eq. (24). The solvents used in the study included water, glycerine, and formamide, for which the glass-transition temperatures are either known or can be determined by extrapolation. For example, the water–glycerine system has been studied extensively, and the T_g of water has been calculated [70] as -146°C. Thus, with T_{g_2} known, T_{g_1} and K can be regarded as calculable parameters and obtained from a plot of $(T_g - T_{g_2})/w_1$ versus w_2 from the slope and intercept, respectively. It should be pointed out that other workers [71, 72] have obtained an extrapolated value of ~ -135°C for the T_g of water. A difference of 12° in the T_g of water, however, will not introduce an appreciable error in the extrapolated glass-transition temperature of the ionenes since the majority of the points used in the extrapolation lie at low water contents.

With glycerine as a plasticizer, where the accessible concentration range is 0 to 80% glycerine, the effect of ion spacing on T_g seems minor. Thus, the 6,4-ionene bromide ($m = 6$, $n = 4$) and the 6,5-ionene bromide both have an extrapolated T_g of -88°C, while that of the 6,8-ionene bromide is ~ -82°C. Using water as a plasticizer, with the accessible concentration range 5 to 50% water, the extrapolated glass transitions come out to be very much higher. For example, the value for the 6,8-ionene is -4°C, in contrast to the -82° obtained with glycerine. The effect of spacing, however, is still quite small.

Since water and glycerine give such drastically different T_g values, several water–glycerine mixtures were used as plasticizers for 6,8-ionene bromide. The values obtained by extrapolation to zero plasticizer content are listed in Table III. It was also found that formamide, with $\varepsilon = 109$ at room temperature (i.e., larger than water) gives a T_g of ~ 0°C, while ethylene glycol ($\varepsilon = 38$) gives a value of ~ -80°C, i.e., very close to that of glycerine. It is highly probable that a conformational transition of the polymer chain occurs in a dielectric constant range corresponding to that encountered in glycerine–water mixtures ($\varepsilon \simeq 90$), and that this conformational effect influences the glass transition profoundly. So far, this effect on the glass transition has not been observed in any other polymer system, but it should

TABLE III

Variation of the T_g of 6,8-ionene with Plasticizer Composition

Glycerine in plasticizer (%):	100	80	60	40	0
T_{g_1} (°C):	−82	−76	−57	−31	−4

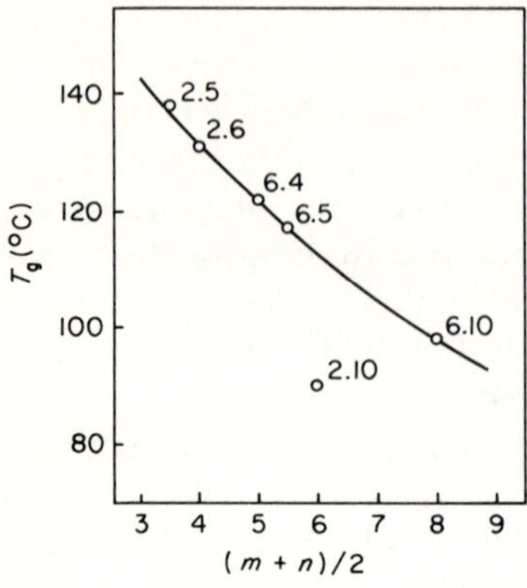

FIG. 2.24. The variation of T_g with the average spacing of the ions in aliphatic ionenes. (m,n) in the figure indicates m,n-ionene. [73].

be recalled that no other ionic polymer-diluent system has been studied as extensively as that of 6,8-ionene with water and glycerine with respect to the glass-transition properties.

An attempt was also made to see whether a q/a relationship existed in the ionenes. Because of the very small range of ion concentrations accessible, the uncertainty was much greater than in any of the previously described systems. However, for the 6,8 series, a plot of T_g versus q/a can be represented by the following linear relationship:

$$T_g = 695(q/a) - 23 \tag{23d}$$

a was taken as 5 Å from crystal structure studies of tetramethylammonium bromide, and the extreme, i.e., the completely ionized polymer, was assumed to be $[N(CH_3)_2{}^+]_x$. The result is shown in Fig. 2.21 and seems to confirm the generality of the q/a approach.

In a subsequent study of the glass transition in aliphatic ionenes, Tsutsui *et al.* [73] determined T_g values for several ionenes using a torsion braid technique. Their T_g values for ionene–water mixtures were consistent with those measured by Eisenberg *et al.* [69]; however, in the absence of any plasticizer at all, the T_g's they found were very much higher than the extrapolated values. Their data for unplasticized ionenes are shown in Fig. 2.24. While no satisfactory explanation for the drastic effect of plasticizer at low levels can be offered at this time, it would seem reasonable to expect appreciable conformational changes to occur for the different plasticizers used.

REFERENCES

1. A. Eisenberg, *Macromolecules* **4,** 125 (1971).
2. J. Brandrup and E. H. Immergut (eds.), "Polymer Handbook." Wiley (Interscience), New York, 1966.
3. L. D. Pettit and S. Bruckenstein, *J. Am. Chem. Soc.* **88,** 4783 (1966).

4. A. Eisenberg, *Macromolecules* **3,** 147 (1970).
5. J. C. Cowan and H. M. Teeter, *Ind. Eng. Chem.* **36,** 148 (1944).
6. C. Paquot, R. Perron, and C. Vassilieres, *Bull. Soc. Chim. Fr.* 317 (1959).
7. C. Paquot, R. Perron, and A. Mathieu, *Bull. Soc. Chim. Fr.* 92 (1960).
8. J. Economy, J. H. Mason, and L. C. Wohrer, *Am. Chem. Soc. Polym. Preprints* **7,** 596 (1966); *J. Polym. Sci. Part A-1* **8,** 2231 (1970).
9. J. Economy and J. H. Mason, *in* "Ionic Polymers" (L. Holliday, ed.), chapter V. Halsted Press, Wiley, New York, 1975.
10. E. P. Otocka, M. Y. Hellman, and L. L. Blyler, *J. Appl. Phys.* **40,** 4221 (1969).
11. G. Kraus and J. T. Gruver, *Rubber Chem. Tech.* **42,** 800 (1969).
12. J. E. Potts, E. G. Hendricks, C. Wu, and A. C. Ashcraft, Jr., paper presented at Middle Atlantic Regional Meeting, Amer. Chem. Soc., Philadelphia, Pennsylvania, February 1968.
13. A. Eisenberg, unpublished.
14. M. Pineri, C. Meyer, A. M. Levelut, and M. Lambert, *J. Polym. Sci. Polym. Phys.* **12,** 115 (1974).
15. E. P. Otocka and F. R. Eirich, *J. Polym. Sci. Part A-2* **6,** 921 (1968).
16. A. Eisenberg and M. Navratil, *Macromolecules* **6,** 604 (1973).
17. A. Guinier and G. Fournet, "Small-Angle Scattering of X-Rays," chapter 2. Wiley, New York, 1955.
18. F. C. Wilson, R. Longworth, and D. J. Vaughan, *Am. Chem. Soc. Polym. Preprints* **9,** 505 (1968).
19. R. Longworth and D. J. Vaughan, *Nature* (*London*) **218,** 85 (1968).
20. S. Bonotto and E. F. Bonner, *Macromolecules* **1,** 510 (1968).
21. B. W. Delf and W. J. MacKnight, *Macromolecules* **2,** 309 (1969).
22. W. J. MacKnight, W. P. Taggart, and L. McKenna, *J. Polym. Sci. Polym. Symp.* **46,** 83 (1974).
23. F. L. Binsbergen and G. F. Kroon, *Macromolecules* **6,** 145 (1973).
24. C. L. Marx, D. F. Caulfield, and S. L. Cooper, *Macromolecules* **6,** 344 (1973).
25. R. J. Roe, *J. Phys. Chem.* **76,** 1311 (1972).
26. J. Kao, R. S. Stein, W. J. MacKnight, W. P. Taggart, and G. S. Cargill, III, *Macromolecules* **7,** 95 (1974).
27. W. J. MacKnight, W. P. Taggart, and R. S. Stein, *J. Polym. Sci. Polym. Symp.* **45,** 113 (1974).
28. A. Eisenberg and M. Navratil, *Macromolecules* **7,** 90 (1974).
29. H. A. Davis, R. Longworth, and D. J. Vaughan, *Am. Chem. Soc. Polym. Preprints* **9,** 515 (1968).
30. R. Longworth, *in* "Ionic Polymers" (L. Holliday, ed.), chapter II. Halstead Press, Wiley, New York, 1975.
31. W. J. MacKnight, *Am. Chem. Soc. Polym. Preprints* **11,** 504 (1970).
32. C. L. Marx, J. A. Koutsky, and S. L. Cooper, *J. Polym. Sci. Part B* **9,** 167 (1971).
33a. M. Pineri, C. Meyer, and A. Bourret, *J. Polym. Sci. Polym. Phys.* **13,** 1881 (1975).
33b. C. T. Meyer, J. M. D. Coey, and M. Pineri. To be published.
34. W. J. MacKnight, L. W. McKenna, and B. E. Read, *J. Appl. Phys.* **38,** 4208 (1968).
35. E. P. Otocka and T. K. Kwei, *Macromolecules* **1,** 401 (1968).
36. K. Sakamoto, W. J. MacKnight, and R. S. Porter, *J. Polym. Sci. Part A-2* **8,** 277 (1970).
37. J. D. Ferry, W. C. Child, Jr., R. Zand, D. M. Stern, M. W. Williams, and R. F. Landel, *J. Colloid Sci.* **12,** 53 (1957).
38. A. Eisenberg, S. Saito, and L. A. Teter, *J. Polym. Sci. Part C* **14,** 323 (1966).
39. D. G. Fesko and N. Tschoegl, *J. Polym. Sci. Part C* **35,** 51 (1971).
40. A. V. Tobolsky, P. F. Lyons, and N. Hata, *Macromolecules* **1,** 515 (1968).
41. H. Matsuura and A. Eisenberg, *J. Polym. Sci. Polym. Phys.* **14,** 773 (1976).
42. E. P. Otocka and D. D. Davis, *Macromolecules* **2,** 437 (1968).

43. O. A. Ponomarev and I. A. Ionova, *Vysokomol. Soyed.* **A16,** 1023 (1974); *Polym. Sci. USSR* **16,** 1181 (1975).
44. J. Moacanin and E. F. Cuddihy, *J. Polym. Sci. Part C* **14,** 313 (1966).
45. M. J. Hannon and K. F. Wissbrun, *J. Polym. Sci. Polym. Phys.* **13,** 113 (1975).
46. K. F. Wissbrun and M. J. Hannon, *J. Polym. Sci. Polym. Phys.* **13,** 223 (1975).
47. R. E. Wetton, D. B. James, and W. Whiting, *J. Polym. Sci. Polym. Phys.* (in press).
48. S. Reich and I. Michaeli, *J. Chem. Phys.* **56,** 2350 (1972).
49. S. Reich and I. Michaeli, *J. Polym. Sci. Polym. Phys.* **13,** 9 (1975).
50. S. Reich, S. Raziel, and I. Michaeli, *J. Phys. Chem.* **77,** 1378 (1973).
51. W. E. Fitzgerald and L. E. Nielsen, *Proc. Roy. Soc.* **A282,** 137 (1964).
52. K. Ogura, H. Sobue, and S. Nakamura, *J. Polym. Sci. Polym. Phys.* **11,** 2079 (1973).
53. T. G. Fox and S. Loshaek, *J. Polym. Sci.* **15,** 371 (1955).
54. M. W. Williams, *J. Polym. Sci. Polym. Symp.* **45,** 129 (1974).
55. T. Tsutsui, T. Yokoyama, and T. Tanaka, *Kobunshi Ronbunshu* **31,** 565 (1974).
56. A. Eisenberg and S. C. Yeo, *J. Appl. Polym. Sci.* (in press).
57. A. Noshay and L. M. Robeson, *J. Appl. Polym. Sci.* **20,** 1885 (1976).
58. A. Eisenberg, H. Farb, and L. G. Cool, *J. Polym. Sci. Part A-2* **4,** 855 (1966).
59. A. Eisenberg, *Adv. Polym. Sci.* **5,** 59 (1967).
60. A. Eisenberg and K. Takahashi, *J. Non-Cryst. Solids* **3,** 279 (1970).
61. A. Eisenberg, H. Matsuura, and T. Yokoyama, *J. Polym. Sci. Part A-2* **9,** 2131 (1971).
62. E. Jenckel and R. Heusch, *Kolloid-Z. Z. Polym.* **130,** 89 (1953).
63. A. Eisenberg and T. Sasada, *in* "Physics of Non-Crystalline Solids" (J. A. Prins, ed.), pp. 99–116. North Holland Publ., Amsterdam, 1966.
64. J. H. Gibbs and E. A. DiMarzio, *J. Chem. Phys.* **28,** 373 (1958).
65. N. H. Ray, *J. Polym. Sci.* (in press).
66. E. A. DiMarzio, *J. Res. Nat. Bur. Std.* **68A,** 611 (1964).
67. N. H. Ray, *in* "Ionic Polymers" (L. Holliday, ed.), chapter VIII. Halsted Press, Wiley, New York, 1975.
68. A. Rembaum, W. Baumgartner, and A. Eisenberg, *J. Polym. Sci. Part B* **6,** 159 (1968).
69. A. Eisenberg, H. Matsuura, and T. Yokoyama, *Polym. J.* **2,** 117 (1971).
70. I. Yannas, *Science* **160,** 298 (1968).
71. C. A. Angell and E. J. Sare, *Science* **168,** 280 (1970).
72. D. H. Rasmussen and A. P. MacKenzie, *J. Phys. Chem.* **75,** 967 (1971).
73. T. Tsutsui, T. Sato, and T. Tanaka, *Polym. J.* **5,** 332 (1973).

Chapter III

Viscoelastic Properties of Homopolymers

Chapter II has dealt with the effects of ionic forces in polymers on the structure and glass transitons. In this chapter, the effect of ionic forces on the viscoelastic properties of homopolmers will be examined. The first group of polmers to be discussed (in Section A) are the inorganic polymers—specifically, the polyphosphate and polysilicate systems. In some ways, these polymers are the simplest materials in which the effect of ions on viscoelasticity can be studied. Unlike ion-containing organic polymers, the inorganic homopolymers can be studied in the unplasticized state. This avoids the complications which result from the presence of diluents. Also, since inorganic polymers are of high dielectric constant, there is less tendency for the ions to aggregate at comparable ion concentrations; thus, microphase separation is less likely. Hence, problems arising from thermodynamic dissimilarities between different elements of the polymer are decreased.

The high values of T_g in the phosphate and silicate systems, as discussed in Chapter II, are due to the hindred segmental mobility which results from ionic interactions. However, since ion aggregation does not occur, the primary relaxation process should be essentially the same as it is in most "simple" polymers, except for the shift in T_g. It will be seen that this is, in fact, the case. The viscoelastic response of these polymers in the vicinity of T_g is very similar to that seen in other rheologically simple polymers; the methods of observation and analysis that apply to polymers such as polystyrene and poly(methyl methacrylate) apply to ionic inorganic polymers as well. The exception to this viscoelastic behavior is seen in the relaxation phenomena that occur in transition metal phosphates. Here, an additional mechanism due to transition-metal-catalyzed bond interchange is observed [1].

A further contribution to the relaxation behavior can be expected to result from the motion of the mobile counterions through the glass. Because of its nature, this relaxation process should occur at temperatures well below T_g or in a much higher frequency range. In the silicate system, it is found to occur in the vicinity of room temperature for frequencies of the order of 1 Hz [2].

The second class of polymers, to be discussed in Section B, are those in which ionic character is introduced by complexation with a low-molecular-weight additive. This could occur by the addition of a Lewis acid to a donor polymer resulting in a polymeric charge-transfer complex, or by the addition of an inorganic salt to a polar polymer to form a charged complex. Regardless of its origin, complexation could produce one of two possible effects on polymer viscoelasticity. If a three-dimensional structure is created, then phenomena associated with time-dependent cross-linking should be seen, the lifetime of the cross-links being temperature dependent. However, if complexation merely hinders segmental motion, the only effects to be expected are those associated with decreased segmental mobility. In the former case, the effects of the structural changes should manifest themselves in the viscoelastic behavior either through a second rubbery plateau, or in an additional loss mechanism, or both. In the latter case, no additional mechanism should result; however, in both instances, complexation should produce an increase in the value of T_g.

Perhaps the most complex ion-containing homopolymer systems are provided by the ionizable organics, to be discussed in Section C. Unlike inorganic polymers, these materials are not amenable to study in the unplasticized state because of decomposition near their T_g values, which tend to be very high. Furthermore, because these polymers are composed of thermodynamically dissimilar elements, namely, hydrocarbon backbones and ionic side groups, there is a strong tendency in them to phase-separate through the formation of ionic aggregates. The presence of plasticizer can contribute to the complexity of the behavior of this type of polymer since hydrophilic diluents could become incorporated into the ionic regions, while hydrophobic diluents preferentially associate with the hydrocarbon portions. It has been found that the extent of ion aggregation does not vary a great deal with the type and amount of plasticizer. However, wide variations in the stability of the polymer superstructure occur with different plasticizers [3]. These variations are reflected dramatically in the viscoelastic response of the polymer system.

The results of an investigation of the viscoelastic properties of plasticized salts of poly(acrylic acid) are presented in Section C1. In the usual range of plasticizer concentration (20–70%), intermolecular associations contribute strongly to the polymer structure, and it is found that the viscoelastic properties reflect this fact. As the concentration of plasticizer increases, the polymer chains become more and more isolated, and consequently, intermolecular associations become relatively less important. In the region of the gelation point, inter- and intramolecular interactions become competitive, and here, a different range of viscoelastic properties is observed. As described in Section C2, in moderately concentrated polymer solutions (1–20% polymer), relaxations due to large-scale segmental motions, detectable by dynamic techniques in the frequency range 1–1000 Hz, are seen.

Relaxation effects due to the motion of the counterions themselves can also be anticipated. Because the relative mobility of a counterion is so much higher than that of a macroion, one would expect to observe the relaxation of the counterion atmosphere in a much higher frequency range than that in which the segmental relaxation of the macroion occurs. Indeed, it will be seen in Section C3 that counterion relaxation in dilute polyelectrolyte solutions can be observed by dynamic techniques in the megahertz range.

In the dilute-solution range, the dominant form of ionic interaction is intramolecular. In Section C3, a theoretical study of the dilute-solution viscoelasticity of ion-containing polymers will be presented. Mechanical models corresponding to simple ion-containing polymers are used to predict the variation in viscoelastic properties with the magnitude of ionic interaction. The development of experimental techniques for the study of the study of the viscoelasticity of very dilute polymer solutions is quite recent. So far, few experimental studies on ion-containing polymers in the dilute-solution range have been reported, and, except for the high-frequency studies, none of these can be considered to represent the behavior where the interactions are exclusively intramolecular.

A. INORGANIC IONICS

1. Chemically Inactive Polyphosphates

The polyphosphates represent perhaps the best explored ionic homopolymer system in the solid state. Although these polymers are classified in current terminology as heteroatomic polymers, because of their alternating P—O backbone, they can also be considered as homopolymers in the sense that they are polymers with a single repeat unit (I) and, with the exception of an occasional branch point or chain end, nothing else.

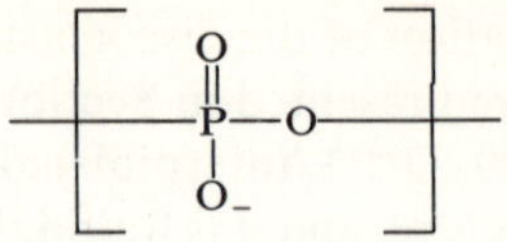

$$\left[\begin{array}{c} \text{O} \\ \| \\ -\text{P}-\text{O}- \\ | \\ \text{O}_- \end{array}\right]$$

(I)

Although the inorganic chemistry of the phosphates is well explored [4], only one series of studies has been performed in which the properties of noncrystalline polyphosphates were treated from a polymeric point of view. These studies were summarized in a review [1]; for this reason, only a brief presentation of the results is given here. Section A1 is devoted to the viscoelastic properties of phosphate polymers which are not subject to bond interchange in the glass-transition region. The viscoelasticity of phosphate polymers subject to bond interchange is discussed in Section A2.

It is known that backbone bonds can be cleaved thermally and be reformed under equilibrium conditions in many inorganic and organic polymer systems [5]; thus, in the glass-transition region, bond interchange is a feasible relaxation mechanism. In the sodium polyphosphates, the possible existence of this form of relaxation must be considered, since it is impossible to predict the relaxation mechanism in these materials from structural considerations alone. From the method of synthesis [4] (a classical condensation polymerization above 650°C) as well as from the very small molecular weight dependence of the viscosity above 600°C [6], it is evident that, in the melt, bond interchange occurs at an appreciable rate. On the other hand, the study of the solution properties of the polyphosphates in aqueous media at room temperature indicates that, aside from cross-link instability, the chain backbones of these materials behave like those of perfectly normal polymers in that temperature region; for instance, the intrinsic viscosity can be correlated with the molecular weight in the usual manner [7]. Since the glass-transition temperature for $NaPO_3$ ($\sim$285°C) is intermediate between the melting point and room temperature, the relaxation mechanism could be either molecular flow or bond interchange and must be ascertained experimentally.

The first part of this section details the elucidation of the primary relaxation mechanism in $NaPO_3$, while the second part is devoted to a discussion of the effect of counterions on the viscoelastic properties. Two possible primary processes are considered: relaxation by molecular (diffusional) flow which leaves the chain backbone intact, referred to as the α mechanism in accordance with the usage of Ferry *et al.* [8], and bond interchange, termed the χ mechanism because of its chemical nature.

Stress relaxation was chosen as the critical experiment in the elucidation of the specific relaxation mechanism. Its use is based on the premise that the viscoelastic properties of a polymer should be dependent on the molecular weight if the α mechanism is operative, but should be independent of it if bond interchange is encountered (at least at high molecular weights), since the chain ends would not participate in the bond interchange process. To make sure that the chain ends did not, in fact, participate and also to ensure that slight changes in the ionic concentration with chain length (due to the presence of two ions at the end of the chain in $—O^-Na^+$-terminated polymers) did not lead to erroneous results, polymers possessing both types of chain ends were studied, i.e., those terminated by —OH groups and those terminated by $—O^-Na^+$ groups.

The stress relaxation curves showed no unusual features and did not differ qualitatively from curves obtained in similar experiments on organic materials except, of course, for the much higher values of T_g. Since stress relaxation runs at a given temperature were found to be reproducible with time, the possibility of significant relaxation by hydrolytic degradation or other similar mechanisms could be discounted [9]. Also, the viscoelastic response was linear, at least for the magnitude of strain employed in this study.

The usual method of holding one curve constant and shifting all the others relative to it was used to construct master curves of modulus versus reduced time for each material. For the sake of uniformity, the reference temperature used in drawing each master curve was the glass-transition temperature of the particular material being tested. This procedure was followed for both sets of materials, the resulting master curves being exceedingly similar. For both forms of sodium polyphosphate, time–temperature superposition was found to hold, at least for moduli greater than 10^8 dyn/cm^2. This provides an indication that a single relaxation mechanism is operative, although it by no means constitutes a proof.

For the —OH-terminated polymers, the master curves are shown in Fig. 3.1, with the degrees of polymerization noted near each curve. A slight indication of a rubbery plateau is observed in the master curve corresponding to the sample of highest molecular weight. Also, it should be noted that the shift factors follow a WLF-type relationship [10]

$$\log a_T = -C_1(T - T_g)/(C_2 + T - T_g) \tag{1}$$

where the WLF parameters C_1 and C_2 are referred to T_g. As might be expected, the parameters differ from the universal values; for the highest-molecular-weight sample, the differences are slight, but they increase with decreasing molecular weight, resulting in smaller shifts for corresponding temperatures. In general, it is clear that the curves are highly molecular weight dependent.

It has been shown by Tobolsky and others [5, 11, 12] that the terminal portion of each master curve can be plotted as $\log E$ versus t, with the result that a straight line is obtained from which the parameters of the equiv-

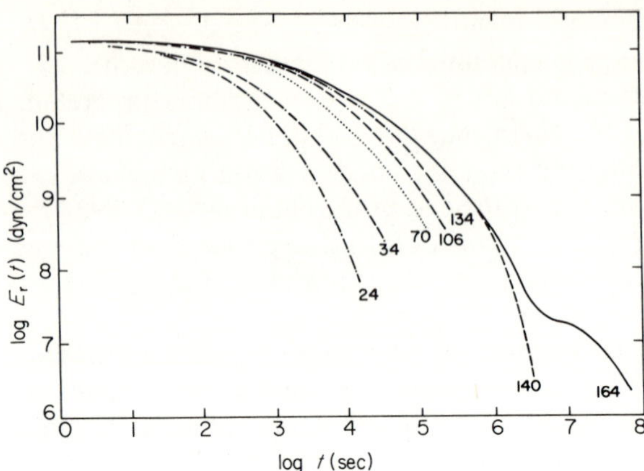

FIG. 3.1. Family of stress relaxation master curves for —OH-terminated $NaPO_3$ at $T = T_g$ for each sample. $\bar{P}_n$ of each sample is indicated [9].

alent ultimate or terminal Maxwell element can be computed, the most important parameter for the purposes of this discussion being the relaxation time. It was further shown that for several representative organic polymers, the relaxation time τ_m varied with temperature and degree of polymerization in the following manner:

$$\log \tau_m = \log A + C_1 \frac{T - T_g}{C_2 + T - T_g} + B \log \bar{P}_w \tag{2}$$

where A is a material-dependent constant, C_1 and C_2 the parameters of the WLF equation, and B a number ranging between 1 and 2 for polymers below the critical entanglement molecular weight and between 3 and 4 for polymers above it.

To ascertain whether the polyphosphates exhibited the same type of τ_m dependence on the molecular weight as the organics, the τ_m values were plotted as a function of chain length (Fig. 3.2) for both the —OH— and —O^-Na^+–terminated materials. The WLF term in Eq. (2) can be eliminated in this instance, since the τ_m values were obtained from master curves drawn with T_g as the reference point. It is evident that in both cases the trends are identical, the constant B being approximately 2.7. This is considerably higher than one would expect it to be for organics of comparable chain length, i.e., below the critical entanglement region; however, because of the polyelectrolyte nature of the chain with its attendant stiffening, the value may not be unreasonable. In any case, it is clear that the viscoelastic properties are strongly molecular weight dependent and that the chain ends to not have an appreciable effect other than displacing the τ_m versus

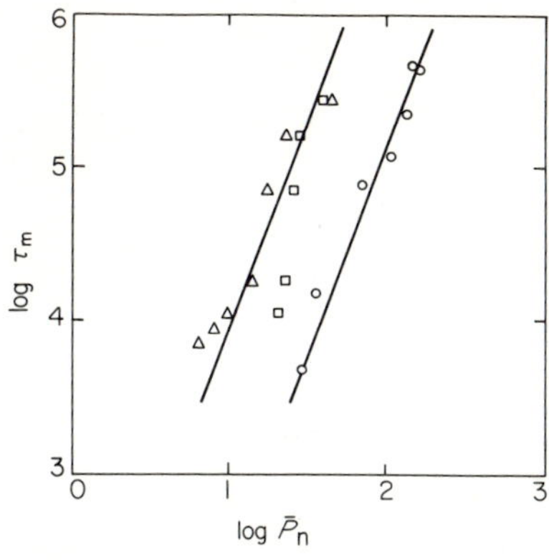

FIG. 3.2. Chain length ($\bar{P}_n$) dependence of the maximum relaxation time τ_m for —OH-terminated (○) and $—O^-Na^+$-terminated (□, △) polyphosphates [9].

$\bar{P}_n$ curve slightly. Therefore, it can be concluded that the primary relaxational process in sodium polyphosphate occurs by the α mechanism. This finding was also confirmed by a radiochemical study [9], which showed that, up to the glass-transition temperature, no detectable bond interchange occurred. It should be pointed out that at very high chain lengths, the two types of materials should yield identical results since the contribution of the chain ends would become negligible. Judging from Fig. 3.2, however, the molecular weight would have to be considerably higher before that identity is reached.

Having established the nature of the primary relaxation mechanism in the sodium polyphosphates, one can now proceed with the study of the effect of counterions on the viscoelastic properties. Although the investigation [13] to be reported here confined itself only to the distribution of relaxation times, the effects to be expected are by no means clear, even for this one function. As discussed in Section B2 of Chapter II, the strength of the intermolecular forces in the polyphosphates depends on the value of q/a, the ratio of the cation charge to the separation between cation and oxygen nuclei. Variations in the q/a value change not only the strength of the intermolecular forces but also the three-dimensional character of the polymer. Thus, according to some investigators [14], an increase in the forces should result in a broadening of the relaxation spectrum, while according to others [15], a narrowing should be observed as a result of increased three-dimensional character.

The latter prediction was based by Tobolsky on a classical three-dimensional oscillator model; it was shown to apply in the case of a molecular glass (dehydroabietic acid), in which a very narrow relaxation spectrum was observed [16]. To test this postulate in the polyphosphates, in which an increase in q/a would clearly increase the three-dimensional character of the polymer, two systems were studied, the $Na^+–K^+$ and the $Na^+–Ca^{2+}$

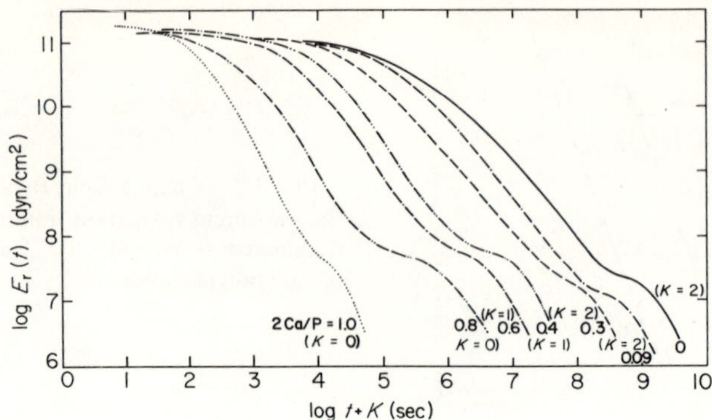

FIG. 3.3. Family of master curves for $NaPO_3$–$Ca(PO_3)_2$ counterion copolymer system [13]. Different abscissa scales are used to avoid overlapping.

counterion copolymers. The original stress relaxation master curves for the second system are shown in Fig. 3.3, while the distribution of relaxation times is shown in Fig. 3.4.

In Fig. 3.3, a pronounced inflection point or incipient plateau can be observed at a modulus of $\sim 10^7$ dyn/cm^2. Since this value lies near the lower limit of detectability for the sample geometries which had to be employed here, the accuracy in the position of this plateau is not very high. It is believed, however, that the plateau itself is real and that it reflects molecular entanglements, just like those encountered in organic materials.

As can be seen from Fig. 3.4, the distribution of relaxation times narrows appreciably with increasing Ca^{2+} content in the Na^+–Ca^{2+} copolymers; a narrowing of the distribution was also seen with increasing Na^+ in the Na^+–K^+ system (not shown). In other words, with increasing value of q/a or increasing three-dimensional character of the material, the distribution

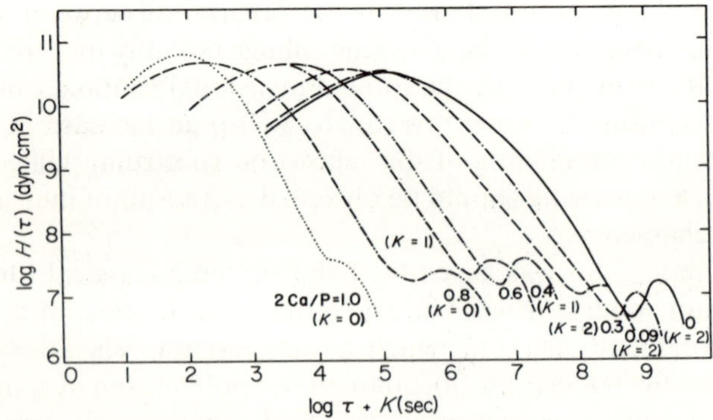

FIG. 3.4. Distribution of relaxation times for the family of curves shown in Fig. 3.3 [13].

narrows. The dependence of the width of the relaxation spectrum on the ionic force can also be seen in Fig. 3.5, in which $\Delta \log \tau$, a quantitative measure of the spectral width [17a], is plotted as a function of q/a for a number of phosphate polymers; it indicates a practically linear relationship between these two variables. The point at $q/a = 0$ was obtained from a study of the polyphosphoryl dimethylamides [17b] and is believed to be at least approximately applicable to the above correlation, the backbone being identical, and the "side chains" being $—O^-Na^+$ in the case of the ionic sodium polyphosphates and $—N(CH_3)_2$ in the nonionic case. The latter is probably somewhat larger than the former, so that the observed linearity down to $q/a = 0$ may be fortuitous; the trend, though, is clear. These results tend to confirm the theoretical prediction of Tobolsky [15]. It should be stressed, however, that superimposed on this q/a effect is the effect of molecular weight, which can also lead to a change in the spectral width. It has been shown, for example, that a decrease in the average chain length below 200 units leads to a decrease in the width of the relaxation spectrum in $NaPO_3$ homopolymers. However, since the effect is not large and since the range of degrees of polymerization used here is small (50–160), the dependence of $\Delta \log \tau$ on molecular weight was not incorporated in the data plotted in Fig. 3.5.

Van Beek *et al.* [18] have recently investigated the dielectric properties of $(NaPO_3)_x$ in solution. Since most of the work on the dielectric behavior of polyelectrolytes has been carried out in connection with organic materials, the development of polyelectrolyte dielectric theory is presented in Section C3. As will be seen, two dielectric relaxation times corresponding to two dispersion mechanisms—one in the kilohertz range and the other at megahertz frequencies—are observed with high-molecular-weight flexible polyelectrolytes. However, for rigid, inflexible polyelectrolytes, only a single

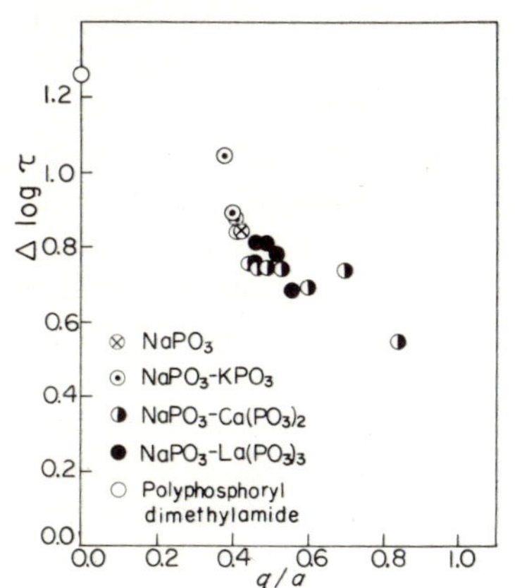

FIG. 3.5. Dependence of $\Delta \log \tau$ on q/a for phosphate polymers (not corrected for molecular weight) [13].

relaxation mechanism is expected. For a sodium polyphosphate of $P = 338$, van Beek *et al.* observed a dielectric response typical of a high-molecular-weight flexible polymer, i.e., two dispersion regions. By contrast, for the $P = 112$ sample, only one dispersion region was seen, implying that at this chain length, the polyion behaves as a rigid rod. For the sample of intermediate chain length ($P = 198$), the dielectric response was intermediate. The results indicate that the borderline chain length for high-molecular-weight dielectric behavior in $NaPO_3$ polymers is about $P = 200$. For the solid-state viscoelastic response, the onset of high-molecular-weight behavior was found to be somewhat lower, at $P = \sim 140$.

2. Polyphosphates Subject to Bond Interchange

In contrast to the behavior of the phosphates with alkali or alkaline earth cations, in which no bond interchange is encountered, some transition metal cations do catalyze bond interchange in the glass-transition region [19]. La^{3+} is one of these ions, not surprisingly, as Bamann [20] showed that lanthanum hydroxide promotes the hydrolysis of α-glyceryl phosphates, while Butcher and Westheimer [21] found that the same cation in a hydroxide gel led to more than a thousandfold increase in the rate of the hydrolysis of phosphate esters.

As in the studies discussed in Section A1, stress relaxation was used to determine the relaxation mechanism. Bond interchange manifests itself in several ways. One of these depends on the fact that the shift factors for the stress relaxation curves a_T do not have the same temperature dependence as the maximum relaxation time τ_m [5]. In normal polymers, these two factors are a reflection of the same mechanism and, therefore, have an identical temperature dependence. Thus, a plot of one against the other yields a straight line of slope 1. By contrast, if more than one mechanism is responsible for the relaxation behavior, and the mechanisms have different temperature dependencies, then a_T and τ_m will reflect differing contributions of the two mechanisms, and a plot of one against the other should no longer yield a straight line of slope 1. This is shown in Fig. 3.6 for various concentrations of La in polymeric $NaPO_3$. It should be understood that the ratio $(La^{3+}/3 + Na^+)/PO_3^-$ is still one. In Fig. 3.6, it is particularly noteworthy that the plot for pure $NaPO_3$ is typical of systems with only one relaxation mechanism, while all the others indicate the presence of more than one mechanism. One other point is worth noting. A very large change in log a_T corresponds to only a small change in τ_m. This suggests that the relaxation process, which predominates at short times (as reflected in a_T), is primarily due to the normal diffusional motions of the chain, whereas the one occurring at long times (as reflected in τ_m) is due to both, with the second mechanism predominating.

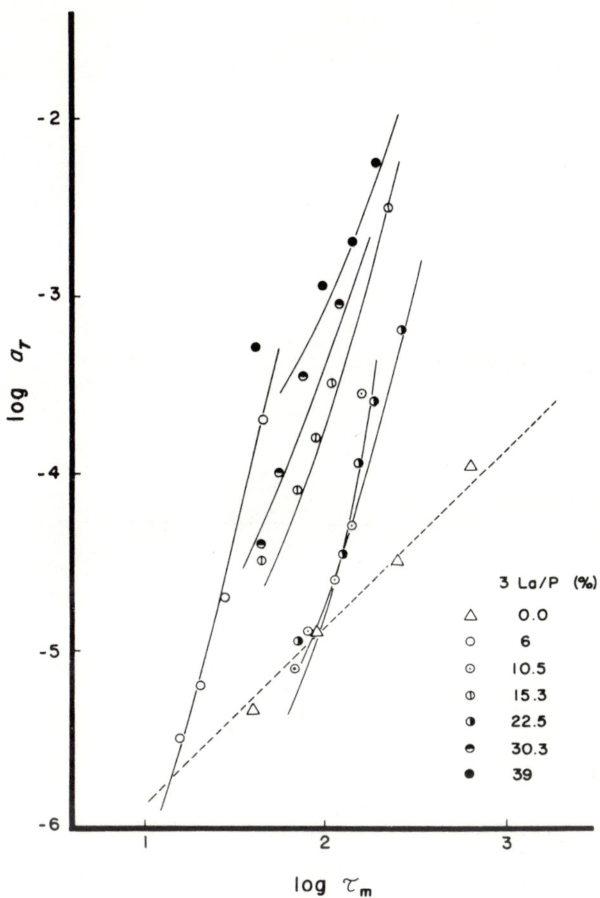

FIG. 3.6. Shift factor versus maximum relaxation time for Na^+–La^{3+} phosphate counterion copolymers. Slope of dashed line is unity [19].

Another manifestation of the presence of a second relaxation mechanism can be found in the breakdown of time–temperature superposition [5]. This can be seen in Fig. 3.7, for both creep compliance and stress relaxation. The graphs in this figure were drawn in such a way as to maximize overlap in the short-time region; most of the experimental points in the region in which time-temperature superposition is applicable have been omitted for simplicity.

The type of viscoelastic response illustrated in Fig. 3.7 suggests the existence of two concurrent relaxation mechanisms, each with a different temperature dependence. It is assumed that the contributions of the two mechanisms to the total compliance are additive [8]. A general method for

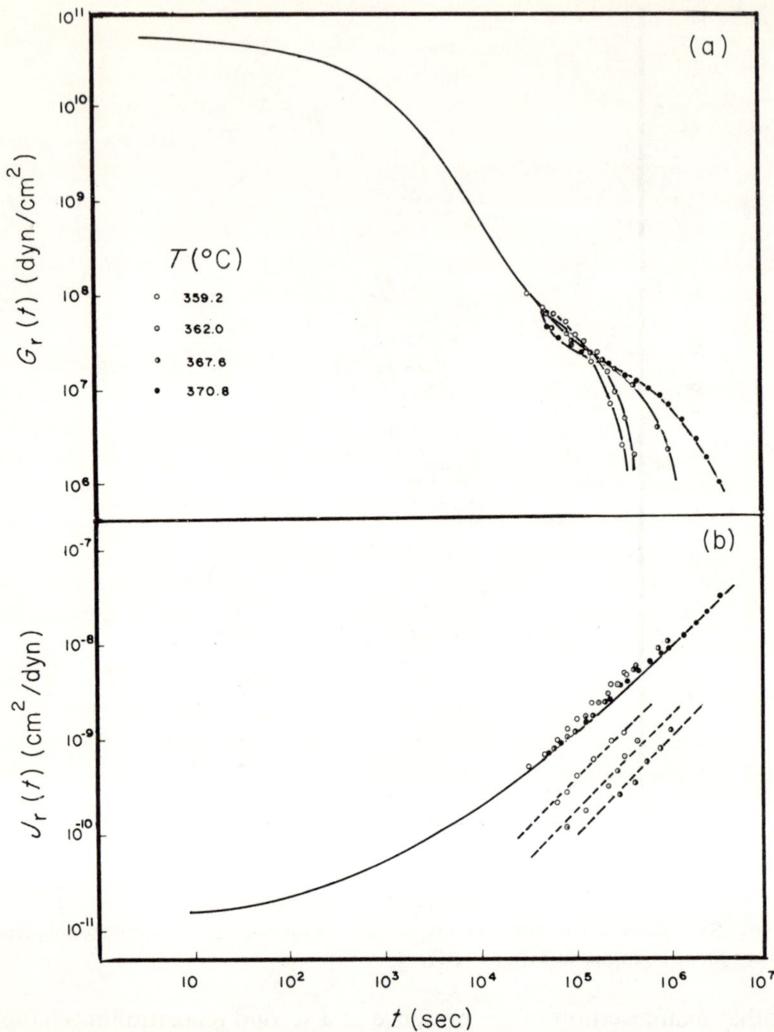

FIG. 3.7. (a) Stress relaxation master curve for phosphate sample with 3La/P = 30.3%. (b) Creep compliance master curve for same material, showing subtracted compliances [19].

the analysis of this type of data from ion-containing polymers has been presented [22], based on the method originally employed by Ferry *et al.* [8] in the analysis of side-chain motion in poly(alkyl methacrylates); a brief summary is given here.

The moduli from the original stress relaxation curves at each temperature, as well as the moduli from the upper envelope of modulus versus reduced time, are converted to compliances [23]. It can be seen in Fig. 3.7 that this procedure results in a pseudo master curve of compliance versus reduced time which is similar to, but not the same as, the mirror image of the original stress relaxation master curve. It is assumed that, at any given temperature, the short-time compliance is mainly due to the primary relaxation mechanism; the upward deviations from the lower envelope of the compliance-time curves are therefore due to the contribution of the second mechanism at that temperature. The separation of the contributions of the two relaxational modes is accomplished by point-by-point subtraction of the compliance values of the lower envelope from the values corresponding to the deviations at long times.

By following this procedure, it is possible to calculate the contribution of each of the modes to the total relaxation and to identify the molecular mechanisms responsible for each of these processes. In this particular case, the mechanism that predominates at very short times (as determined from a_T) has activation energies at $T_g + 30°$ ranging from ~ 150 to ~ 210 kcal (depending on the La content). The mechanism that dominates the long-time behavior (as reflected in the T_m values) has activation energies of 45 ± 10 kcal. As might be expected, the pure $(NaPO_3)_x$ shows an activation energy at T_g determined from either a_T or τ_m of 230 kcal, reflecting the presence of only one mechanism. Since the high value of the activation energy determined from a_T is characteristic of molecular chain diffusion, the mechanism responsible for it is taken to be the same as that encountered in the normal organics. By contrast, the mechanism that gives rise to the low activation energy is taken to reflect bond interchange. It is worth noting that a plot of log T_m versus $(1/T - 1/T_g)$ gives essentially a single curve independent of the La content within wide concentration ranges.

From a correlation of the La concentration with the rate of stress relaxation, it was shown that the bond interchange occurs at the La ions, and not at random. Finally, from a consideration of the kinetics of bond interchange as measured by stress relaxation [24–27], the first-order rate constant for the bond interchange reaction was calculated. At T_g, the value of k is $\sim 1 \times 10^{-3}$ sec^{-1} independent of the concentration of La, the activation energy being 47 ± 10 kcal. While no analogs to the above phenomena have, as yet, been found in organic polymers, it can be anticipated that some polyesters or other materials containing labile bonds will be subject to catalyzed bond interchange in the presence of suitable ions.

3. Silicates

The viscoelastic properties of the silicates, like those of the phosphates, have received considerable attention. Because of the enormous differences in properties between the silicates and most common organic polymers, in most of the studies dealing with the silicates, viscoelasticity has not been approached from the point of view that polymer chemists would normally take. In general, three major contributions to mechanical relaxation in silicates may be considered: bond interchange and molecular diffusion in the region of T_g and above, and the motion of ions or small segments at lower temperatures. The viscoelastic properties of silicates in the region of their glass transitions can be described in the same terms as the viscoelasticity of normal linear polymers, i.e., through the preparation of master curves by time–temperature superposition, the reduction of master curves to T_g and the determination of WLF constants. The viscoelastic properties of silicates in the vitreous state are only briefly described here, and only insofar as they relate to their properties as ion-containing polymers. Mechanical-properties studies on silicate glasses have received extensive treatment in the literature, and the interested reader is referred to the various treatises which deal with this subject [28–30].

One of the major topics of interest in the study of the viscoelastic behavior of silicate glasses has concerned the investigation of the ionic relaxation mechanism below the glass-transition region. This research is of considerable commercial interest since the motion of ions in silicate glasses affects not only their mechanical properties but other properties such as electrical conductivity. The most widely used experimental approach in these studies is the measurement of internal friction, generally by means of a torsion pendulum. Although the analysis of data has not proceeded from a polymeric point of view, the same principles of linear viscoelasticity have still been applied.

The temperature dependence of the internal friction has been found to vary with a number of factors, including thermal history, frequency, and composition. It is worthwhile to summarize the main features of the composition dependence of internal friction in the silicate glasses. Except for a peak at very low temperature [31], no maxima are seen (below 500°C) in the internal friction of fused silica. In binary alkali metal silicates, two maxima are typically seen: one in the region of room temperature and one about 200° higher [32], as shown in Fig. 3.8. It can be seen that an increase in the cation content of the silicate glass results in increased damping as well as a shift to lower temperatures, corresponding to a loosening of the silica structure. The low-temperature peak is attributed to a relaxation mechanism involving the cation [2]. Similar peaks are seen in the alkaline earth silicates as well as in the phosphate glasses [33]. Several possible mechanisms have been proposed to account for the higher peak. Its most probable origin is in the migration of the nonbridging oxygens, as suggested by Fitzgerald and Laing [34] and confirmed by others.

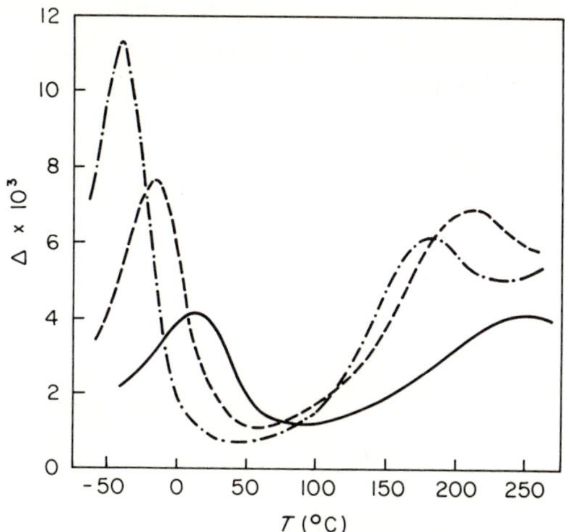

FIG. 3.8. Comparison of internal friction versus temperature curves for three annealed soda silicate glasses: 34% Na_2O (–·–), 25.6% (– – –), 17% (———) [32].

The viscoelastic behavior of silicate glasses in the glass-transition region was used to elucidate the relaxation mechanism of silicates in the composition range $M_2O:SiO_2 \approx 1$, i.e., in the region of maximum linearity [35]. Stress relaxation measurements were made on samples in which the composition ratio ranged from 1.2:1 to 1:1.2, i.e., from very short chains ($\bar{P}_n \approx 5$) to the high polymer (1:1) all the way to highly cross-linked systems (1:1.2). A mixture of counterions had to be used, since the presence of only one cation led to rapid crystallization. As can be seen in Fig. 3.9, all the master curves are identical within experimental error, indicating that bond interchange rather than molecular diffusion must be the predominant relaxation mechanism, since the structures of the samples are drastically different.

B. ORGANIC NONIONIC HOMOPOLYMERS

1. LOW-MOLECULAR-WEIGHT SALTS IN POLAR POLYMERS

Several studies have been made with regard to the interaction of polymers and low-molecular-weight salts, but little work has been done in the area of the mechanical properties of undiluted polymer–salt mixtures. It is known that the transport properties of poly(ethylene oxide)—PEO—solutions are affected by neutral salts [36–38]. Liu and Anderson studied the interaction of PEO and potassium salts in solution by means of NMR spectroscopy [39].

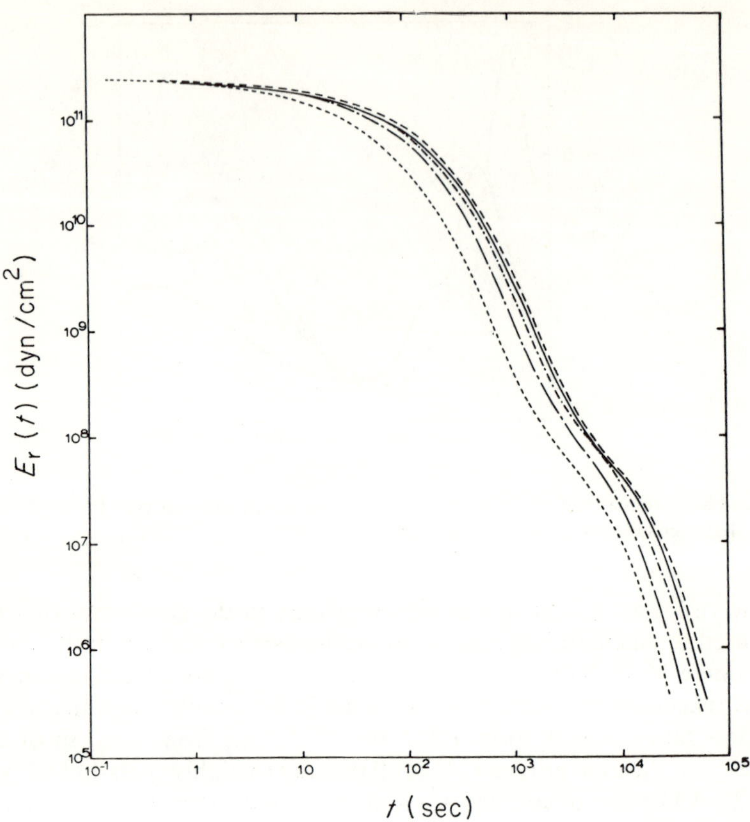

FIG. 3.9. Stress relaxation master curves for five silicate samples: $M_2O:SiO_2$ = 1.0:1.0 (——), 1.1:1.0 (–––), 1.0:1.1 (----), 1.0:1.2 (–––), 1.2:1.0 (–·–). All curves drawn with respect to T_g [35].

In a prior study [40], it had been shown that many factors that influence the macroscopic transport properties profoundly (such as molecular weight or concentration of polymer) have no effect on T_1, the spin-lattice relaxation time in NMR, which is sensitive only to localized molecular motions.

Liu and Anderson investigated two PEO samples of different molecular weights (6×10^3 and 3×10^5) in D_2O and CD_3OD with dissolved KI and K_2SO_4. They first showed that T_1 for the low-molecular-weight sample in D_2O did not change as a function of concentration of either salt, indicating that in that solvent no unusual association effects were present. The behavior of the sample of higher molecular weight was identical. By contrast, studies in CD_3OD indicated that T_1 decreased profoundly both as a function of the salt concentration at constant PEO concentration and as a function of the PEO concentration with a constant amount of salt.

It has been shown [36, 37] that a definite association complex is formed between K^+ and PEO in methanol, and the formation constant for this association is known. Liu and Anderson [39] assumed that two species existed in solution, one of these involving PEO with bound K^+ and the other free PEO. By choosing different values of T_1 for these two species along with the known values of the association constant, the authors could duplicate the experimental T_1 curves. For the free PEO, the value of T_1 is 1.92 sec, while for the bound species it is 0.089 sec, indicating that the mobility of the bound species is considerably smaller than that of the free species. The authors demonstrated, furthermore, that the observed effect could not be interpreted in terms of the influence of the dissolved salt on the transport properties of the solvent, which can be appreciable.

In view of the above, it is reasonable to expect that strong interaction between polyethers and salts will also be observed in the absence of solvent in cases where the salt is directly soluble in the polyether. The first study that dealt with the mechanical properties of polyether–salt mixtures is that of Moacanin and Cuddihy who investigated the dynamic mechanical properties (at ~1 Hz) and the modulus-temperature behavior of poly-(propylene oxide)—PPO—containing dissolved lithium perchlorate [41]. Two samples of PPO (molecular weight 2×10^3 and 7×10^5) were employed in the investigation. Volumetric studies on both samples showed a pronounced negative deviation from volume additivity, indicating a strong interaction between the two components, i.e., the polymer and the salt.

The damping constant as well as the modulus–temperature curves of the low-molecular-weight sample for various lithium perchlorate concentrations are shown in Fig. 3.10. It is evident that initially, as the salt concentration increases, the glass temperature does also (see Section B of Chapter II), and that the damping peak itself broadens. As the salt concentration increases beyond ~10%, the peak narrows again, but the rate of change of T_g with salt concentration increases once the salt concentration exceeds ~20%. The authors hold that the broadening of the peak is suggestive of the superposition of two relaxation spectra, one corresponding to the polyether and the other to an ether–salt complex in which each oxide unit is associated with one lithium perchlorate nearest neighbor. Since the material exists in one phase, however, it exhibits only one glass-transition temperature.

The high-molecular-weight PPO–salt mixtures, by contrast, yield a completely different set of damping curves, as shown in Fig. 3.11. It is seen here that, as the concentration of salt increases, the intensity of the peak at ~ −65°C decreases. At the same time, a second peak at higher temperatures makes its appearance; both its intensity and its position in temperature increase with increasing salt concentration. Between 12 and 15% salt, this new peak is located at ~ −10°C. Above 15%, a completely new peak begins to appear, the low-temperature peak having disappeared completely.

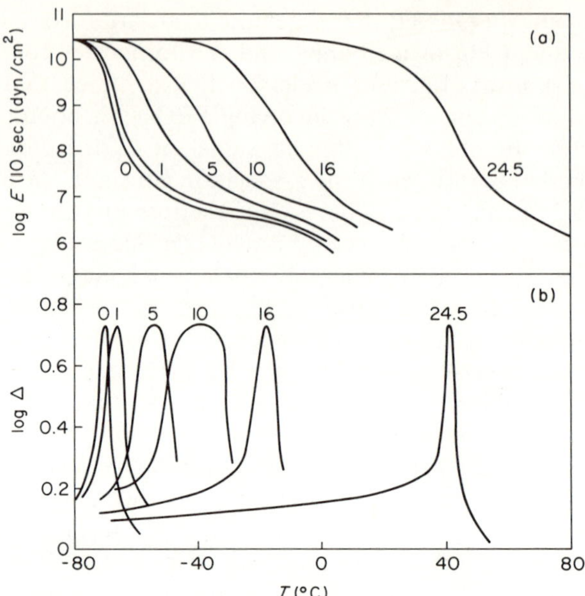

FIG. 3.10. (a) Ten-second Young's modulus and (b) damping constant at ~1 Hz for low-molecular-weight poly(propylene oxide) containing dissolved $LiClO_4$. Numbers on curves indicate weight percent $LiClO_4$ [41].

These phenomena can be explained on the basis of phase separation; in the polymer of high molecular weight the first peak is due to the polymer, while the high-temperature one is due to the adduct or complex of salt and polymer. The phase separation in this case is due to the increased molecular weight, which, as in all other cases, decreases the solubility of the material in solvents. The phase-separation phenomenon is confirmed by the modulus–temperature plots for these materials, also shown in Fig. 3.11, which are typical of two-phase systems, in contrast to those for the low-molecular-weight polymer.

Moacanin and Cuddihy point out that a composition of ~20% (wt) of $LiClO_4$ corresponds to the situation in which every polymer repeat unit is associated with one lithium perchlorate moiety, assuming a coordination number of 12 for the salt. Above this salt concentration the glass transition

FIG. 3.11. (a) Ten-second Young's modulus and (b) damping constant at ~1 Hz for high-molecular-weight poly(propylene oxide) containing dissolved $LiClO_4$ [41]. (c) Dielectric loss versus temperature at 1000 Hz for pure poly(propylene oxide) and for poly(propylene oxide)–$ZnCl_2$ mixtures; 1 = poly(propylene oxide), 2 = poly(propylene oxide) + 12.2 mole % $ZnCl_2$, 3 = poly(propylene oxide) + 16.6 mole % $ZnCl_2$ [43].

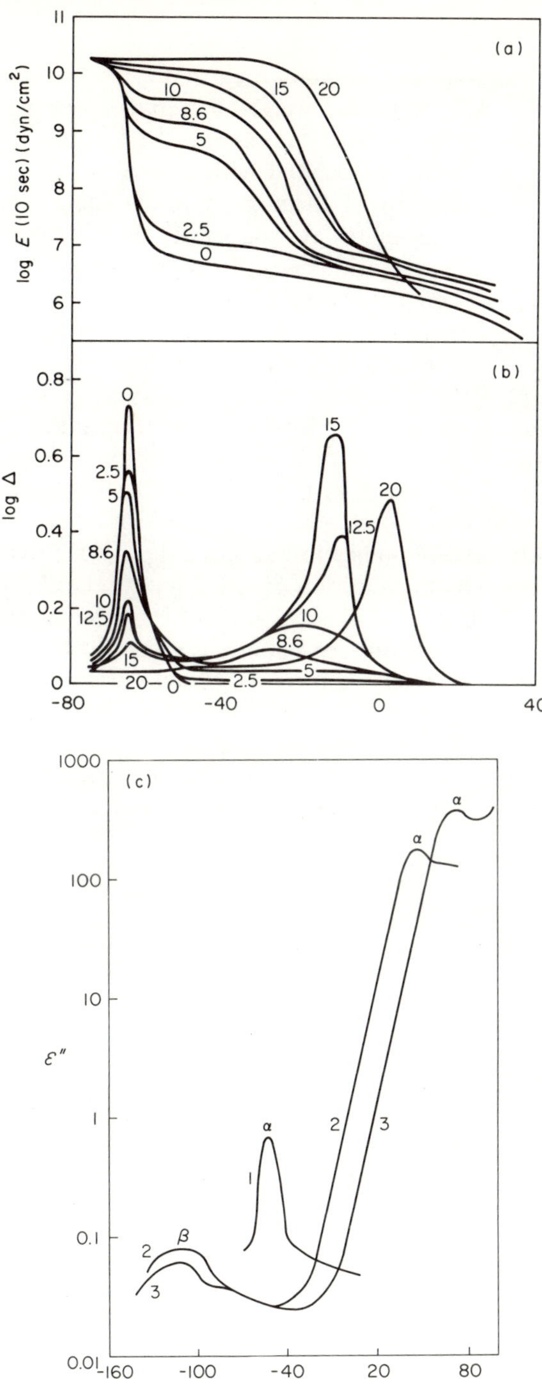

(a)
log E (10 sec) (dyn/cm²)
11
10
9
8
7
6
10
15
20
8.6
5
2.5
0
(b)
log Δ
0.8
0.6
0.4
0.2
0
15
20
12.5
10
8.6
5
2.5
-80
-40
0
40
(c)
1000
100
10
1
0.1
0.01
ε″
α
β
1
2
3
-160
-100
-40
20
80
T (°C)

of the low-molecular-weight polymer begins to rise more steeply than before, and in the high-molecular-weight material a new peak makes its appearance. This is taken to imply that above ~20% $LiClO_4$ not only are all the polymer repeat units associated with salt species but that multiple associations are beginning to be formed, in which some repeat units are associated with more than one ionic species. It should be noted that the occurrence of multiple associations could result from ion clustering, and that the 20% salt level might very well represent the onset of clustering in this system. This concentration corresponds to 14 mole % salt, and in the PAA system, as will be seen in Section Cl, the onset of clustering occurs between 10 and 22% neutralization, i.e., at roughly similar ion content. In other ion-containing systems, as pointed out in Section A2 of Chapter II, the onset of clustering is seen at various values of ion concentration.

Hannon and Wissbrun [42a], in the study of Phenoxy –$Ca(SCN)_2$ mixtures discussed in Section B1 of Chapter II, measured melt viscosities on these polymer–salt mixtures. They found that although melt viscosities were considerably elevated by the addition of salt (e.g., by a factor of 10 at 190°C for 5 wt % salt), the effect could be accounted for by the increase in T_g; i.e., the effect of the added salt on free volume is sufficient to account for the observed increases in viscosity. This is in contrast with the effect seen in a polyoxymethylene copolymer [42b], and also with the effects observed with other ionomer melts, to be discussed in Chapter IV. It is, however, consistent with their finding that the strength of polymer–salt interaction in the Phenoxy system is considerably less than that seen in polyether–salt mixtures.

The dielectric behavior of polyether–salt complexes was investigated by James and Wetton [43], who studied mixtures of polypropylene oxide and $ZnCl_2$. On cursory inspection, plots of ε'' versus frequency at various temperatures appear perfectly normal in that a peak (designated α) is visible, that moves to higher frequency with increasing temperature and gives activation energies typical of a glass transition. The height of the α peak, however, is most unusual in that it is two orders of magnitude larger than that seen in normal amorphous polymers. This can be seen in Fig. 3.11c which illustrates the effect of added $ZnCl_2$ on the height and position of the α transition at 1000 Hz. ε' versus temperature plots show parallel increases in height above those of normal polymers, the α peak appearing again in the range $100 < \varepsilon' < 1000$.

In pure PPO, the α peak at 1000 Hz appears at ~20° to 30° above the T_g as measured by DSC. However, with the addition of $ZnCl_2$, this temperature difference increases by a further 30 to 40°. The authors suggest that an additional hgh-temperature ($T > T_g$) process overwhelms what would otherwise be a normal α transition. One possibility is the displacement of

Cl by ether oxygens in the coordination sphere of Zn, which would lead to the formation of ionic species. As pointed out, the values of E_{act} for the α transition are typical for the glass transition (pure PPO: $E_{act} > 115$ kcal; 12.2% $ZnCl_2$: ~80 kcal; 16.6% $ZnCl_2$: ~115 kcal). The position and activation energy of the β peak are independent of salt content in the region studied.

2. Charge-Transfer Complexes

Charge-transfer bonding in polymers has been a subject of interest for a number of years. Several physical aspects of these complexes have received attention, the most important from the point of view of potential applications being their electrical properties [44–46]. Most of the polymeric charge-transfer complexes that have been prepared have involved polymers of comparatively low molecular weight. As a result, their mechanical properties have been of limited value and consequently not well explored. However, the few studies that have involved high-molecular-weight complexes indicate that they may have useful mechanical properties in conjunction with the electrical applications for which they are being considered.

A recent investigation by Kumanotani *et al.* [47] found that complexation between polymeric donor molecules and low-molecular-weight acceptor molecules leads to increased T_g of the polymer and increased mechanical strength. The complexes were formed by adding chloranil, a Lewis acid, to various donor polymers of the addition type, such as poly (*N,N*-dimethylaminophenylethyl acrylate), and of the polyurethane type, containing the moiety

$$-\phi-N-(CH_2CH_2OH)_2$$

With the addition-type polymers, the T_g was found to increase with increasing concentration of chloranil at the rate of ~1.3°/mole % chloranil. The modulus-temperature curves of these materials, represented by a typical example in Fig. 3.12, show that the level of the rubbery plateau increases with increasing mole % chloranil, suggesting a type of cross-linking effect. With the polyurethane complexes, two inflection points were observed in the modulus–temperature curves, and corresponding to this, two maxima in the loss tangent. It is assumed that one corresponds to T_g and the second to a new transition above T_g.

The mechanical properties, which are observed to result from charge-transfer complexation, are entirely analogous to those that result from polymer–salt interaction, as discussed in Section B1. It will be seen that this behavior is also exceedingly similar to that seen in the styrene ionomers to be discussed in Section A of Chapter IV. The mechanical effects of each type of interaction are very similar—increased T_g, the formation of a

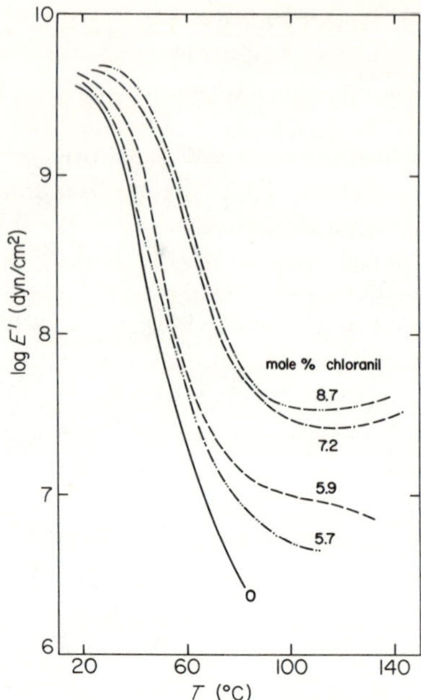

FIG. 3.12. Dynamic storage modulus at 11 Hz as a function of temperature for poly(*N*,*N*-dimethylaminophenylethyl acrylate) complexed with chloranil (mole % shown) [47].

second rubbery plateau, and the appearance of a new loss tangent maximum above T_g.

In addition to polymer–monomer charge-transfer formation, polymer–polymer complexation is also possible, and the development of suitable acceptor polymers has received some attention. A charge-transfer complex was prepared from two condensation polymers of the polycarbonate type by Sulzberg and Cotter [48]. They found that a 50:50 mixture of donor and acceptor polymer had a higher tensile modulus than either species by itself, indicating that the effect of polymer–polymer complexation is network formation, involving at least some of the active units in each chain.

Charge-transfer complexes of the polymer–polymer type are analogous to the polysalt complexes, which will be discussed in Section E of Chapter IV. It is to be expected that there are many qualitative similarities between the properties of charge-transfer complexes and polysalt complexes. Differences between the two types of complex result from differences in the strengths of the interactions, which are presumably greater in the case of the polysalts.

C. IONIZABLE ORGANICS

1. Polyelectrolytes in Bulk and Concentrated Solution

Ionizable organic homopolymers are generally termed polyelectrolytes when they are aqueous solution. Their thermodynamic and conformational properties have received considerable attention in other works [49–51] and are treated only briefly in this review. In this section, the viscoelastic properties of polyelectrolytes are described. Section C1 deals with their solid-state viscoelasticity; Sections C2 and C3 are concerned with their viscoelastic behavior in solution. Chapter V summarizes their conformational properties.

The polyelectrolyte solid state is operationally defined as polyelectrolyte materials, which are self-supporting and includes both unplasticized polymers and highly concentrated solutions. It will be seen that the properties of plasticized polyelectrolytes are similar to those of unplasticized polymers in so far as they can be determined, even at high plasticizer contents. Studies on the mechanical properties of unplasticized polyelectrolytes have been limited by the difficulties involved in the preparation of well-defined samples; many of these difficulties can be overcome by studying plasticized materials, although other complications are then introduced.

The only investigations of note on the viscoelasticity of unplasticized polyelectrolytes are those reported by Nielson and co-workers [52–54]. They obtained viscoelastic data for several ion-containing polymers, including salts of poly(methacrylic acid), poly(acrylic acid), and an acrylic acid copolymer containing 6% 2-ethylhexyl acrylate (hereafter referred to as PMAA, PAA, and AA–0.06EHA, respectively). The studies showed that ionization produces drastic changes in the mechanical properties of these materials. Young's moduli greater than 1×10^{11} dyn/cm^2 were reported for the sodium salts of PMA and PAA, while shear moduli of various divalent salts (Zn, Pb, Ca) of AA–0.06EHA were found to range between 5.5×10^{10} and 6.5×10^{10} dyn/cm^2. These values represent increases by a factor of about 1.5 for the sodium salts and about 2 for the divalent salts when compared with the corresponding moduli of the unionized polyacids. The moduli of the polyacids themselves are greater by a factor of 2 than those of typical nonionic, non-hydrogen-bonded polymers like polystyrene. The modulus values of the divalent salts may in fact be too low [54] because the method of preparation—simultaneous molding and neutralization with metallic oxide—would likely produce an inhomogeneous product.

In addition to the moduli, the glass-transition temperatures of these ionized polyacids are greatly increased; they were not measured specifically, but no significant softening was observed up to 275°C. In fact, other tests indicated that they neither soften nor lose shape before their decomposition temperatures. Another notable change in the physical properties of polyelectrolytes is in the (glassy) linear expansion coefficient α_g. The salts of the polyacids were found to have values of α_g as low as one-fourth of those of the unionized polymers. Indeed, in this property, as well as in many others, solid-state polyelectrolytes are more like inorganic glasses than organic polymers.

The dramatic effects of ionization on the viscoelastic behavior of polymers can still be observed in plasticized systems. Eisenberg *et al.* [3] studied the viscoelasticity of plasticized sodium salts of PAA, using a number of aqueous and nonaqueous plasticizers. X-ray diffraction studies, similar to those discussed in Section A of Chapter II, indicate that these materials undergo microphase separation through the phenomenon of ion aggregation, as do most other ion-containing organic polymers, at least in some concentration regions.

As a consequence of their structure, the viscoelastic response in the plasticized poly(sodium acrylates) is complex [3]; this is demonstrated by the breakdown of time–temperature superposition over a wide range of composition. Stress relaxation measurements taken over ~3.5 decades of time indicate that this occurs for a variety of both polymer and plasticizer compositions [55]. An example of the type of stress relaxation behavior in this system can be seen in Fig. 3.13, for a PNaA sample of high molecular weight containing 35% (wt) H_2O. The high glassy moduli ($\sim 1 \times 10^{11}$)

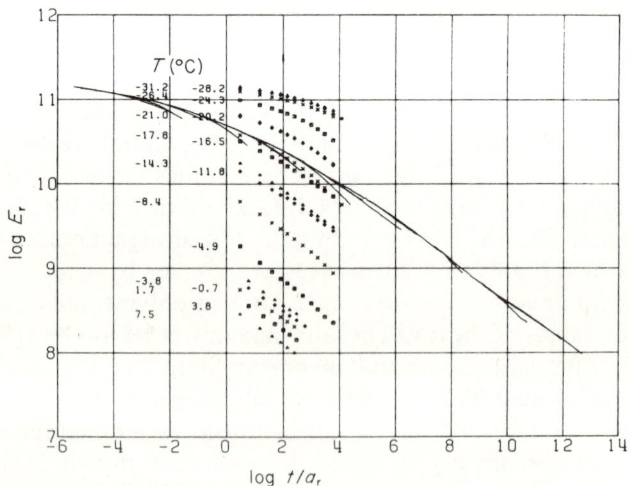

FIG. 3.13. Reduced Young's modulus versus time as a function of temperature for PNaA–H_2O (35%). Solid lines represent pseudo master curve [55].

seen in the unplasticized polymer are still apparent in this highly plasticized material.

Figure 3.13 shows the individual curves of Young's modulus versus time for the various constant temperature runs, as well as an attempt to construct a master curve in which the overlap between successive curves is maximized in their short-time regions. It is apparent that the result of this procedure is not a true master curve, but what may be termed a "pseudo master curve," which describes only the short-time relaxation behavior at any given temperature.

The failure of time–temperature superposition in PNaA was found to be quite general, occurring with each plasticizer studied (water, glycerine, ethylene glycol, and formamide); it was also observed in partially neutralized PAA (above 10% neutralization). It was seen that the magnitude of the long-time deviations from the upper envelope of short-time moduli varied with the amount and type of plasticizer and with the degree of neutralization of the PAA; however, in every case, the deviations were of the same nature as those seen in Fig. 3.13, i.e., toward values of moduli lower than those on the upper envelope of short-time moduli. In this latter respect, the similarity between the pseudo master curves in Figs. 3.7 and 3.13 should be noted, showing the similarity in viscoelastic behavior for polymers of completely different backbone structures. In Chapter IV, it will be seen that styrene ionomers containing more than 6 mole % ions also exhibit the same type of viscoelastic response.

Because the pseudo master curves determine short-time behavior only, it is not surprising that the variations in their shapes are closely paralleled by the variations in the shapes of curves of 10-sec modulus versus temperature. Because of their greater ease of presentation and simplicity in interpretation, modulus-temperature curves are shown here to illustrate the variation in mechanical behavior of these polymers with the percentage and type of plasticizer and with the degree of neutralization. In Fig. 3.14, the 10-sec

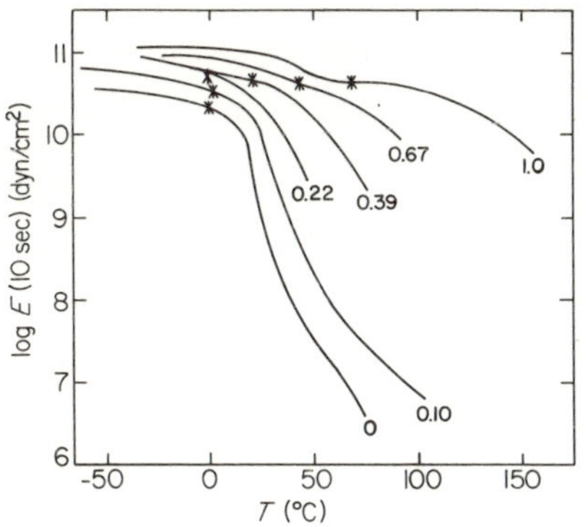

FIG. 3.14. 10-sec modulus of PAA versus temperature as a function of the degree of neutralization. The stars on each curve indicate the position of T_g; degree of neutralization is given on each curve; weight fraction formamide $\simeq$ 0.2 [3].

Young's modulus is given as a function of the temperature for PAA samples with degrees of neutralization between 0 and 100% and with a common plasticizer content (~20% formamide). The stars on each curve indicate the position of T_g, as measured by a thermal-expansion technique [56]; the values correspond reasonably well with those determined from DSC measurements [57]. It is worthwhile to note the rise in glassy modulus (by a factor of ~3 at 0°C, for example) and the great increase in the breadth of the transition as the ion content increases. It is clear that the high glassy moduli must be due to additional intermolecular bonding which results from the presence of ions.

Figure 3.15 illustrates the variation in the modulus of the fully neutralized polymer as a function of the percentage of formamide. One can see that the breadth of the transition and the glassy modulus decrease as the plasticizer content increases, as one might expect. In Figs. 3.14 and 3.15, an inflection point below T_g appears in the modulus–temperature curves for low percentages of formamide and high degrees of neutralization. This inflection point seems to correlate with the occurrence of a broad shoulder in the loss tangent below T_g, as indicated by the dynamic mechanical measurements to be discussed in more detail later.

The shapes of some of the modulus–temperature curves presented thus far are reminiscent of those obtained for partially crystalline polymers such as low-density polyethylene [58]. However, X-ray diffraction patterns obtained from a variety of PAA salts showed no evidence of crystallinity.

With other plasticizers, and most notably with water, the decrease in modulus with temperature (which parallels the rate of relaxation) was found to be considerably greater than it was for formamide. A comparison of the effect of various plasticizers on the modulus–temperature behavior of PNaA is shown in Fig. 3.16. The curves drawn in this figure were all obtained for

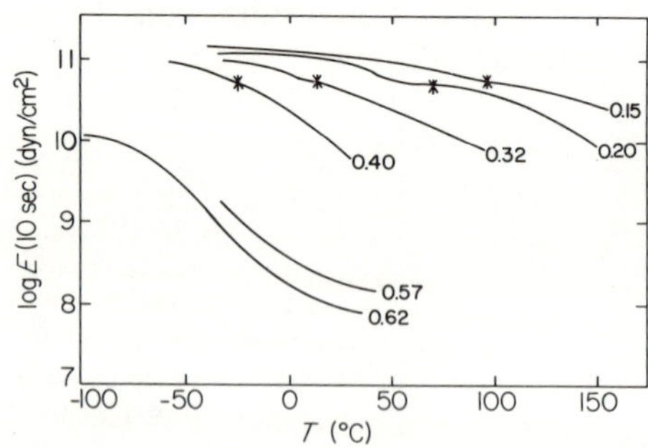

FIG. 3.15. 10-sec modulus of PNaA versus temperature as a function of the formamide content. The stars on each curve indicate the position of T_g; the weight fraction of formamide is given on each curve [3].

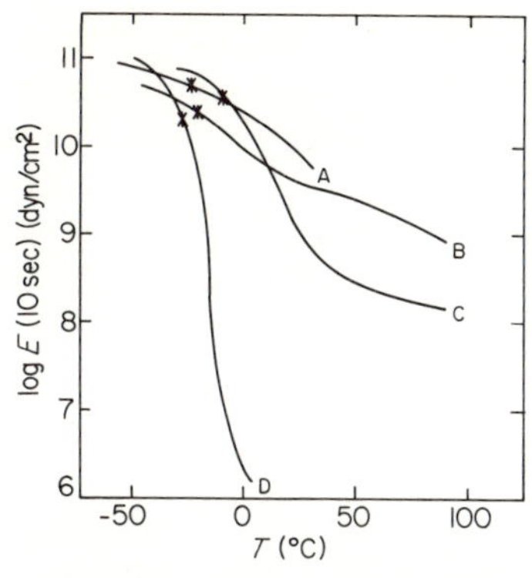

FIG. 3.16. Modulus–temperature curves for PNaA with four different plasticizers (weight fraction $\simeq$ 0.4): (A) formamide, (B) ethylene glycol, (C) glycerine, (D) water [3].

approximately the same volume fraction of plasticizer. The glassy moduli and the glass-transition temperatures of these polymers are all similar; they also contain clusters, as seen by X-ray diffraction studies in the formamide and water-plasticized systems. Nevertheless, above T_g, considerable variation in the efficiency of these plasticizers is seen. This variation is a good indication that the plasticizer plays a very important role in influencing the lifetimes of the ionic clusters.

The same method of analysis that was used in the La^{3+} polyphosphate system described in Section A2 was applied to nonsuperposable stress relaxation data obtained for the PAA salts [22]. This procedure enabled one to determine the contribution of the second (long-time) mechanism to the total compliance. The results of this procedure for PNaA–H_2O (35%) are shown in Fig. 3.17. In addition to the curves of secondary compliance versus time, this figure shows an attempt to prepare a second master curve by shifting the individual compliance curves horizontally relative to the curve at the highest temperature. In view of the errors involved in the subtractive procedure, the degree of overlap obtained seems quite reasonable. The significance of the dashed line in Fig. 3.17, representing a theoretical slope of unity, will be discussed later.

The natures of the two relaxation processes and their relationships with the polymer structure are not immediately clear; however, an analysis of their temperature dependencies will be useful in relating them to physical processes. The mechanism whose contribution dominates the short-time behavior above T_g can be investigated by an analysis of the primary shift factors a_T, which contribute to the primary master curves.

It is of interest to determine whether the shift factors corresponding to the primary mechanism follow an Arrhenius-type temperature dependence, or whether they are of WLF type, as given in Eq. (1). In the system PNaA–glycerine, where a wide range of temperature above T_g is accessible for stress

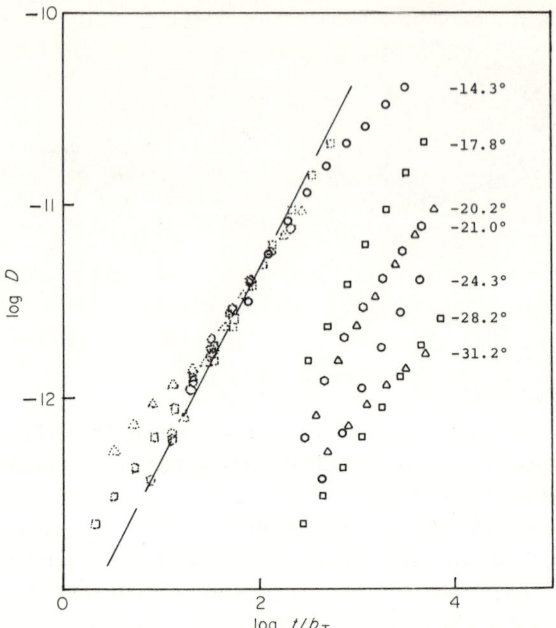

FIG. 3.17. Master curve of subtracted compliance for PNaA–H_2O (35%). Individual curves and corresponding temperatures are shown. The dashed line represents a theoretical slope of unity [55].

relaxation measurements, the WLF-type temperature dependence of the short-time shift factors is clearly observed [55].

With volatile plasticizers, notably formamide and water, the accessible temperature range above T_g is small because of plasticizer loss. Thus, the establishment of WLF parameters from stress relaxation measurements is impossible in these plasticized systems. However, indirect evidence was found to support the applicability of the WLF equation to the primary relaxation process even in these cases. The WLF equation can be used to predict the apparent activation energy at T_g, $(\Delta H_a)_{T_g}$; the relationship, which can be derived from Eq. (1), is given by

$$(\Delta H_a)_{T_g} = 2.303R(C_1/C_2)T_g^2 \tag{3}$$

The constants C_1 and C_2 could not be established for the PNaA–formamide series. With some plasticizer–polymer systems, the distribution of relaxation times changes considerably from that of the pure polymer [59]; however, no information is available regarding the variation of WLF constants with plasticizer content, and in the absence of such information, it was assumed that within a single plasticizer–polymer system, C_1/C_2 is either constant or, at worst, a slowly varying function of the plasticizer content, as it is in the glycerine series [55]. Thus, a linear relationship between $(\Delta H_a)_{T_g}$ and T_g^2 can be anticipated for the PNaA–formamide series, according to Eq. (3). This was found to be the case, as can be seen from Fig. 3.18.

In view of the evidence to support the applicability of the WLF equation to the primary relaxation process in PNaA, it is reasonable to attribute its mechanism to the same type of diffusional process that characterizes the glass transition in most nonionic polymers. This assignment is also reasonable in view of the magnitude of the activation energy associated with the primary mechanism. By extrapolating the linear relationship obtained in Fig. 3.18 to the T_g of undiluted PNaA ($T_g = 250°C$, as reported previously [57]), $(\Delta H_a)_{T_g} = 175$ kcal is obtained. This value of $(\Delta H_a)_{T_g}$ is in general agreement with values that can be obtained for a wide variety of thermorheologically simple polymers by the application of Eq. (3) [10].

The activation energy associated with the secondary process can be calculated from the shift factors used in obtaining master curves of the type shown in Fig. 3.17. The activation energies obtained varied from 32 kcal for PNaA–H_2O (35%) to 17 kcal for both PNaA–glycerine (52%) and PNaA–formamide (37%). The variations in these values with the percent plasticizer were not tested because of the insufficient number of long-time runs and the tediousness of the method. It is difficult to define specifically the relaxation process involved here, but it probably involves ionic bond interchange. The energy required would depend on the size and geometry of the aggregate and on a number of other factors, including the solvating effect of the plasticizer. These values of activation energy, which lie between 17 and 32 kcal, seem reasonable for such a mechanism.

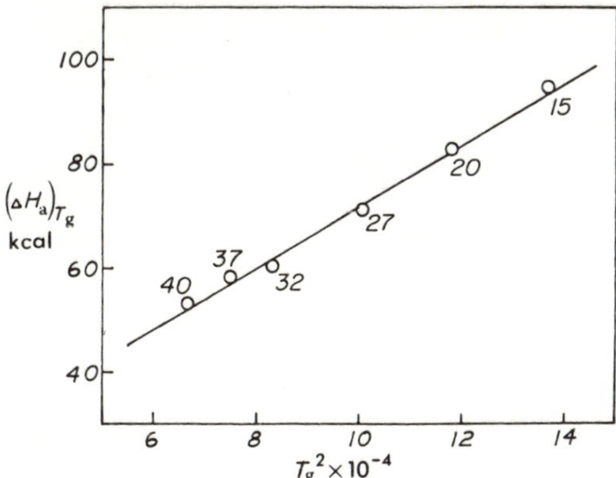

FIG. 3.18. $(\Delta H_a)_{T_g}$ versus T_g^2 for PNaA–formamide series. Numbers on the points indicate weight % formamide [3].

Whatever its form, if the secondary relaxation process were due to the yielding of an ionic phase, its contribution to the compliance would correspond to a pure viscosity in the same fashion as simple bond interchange [5]. In this case, the relationship

$$D(t) \simeq t/\eta \tag{4}$$

should hold. It can be seen in Fig. 3.17, by comparison with the doubly logarithmic slope of unity, that this is a reasonable approximation. Similar slopes were found in the other plasticized systems studied.

Measurements of the loss tangent, tan δ, at ~1 Hz as a function of temperature were made on three samples of PNaA, plasticized with formamide (37%), glycerine (46%), and water (35%). The data are shown in Fig. 3.19, normalized with respect to the primary transition, for the sake of clarity. The dotted lines in Fig. 3.19 indicate the regions that were inaccessible to measurement of tan δ. Although the relative intensities are different, each curve shows a primary maximum and a broad shoulder just below it in temperature. It should be recalled that inflection points below T_g were observed in the modulus–temperature curves of PNaA at low formamide contents and high degrees of ionization (see Fig. 3.14 and 3.15). The shoulder in the loss tangent thus seems to correlate with the inflection points in the modulus–temperature curves of the PNaA–formamide series, as might be expected. The fact that no inflection point is seen in the PNaA–water or PNaA–glycerine series is likely due to the fact that, since the relative intensities of the primary maximum and the shoulder are quite different in these two cases,

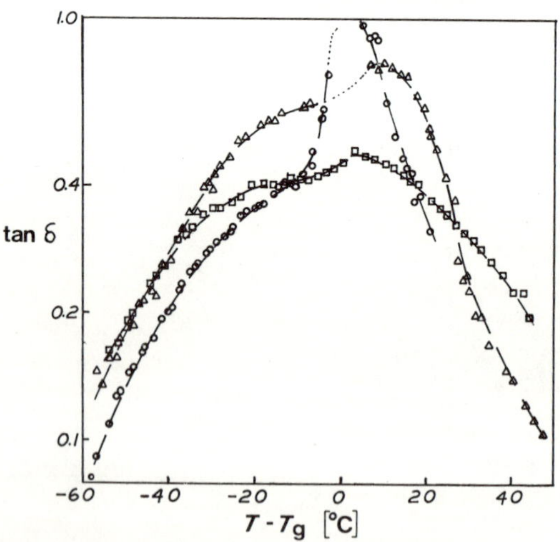

FIG. 3.19. Tan δ versus ($T - T_g$) for PNaA with three different plasticizers: water (○), glycerine (△), and formamide (□) [3].

the effect of the low-temperature shoulder on the modulus would be relatively small.

The correlation of the sub-T_g shoulder observed in the dynamic mechanical study with one of the relaxation processes observed in the stress relaxation experiments should be possible through an analysis of the activation energies of the two dynamic mechanical peaks. This would involve dynamic mechanical measurements in a frequency range in which the peaks would be separated. Assuming that sub-T_g dispersion is of lower activation energy than the primary transition, as might be expected, the separation of the peaks would require lower frequencies than those available on a torsion pendulum.

One method of investigating dynamic mechanical behavior at low frequencies is by the conversion of stress relaxation data to a dynamic mechanical function. This was done for some members of the PNaA–formamide series. The static creep compliance $D(t)$, which had been obtained from the stress relaxation data in the manner described in Section A2, was converted to dynamic compliance by the method of Schwarzl and Struik [60], assuming a semilogarithmic approximation of the third order for $D(t)$. Although the conversion to dynamic compliance is highly sensitive to experimental errors and is subject to large uncertainties because of truncation errors, it was nevertheless possible to reproduce qualitatively the pattern in tan δ that can be obtained from direct dynamic mechanical measurements using a torsion pendulum [55].

2. Polyelectrolytes at Intermediate Concentrations

Any quantitative definition of a concentrated solution must necessarily be rather arbitrary. For the purposes of the present discussion, a polymer solution may be considered to be of intermediate concentration if it lies in the range of polymer concentrations which extends from the lowest concentration in which significant intermolecular contact occurs up to the point of gelation. In general, the boundaries of this region will depend on the choice of variable used to define them, as well as on the particular polymer–solvent system. The properties that define the upper boundary, the gelation point, will be discussed in detail below; the study of reversible gelation discussed in this section is, in fact, an investigation of the effect of a number of physical variables on the concentration dependence of the upper boundary of this region. The lower boundary is considerably less well defined; since this chapter deals with viscoelasticity, the lower boundary is considered here to be the concentration at which the deviations from intrinsic dilute solution viscoelastic behavior become significant. The level of significance is itself arbitrary, and for some of the studies to be presented here, arguments could also be made for their inclusion in Section C3 on dilute solution viscoelasticity.

One of the most valuable investigations in this area is the study of reversible gelation in partially neutralized poly(methacrylic acid)—PMAA—by Silberberg and Mijnlieff [61]. These authors employed dynamic mechanical measurements in the region of 1 Hz to study the sol–gel transition as a function of polymer concentration and the degree of neutralization.

A reversibly cross-linked gel results from the formation of intermolecular bonds of finite lifetime in a moderately concentrated solution of macromolecules. In general, a system will show the mechanical behavior of a gel if there is a minimum of one cross-link per macromolecule whose lifetime exceeds the experimental time scale. The choice of an experimental definition of gelation is somewhat arbitrary. It is possible that a given system will exhibit the properties of a gel in one experiment and those of a solution in another. If the sol–gel transition is sufficiently sharp, as it is in PMAA, the choice of experiment is not so critical.

A workable definition of a sol–gel transition can be obtained by consideration of some of the parameters determining the mechanical behavior of macromolecules in solution. The relaxation spectrum of a solution of macromolecules is characterized by a maximum relaxation time which is dependent on the configuration and size of the macromolecules and on the solvent viscosity. In dilute solution, where no intermolecular interaction occurs, this time is typically of the order of 10^{-5} sec. Increasing the size of the polymer molecules by aggregation causes the maximum relaxation time to increase. Aggregation on a macroscopic scale, as in gelation, results in relaxation times of the order of seconds or minutes. Hence, the onset of gelation should be observable in an experimental time scale of the order of 1 sec, or correspondingly, at a frequency of 1 Hz.

The particular rheological quantity monitored by Silberberg and Mijnlieff was the dynamic shear modulus, $G'(\omega)$, at $\omega \approx 1$ Hz. Simple considerations of rubber elasticity predict that the modulus at the point of gelation in a typical polymer solution (high molecular weight, $\sim 10\%$ polymer) will be of the order of several hundred dynes per square centimeter. The actual value chosen to describe the gel point was $G'(\omega_0) = 500$ dyn/cm^2, where the resonant frequency ω_0 was in the range 1–10 Hz. It was shown that the location of the transition in the PMAA system is quite insensitive to the specific values of either G' or ω_0, since the buildup of relaxation times, corresponding to the increase in aggregate sizes from the colloidal to the macroscopic, is quite rapid.

Experimentally, the study by Silberberg and Mijnlieff involved the determination of the resonant shear modulus $G'(\omega_0)$ and the calculation of the relaxation spectrum $H(\tau)$ in the vicinity of resonance for solutions of PMAA ($M_v = 3.35 \times 10^5$) neutralized to 0, 4, and 8%, respectively. The viscoelastic parameters were determined as a function of temperature for several concentrations in the vicinity of the gel point.

The variation in $G'(\omega_0)$ at 25°C with polymer concentration is illustrated in Fig. 3.20. The sharpness of the sol–gel transition is apparent from these curves, and based on the criterion for gelation of $G'(\omega_0) = 500$ dyn/cm^2, the gel point concentration can be estimated. Increasing the level of ionization at these low degrees of neutralization was found to result in a sharp decrease in the height of the relaxation spectrum at any given PMAA concentration, and corresponding shift in the gel point to higher concentrations. This indicates that under these conditions, ionization tends to inhibit the development of three-dimensional structure instead of promoting it, as one might intuitively expect from observations on ion-containing polymers at higher concentrations or in nonaqueous media.

The relaxation spectrum was used to estimate the dynamic viscosity $\eta'(\omega)$ from the relationship

$$\eta'(\omega) = (1/\omega)G''(\omega) \approx (\pi/2\omega)H(1/\omega) \tag{5}$$

and from this, values of reduced specific viscosity $\eta_{sp}(\omega)/c$ were obtained; these are plotted in Fig. 3.21 as a function of PMAA concentration. The dramatic rise in $\eta_{sp}(\omega)/c$ with concentration in the gelation region, as seen in this figure, parallels the change in $G'(\omega_0)$ with concentration in the same region, as seen in Fig. 3.20. Also, it was shown that the sign of the temperature dependence of $\eta_{sp}(\omega)/c$ changes from negative to positive as the system passes into the gelation zone. This effect is to be expected since it is known that unionized PMAA in dilute solution is highly intramolecularly bonded, and

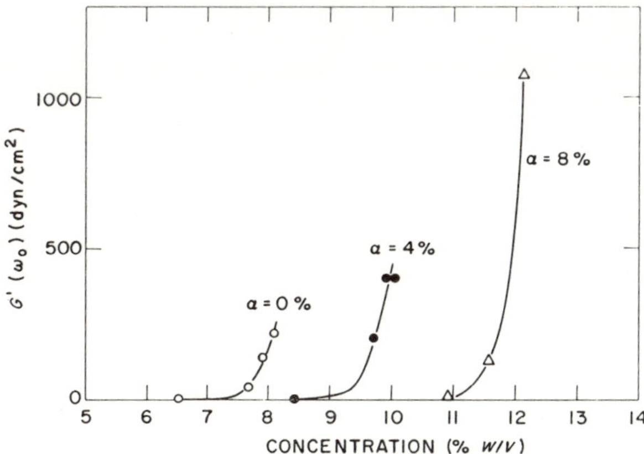

FIG. 3.20. Storage modulus of PMAA at resonance frequency as a function of concentration for various degrees of neutralization. Temperature = 25°C, $M_v = 3.35 \times 10^5$ [61].

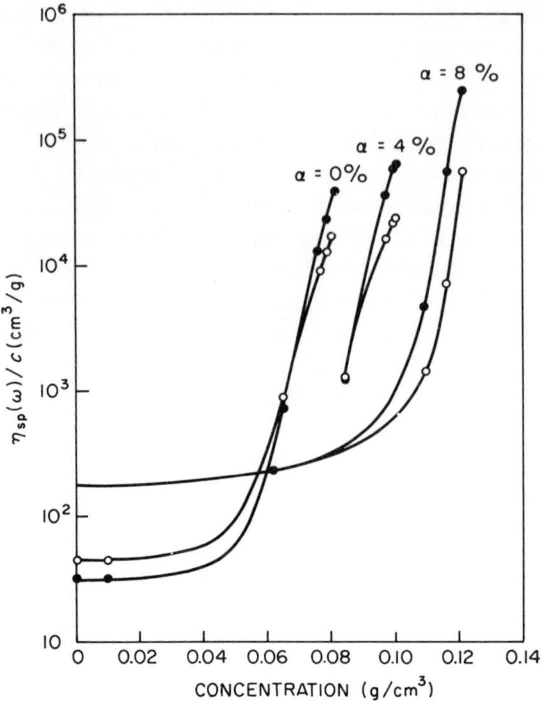

FIG. 3.21. Reduced specific viscosity of PMAA at $\omega = 2\pi$ rad/sec as a function of concentration at 15°C (○) and at 35°C (●) [61].

that bond formation in this system is enhanced at higher temperatures [62]. At these low levels of neutralization, ionization is seen to produce effects which operate in opposite directions in the dilute- and concentrated-solution ranges. In the dilute range electrostatic repulsion leads to a reduction of intramolecular bonding and an increase of the polymer dimensions, while at higher concentrations it leads to an inhibition of intermolecular bonding and a decrease of $\eta_{sp}(\omega)/c$.

The phenomena in the PMAA study were rationalized by Silberberg and Mijnlieff in the following manner: The number of bonds formed between chain segments must be a function of both the inherent bonding tendency of the chain segments and the local-segment concentration. In dilute solution, where the macromolecules are isolated from each other, the local-segment concentration derives mainly from segments of the same macromolecule, as determined by chain statistics; thus, in this case, the number of intermolecular bonds per molecule is extremely small.

At higher concentrations, macromolecular interpenetration increases, and the number of intermolecular contacts (but not necessarily bonds) becomes significant. In this case, the local-segment concentration is composed of contributions from the macromolecule being considered (self-segment concentration) and the contributions from neighboring macromolecules (foreign-segment concentration). The ratio of these two contributions determines the degree of intermolecular bonding. This ratio, and thus the number of intermolecular bonds, is dependent not only on the overall polymer concentration but also on changes in polymer conformation, since a contraction of the coils would decrease, and an expansion would increase the number of intermolecular contacts. If the number of bonding sites is limited and the available sites are already nearly saturated with intramolecular bonds in dilute solution, the cross-linking process is a transfer from intramolecular to intermolecular associations as the concentration of foreign segments increases. If the bonding capacity is far from saturated, there will be little competition for bonding sites, and the number of intramolecular bonds will change only to the extent required by changes in conformation.

In PMAA, which is highly intramolecularly bonded in dilute solution, the former is indicated. Thus, the sharpness of the sol–gel transition derives from the shift from predominantly intramolecular to predominately intermolecular bonding as the concentration increases. With ionization, the antibonding tendency of the electrostatic repulsion causes the gelation point to shift to higher concentrations. However, the transition remains sharp at low levels of ionization, since partially neutralized PMAA in dilute solution is still intramolecularly bonded. At higher degrees of neutralization, electrostatic interaction predominates in both dilute and concentrated solutions, and the sharpness of the transition is lost. In fact, at high levels of ionization, the behavior of PMAA is very similar to that of other linear flexible polyelectrolytes, as might be expected.

Various other aspects of the viscoelasticity of synthetic polyelectrolyte solutions have been studied. In a study by Nishida [63], the dynamic viscosity and rigidity at 500 Hz were measured for aqueous solutions (10 and 20% by weight) of poly(sodium acrylate)—PNaA—as a function of the temperature and of the concentration of a number of low-molecular-weight additives. Several interesting trends were observed. The rigidity was found to decrease and the viscosity to increase with increasing temperature. With increasing concentration of added sodium chloride or urea, the rigidity at 0°C was found to pass through a maximum and the viscosity through a minimum; however, no optima were observed at 35°C. Unfortunately, because of the nature of the treatment, the value of this study in elucidating the molecular basis of the phenomena is limited.

Konno and Kaneko [64] examined the behavior of the dynamic viscosity and rigidity at a fixed frequency (100 Hz) of aqueous solutions of PAA as a function of the degree of neutralization in a lower concentration range

(~0.5% by weight). Although these polymer solutions were quite dilute, nevertheless, even at this low concentration the mechanical properties are still influenced by the effects of intermolecular bonding. The authors determined G' and η' as a function of pH and of temperature for PAA solutions of various concentrations ranging from 2.1×10^{-2} to 8.1×10^{-2} monomol/liter. They also determined G' and η' as a function of temperature in PNaA solutions with various concentrations of added salts.

The most interesting facet of this study is the variation of G' and η' with pH at 30°, as shown in Fig. 3.22. It can be seen that the rigidity passes through a maximum at pH ≈ 6.4, which corresponds to approximately 50% neutralization; the position of the maximum does not change with concentration. The viscosity, on the other hand, shows only a monotonous increase throughout the interval. The variation in dynamic intrinsic viscosity with pH is very similar to the variation in the steady-flow intrinsic viscosity

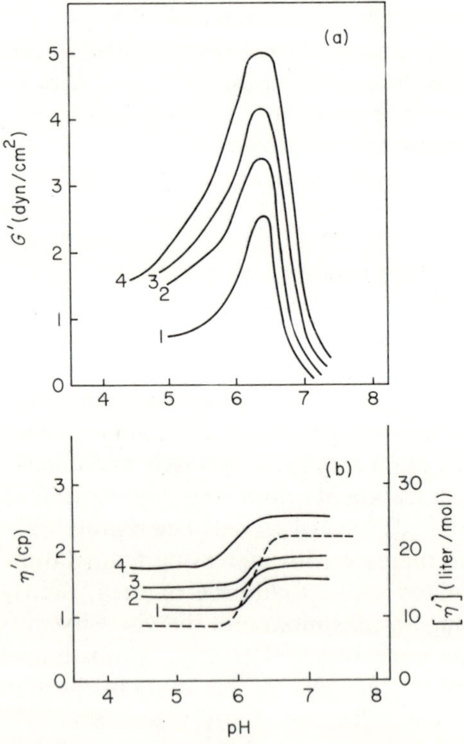

FIG. 3.22. The pH dependence of the rigidity, the dynamic viscosity, and the dynamic intrinsic viscosity of PAA at 30°C (ω = 100 Hz). Concentrations in monomoles per liter; (1) 3.7×10^{-2}; (2) 4.4×10^{-2}; (3) 5.8×10^{-2}; (4) 8.1×10^{-2} [64].

with increasing degree of neutralization seen in PAA and in other synthetic polyelectrolytes [65, 66]. This variation can be considered to reflect the continuous conformational change which accompanies ionization in dilute aqueous solution. Thus the anomalous behavior in G' cannot be explained in terms of conformational change alone.

G' and η' were also measured as a function of temperature at three different pH values—3.95, 6.45, and 8.50—corresponding to 0, 50, and 100% neutralization, respectively. The variation in G' with temperature at these values of pH is shown in Fig. 3.23. It can be seen that maxima with temperature occur in the curves corresponding to 50 and 100% neutralized PAA. For the fully neutralized polymer, this maximum is diminished and shifted slightly in temperature by the addition of NaCl or $MgCl_2$, but it is still apparent. In contrast, the variation in η' was featureless in each case, a slight decrease with increasing temperature being observed. This was attributed mainly to the decrease in solvent viscosity, offsetting the likely increase in polymer dimensions with temperature. Reduced viscosities were not reported, but sample calculations indicate that $\eta'_{sp}[= (\eta' - \eta_s)/\eta_s]$ increases with temperature in this system; this appears to be in line with the increase in polymer coil size with temperature observed in other polyelectrolyte systems [67].

Several explanations may be offered to explain the anomalous variation in G' with pH and temperature. Kanno and Kaneko suggest that it might

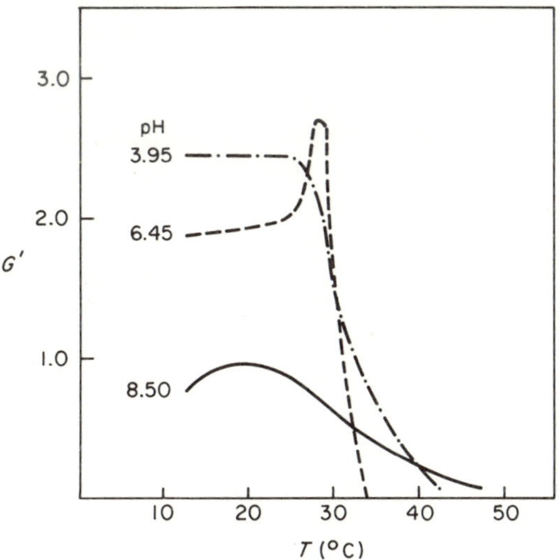

FIG. 3.23. G' of PAA as a function of temperature for three pH values: 3.95 (— - — -) ($C_p = 5.3 \times 10^{-2}$), 6.45 (— — —) ($C_p = 3.7 \times 10^{-2}$), 8.50 (———) ($C_p = 3.7 \times 10^{-2}$) [64].

be due to intermolecular charge–dipole interactions, which would presumably reach a maximum at about 50% neutralization. Such interactions would lead to a buildup of three-dimensional structure, which would be reflected by the increase in G' with pH. Above 50% neutralization, the increased electrostatic repulsion would decrease the intermolecular bonding and result in the drastic reduction in G' at high pH. This would also explain the fact that the position of the maximum in G' does not vary with concentration. However, it does not adequately explain the variation of G' with temperature at pH = 6.45 or the anomalous behavior of G' in fully neutralized PAA. Furthermore, since the energy of charge–dipole interactions in aqueous solution is generally rather low, it seems doubtful that the energy gained in this process would be sufficient to offset the loss in entropy associated with network formation. The energy gain from ion multiplet formation in aqueous systems is even lower, and thus the possibility of network formation through this mechanism is even more unlikely.

An alternative explanation of the anomalous behavior of G' may be found in the fact that the antibonding effect of ionic repulsion is both intra- and intermolecular. In the concentration range of this study ($\sim$0.5% by weight, $M_v = 7.0 \times 10^5$), the polymer coils are not completely isolated, especially not in their most highly expanded conformations (see Table I). Upon ionization of the PAA, both the polymer dimensions and the degree of molecular overlap increase. This is reflected in the variation in viscosity with pH. Initially, at low degrees of ionization, the increase in overlap is sufficient to cause an increase in the number of effective entanglements, since charge repulsion in the overlapped regions is still fairly low. This would account for the initial rise in G' at low pH. However, in spite of the continuous increase in the degree of molecular overlap with increasing pH (corresponding to increasing degrees of neutralization), the number of effective entanglements begins to decrease because of the increase in the magnitude of intermolecular repulsion. Thus the rigidity reaches a maximum with pH (at about 50% neutralization, in the region of most rapid coil expansion) and then falls off in value. The temperature dependence of G' can likely be interpreted on a similar basis. In the PAA system, the increase in coil size with temperature at constant pH indicates that the bonding tendency decreases with temperature. Hence the shape of the curves of G' versus temperature can be considered to reflect a balance between a generally increasing coil size and a generally decreasing bonding tendency.

In a study by Sakai *et al.* [68], the steady-flow mechanical properties of PNaA solutions (1–4% by weight) were investigated. The study involved the determination of the shear rate dependence of the steady-flow viscosity function $\check{\eta}$, the primary normal-stress function θ, and the compliance function J_s, defined by the ratio $J_s = \theta/2\check{\eta}^2$. Following the notation of Ferry

[69], $\breve{\eta}$ and θ are defined in terms of the shear stress σ_{21}, the normal-stress difference $(\sigma_{11} - \sigma_{22})$, and the rate of shear $\dot{\gamma}$ by the relations

$$\sigma_{21} = \breve{\eta}\dot{\gamma}$$

$$\sigma_{11} - \sigma_{22} = \theta\dot{\gamma}^2 \tag{6}$$

In this study, the secondary normal stress function was assumed to be negligible.

The variation in σ_{21} and $(\sigma_{11} - \sigma_{22})$ with the rate of shear in PNaA solutions was determined for a range of polymer molecular weights and for various concentrations of added NaCl (0.01–1 N). The solvent used was 30% aqueous glycerine, in order to obtain solutions with a workable viscosity range. As might be expected, the material functions $\breve{\eta}$, θ, and J_s were all found to vary with shear rate, as well as with molecular weight, polymer concentration, and added salt concentration. It was found that the curves of the material functions versus $\dot{\gamma}$ at constant concentration of polymer and added neutral salt could be superposed with respect to molecular weight by horizontal and vertical shifts. A set of master curves (log J_s versus log $\dot{\gamma}$) for various salt concentrations and a single polymer concentration (4%) is shown in Fig. 3.24, and reduced plots of log J_s versus log $\dot{\gamma}$ as a function of polymer concentration at a fixed salt concentration (0.025 N NaCl) are presented in Fig. 3.25. All of the data shown in these two figures are shifted

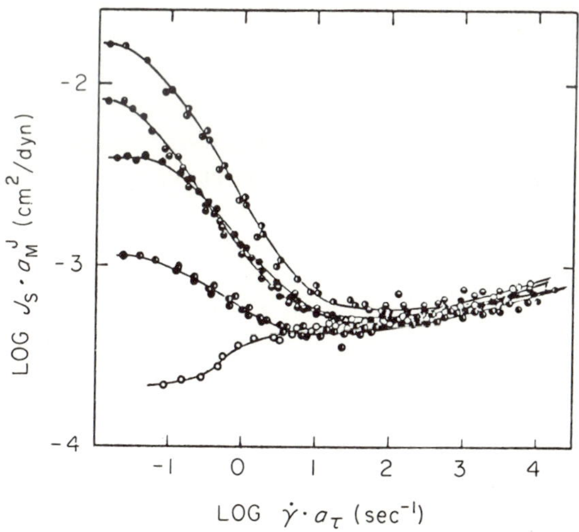

FIG. 3.24. Master curves of J_s versus $\dot{\gamma}$ for 4% PNaA at various NaCl concentrations. (○) 0.01 N; (◐) 0.025 N; (●) 0.05 N; (◒) 0.1 N; (◑) 0.25 N. Data superposed on curves for $M_v = 5.4 \times 10^6$ [68].

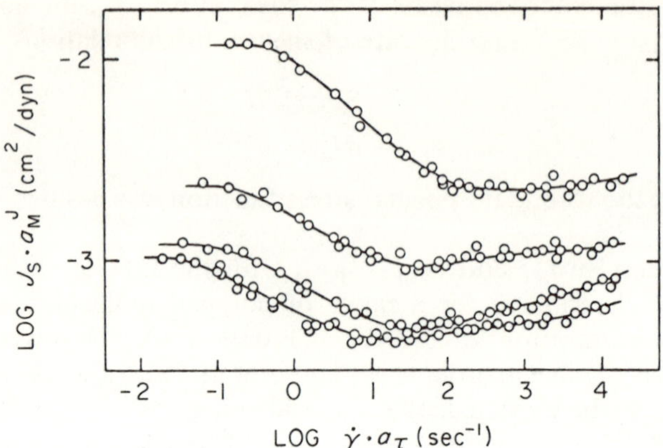

FIG. 3.25. Master curves of J_s versus $\dot{\gamma}$ for various PNaA concentrations. Data superposed on curves for $M_v = 5.4 \times 10^6$ NaCl concentration 0.025 N; polymer concentrations 1%, 2%, 3%, and 4% from top to bottom [68].

with respect to data for an intermediate-molecular-weight sample ($M_v = 5.4 \times 10^6$), the amount of displacement along the ordinate being approximately proportional to $1/M$.

Of particular interest in terms of structure are the zero-shear values of the material functions $\eta(=\eta^0)$, θ^0, and J_e $(=J_s^0)$. The molecular-weight dependences of the zero-shear functions at two polymer concentrations are shown in Fig. 3.26. The slope of the log η versus log M plot increases from 1.9 to 2.6 as the polymer concentration is increased from 1 to 4%, while the slope of log θ^0 versus log M varies from 4.8 to 6.1 over the same range. However, the slope of log J_e versus log M remains at 1.0 as the concentration is increased, and the relationship

$$\theta^0/(\eta - \eta_s)^2 \propto M \tag{7}$$

predicted from dilute-solution theory remains valid even at these polymer concentrations. (η_s, the solvent viscosity, is negligible with respect to the solution viscosity in this concentration range.)

In dilute polymer solutions, viscoelasticity is due to the storage and dissipative capacities of noninteracting polymer coils. The viscoelastic properperties of a dilute polymer solution can be predicted from the bead–spring model of Rouse [70]. In the theories based on this model, the viscosity is calculated from the energy dissipation due to the friction between the Rouse beads and the solvent, while the steady-state compliance is determined by

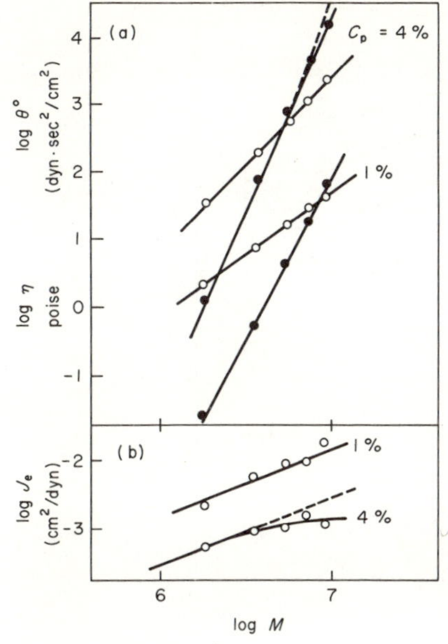

FIG. 3.26. (a) Molecular-weight dependence of (○) zero-shear viscosity η and (●) zero-shear normal-stress coefficient θ^0. (b) Molecular-weight dependence of the steady-state compliance J_e. PNaA concentrations are denoted on the figure. NaCl concentration 0.025 *N* [68].

the spring constant of the Rouse spring. At higher polymer concentrations, the polymer domains overlap, and the viscoelastic properties are determined mainly by entanglements between polymer coils [71].

In the concentration range 1–4%, it is apparent that the polymer concentration is intermediate between the state in which the polymer coils are completely isolated and the state in which complete entanglement has occurred. If the only significant dissipative mechanism were that due to entanglements, then a 3.4-power dependence of molecular weight on viscosity should be observed [71]. The fact that the slope of log η versus log M lies between the values 1.0 (as predicted by Rouse) and 3.4 probably indicates that the viscoelastic properties in this concentration range cannot be interpreted on the basis of entanglements alone.

It is interesting to recall that a slope of 2.7 was obtained for log τ versus log P (equivalent to log η versus log M) in the polyphosphate system described in Section A1. Although the polyphosphates were of low molecular weight and unplasticized and thus not directly comparable to the present system, nevertheless in both cases, the ionic interaction imparts a sufficient degree of three-dimensional character to the material to result in flow behavior intermediate between unentangled and entangled systems.

The dependence of the zero-shear material functions on the concentration of added salt C_s is shown in Fig. 3.27. As might be expected, the values of η at any given polymer concentration increase with decreasing ionic strength.

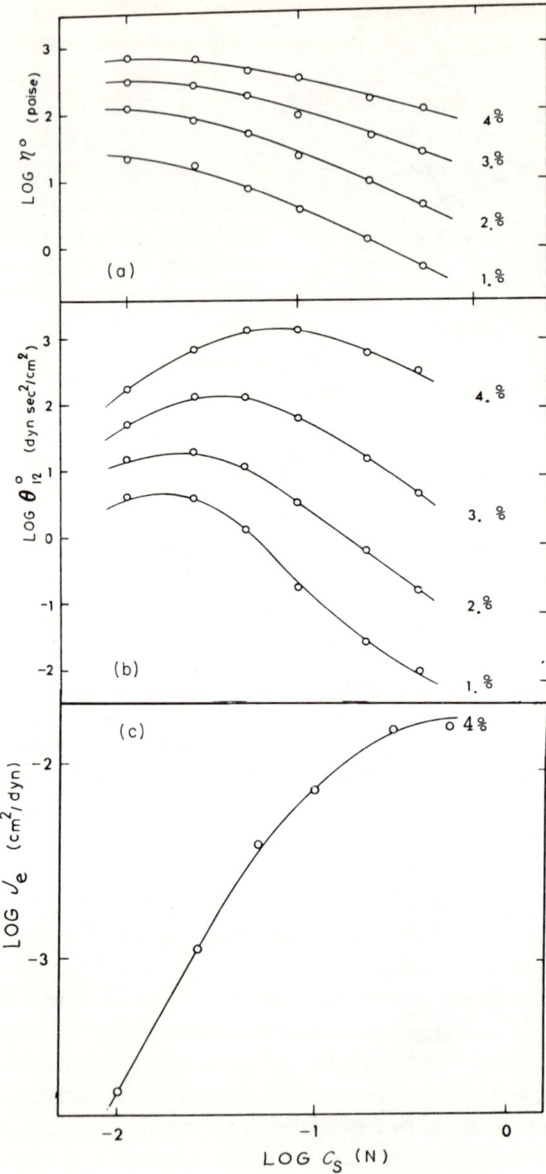

FIG. 3.27. Dependence of (a) zero-shear viscosity η, (b) normal-stress coefficient θ^0, and (c) steady-state compliance J_e of $M_v = 5.4 \times 10^6$ sample on concentration of added salt at PNaA concentrations denoted on the figure [68].

This increased viscosity is undoubtedly due to the expanded polymer conformation which results from decreased ion shielding at low salt concentrations, but whether it reflects an increase in the number of effective entanglements or an increase in the free-draining nature of the coils or both remains to be seen. It was pointed out in connection with the study of Konno and Kaneko [64] that an increase in polymer coil size in a concentration region in which partial overlap already occurs will not necessarily lead to greater numbers of effective entanglements. However, in view of the molecular-weight dependence of the viscosity function, it is reasonable to assume that the increased viscosity with decreasing salt content reflects both an expanding conformation and an increase in the number of effective entanglement points.

The same argument may be used to explain the decrease in J_e with decreasing ionic strength. In this concentration range, it may be assumed that J_e is determined in part by the spring constant of the polymer segment between entanglement points and in part by the spring constant of an isolated polymer coil. Decreasing the ionic strength increases the chain stiffness and, as before, increases the number of entanglements (i.e., decreases the chain length between entanglements). Both of these effects lead to a decreased compliance, hence, the results shown in Fig. 3.27.

It is seen in Fig. 3.27 that the curve of θ^0 versus C_s goes through a maximum in this concentration range. This fact, in itself, is not surprising since θ^0 $[= 2\eta^2 J_e]$ is the product of two functions, one continuously decreasing and the other continuously increasing. However, $\theta^0\dot{\gamma}^2$, which represents the energy storage in the material at low shear rates, can be correlated, at least at low frequencies, with $G'(\omega)$, the dynamic rigidity [69]. Thus, the maxima in θ^0 with added salt concentration, as observed in the study by Sakai *et al.* [68], seem to correspond with the maxima in $G'(\omega)$ observed by Nishida, in the study mentioned previously [63]. The correspondence between the two studies can only be considered qualitative, however, since the types of measurement and the concentration ranges were quite different.

Recently Okamoto *et al.* [72, 73] have reported viscoelastic studies on ion-containing polymer solutions. They measured the complex shear viscosity, $\eta^*(\omega)$ $[= \eta'(\omega) - i\eta''(\omega)]$, at frequencies from 2 to 500 kHz, and the steady-flow viscosity of dilute ($\sim$0.15% by weight) aqueous solutions of PAA and PMAA. For each polymer, viscosity–frequency curves at various degrees of neutralization were obtained, and for PAA dynamic viscosity measurements were also made at different degrees of polymerization and concentrations of added salt (NaCl).

The variation in dynamic viscosity as a function of pH (or degree of neutralization) was found to parallel the change in steady-flow viscosity;

however, the transition became much less accentuated as the frequency increased. This is illustrated in Fig. 3.28, which shows dynamic and steady-flow viscosity for an aqueous solution of high-molecular-weight PAA. The variation in $\eta'(\omega)$ with pH is very similar to that obtained by Konno and Kaneko (Fig. 3.22); however, in contrast to the latter group's finding of a sharp maximum in the dynamic rigidity $G'(\omega)$ at pH $= 6.4$, Okamoto *et al.* observed no maxima in their $\eta''(\omega)$ $(= G'/\omega)$ versus pH curves.

For the PMAA system, the steady-flow viscosity transition is considerably more accentuated than that of PAA because of the hydrophobic bonding at low levels of ionization, which tends to constrict the PMAA coil, and the augmented chain stiffness at higher levels due to the bulky methyl group. However, the basic pattern—decreasing sharpness of transition with increasing frequency—obtained by Okamoto and Wada for PMAA was the same as that shown in Fig. 3.28 for PAA.

The variation in the viscosity–frequency curves of fully neutralized PAA with degree of polymerization can be seen in Fig. 3.29. The steady-flow viscosity values are also plotted in this figure. They indicate that for the higher degrees of polymerization, considerable relaxation must take place at frequencies lower than the experimental limit of 2 kHz. The extent of this low-frequency relaxation is at its maximum in fully neutralized species of

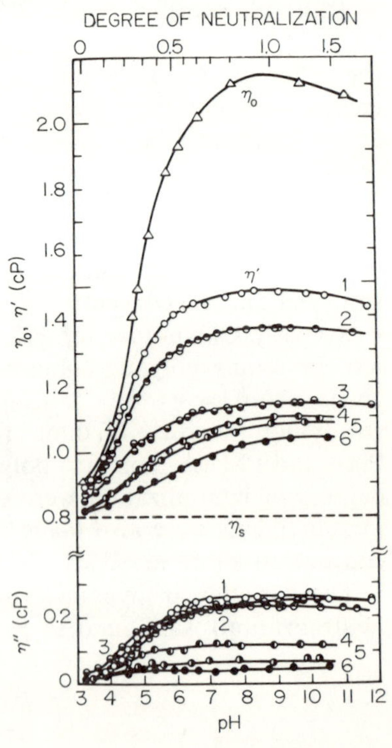

FIG. 3.28. Steady-flow viscosity η_0 and real and imaginary parts of complex viscosity η' and η'', respectively, for aqueous PAA solution at 30°C plotted against degree of neutralization and pH: $P = 3300$, $C_p = 0.14$ g/dl, $C_s = 0.021$ N; (○) 2.2 kHz; (◓) 6.6 kHz; (◒) 19.9 kHz; (◐) 53.9 kHz; (◑) 157 kHz; (●) 448 kHz [72].

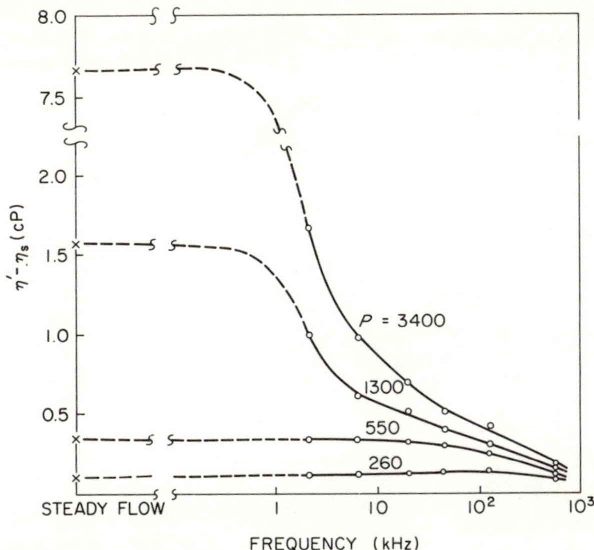

FIG. 3.29. Plots of $(\eta' - \eta_s)$ versus frequency for aqueous PAA at different degrees of polymerization. Steady-flow viscosity shown extrapolated to zero frequency: $\alpha = 1$, $C_p = 0.16$ g/dl, $C_s = 0.005$ N, $T = 30°C$ [72].

high molecular weight. As the degree of polymerization increases, the relaxation region shifts to higher frequencies and lower intensities. A similar shift was seen with decreasing degree of neutralization and with increasing concentration of added salt.

Okamoto *et al.* computed relaxation spectra from their dynamic viscosity data. Examples of their relaxation spectra are shown in Figs. 3.30 and 3.31,

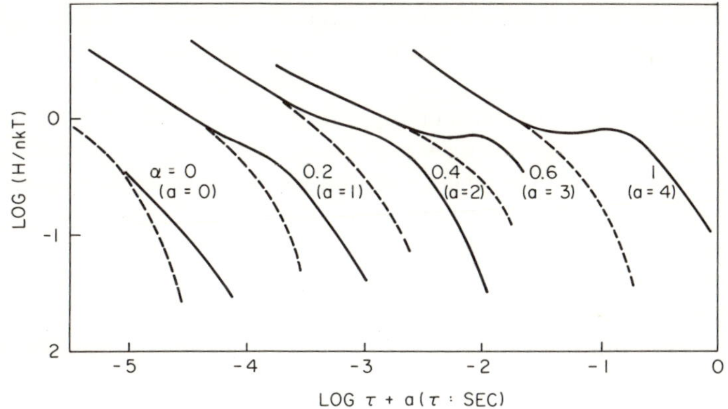

FIG. 3.30. Relaxation spectra for aqueous PAA at various degrees of neutralization. Different abscissa scales are used to avoid overlapping. Calculated from data in Fig. 3.28. Dashed curves represent Zimm spectra [72].

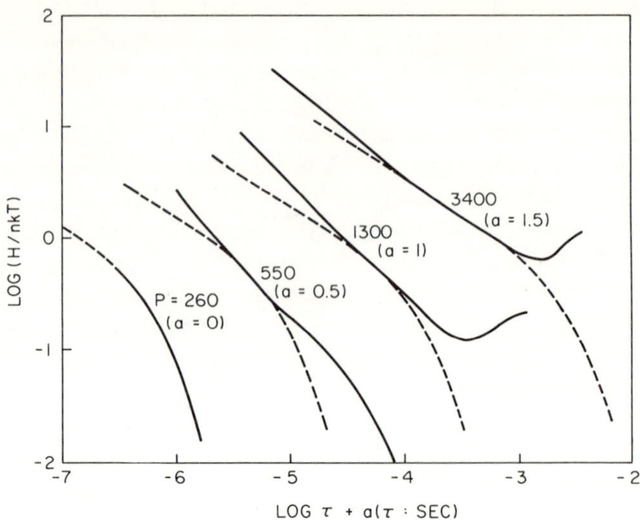

FIG. 3.31. Relaxation spectra for aqueous PAA at various degrees of polymerization. Calculated from data in Fig. 3.29. Dashed curves represent Zimm spectra [72].

illustrating, respectively, the variation in spectral shape with degree of neutralization and degree of polymerization. At short relaxation times, the spectral shapes were found to conform reasonably well with the relaxation spectrum predicted by Zimm [74] for a bead–spring model with dominant hydrodynamic interaction (dashed lines in Figs. 3.30 and 3.31). The relaxation mechanism at short times was therefore associated solely with Zimm-type conformational modes of relaxation.

At longer relaxation times, the excess spectrum was computed as the difference between the observed spectrum and the best-fitting Zimm spectrum. The Zimm treatment of the bead–spring model accounts for rotational relaxation of spherically distributed beads. However, if rotational movement occurred at a slower rate than changes in internal conformation and if the time-average conformation were aspherical, then one would anticipate an extra contribution to the relaxation due to rotation. Okamoto *et al.* [72] interpreted their excess spectra in terms of this type of aspherical rotation, assuming that the polyelectrolyte took the shape of a prolate ellipsoid at high degrees of neutralization and low salt concentrations. Their analysis enabled them to describe the relaxation in terms of two characteristic times τ_c and τ_r—the conformational and rotational relaxation times, respectively.

The conformational relaxation time τ_c was found to be proportional to P^2 (where P is the degree of polymerization), in both unneutralized and fully neutralized PAA. It should be noted that this result more correctly agrees with the Rouse [70] theory (no hydrodynamic interaction) than with the Zimm theory, which predicts $\tau \propto P^{3/2}$. τ_c was observed to pass through a maximum with increasing degree of neutralization and to increase marginally with increasing salt content. In the Zimm treatment $\tau \propto P^{3/2}/f^{3/2}$, where f is the force constant of the Hookean spring of the submolecule. f decreases with increasing degree of neutralization or decreasing small ion concentration, since increased ionic repulsion facilitates extension of the submolecule. Thus at constant degree of polymerization, the Zimm theory predicts that τ_c should increase monotonously with both increasing α and decreasing C_s. The Rouse theory predicts a similar variation in τ_c, C_s, and α. The observations are not in accord with these predictions; however, they do agree, at least qualitatively, with the results of a computer simulation study to be described later.

The excess spectra at long relaxation times were attributed to aspherical rotation. The computed relaxation times increased sharply with increasing degree of polymerization; in fact, τ_r was found to be roughly proportional to P^3, for both ionized and unionized PAA. Since the relaxation time of a prolate ellipsoid is proportional to L^3, the cube of its major axis, these results could be interpreted by assuming that L increases linearly with P. However, it seems doubtful that this could apply even in the ionic case, and with unneutralized PAA, the degree of asphericity is surely negligible.

An alternate explanation for the origin of the excess spectra is their assignment to intermolecular interaction, i.e., chain entanglement. In Table I, average polymer dimensions computed for the polyelectrolyte solutions studied by Okamoto *et al.* [72, 73] and Konno and Kaneko [64] are compared with the average volume available per macromolecule. The values of $\langle h^2 \rangle^{1/2}$, the rms end-to-end displacement of the macromolecule, were estimated from the following equation [75]. The value of Φ, the so-called

$$\langle h^2 \rangle^{1/2} = ([\eta]M/\Phi)^{1/3} \tag{8}$$

universal parameter, was taken as 2.87×10^{21} dl. This represents an upper limit to the usual estimates of Φ; another common estimate of Φ is 2.1×10^{21} dl. The use of the latter figure would lead to only a slightly higher estimate of $\langle h^2 \rangle^{1/2}$. Values of $[\eta]$ were not available for the data of Okamoto *et al.* [72, 73], but were estimated from the reduced specific steady-flow viscosity, $(\eta_0 - \eta_s)/\eta_s C_p$. For the data of Konno and Kaneko [64], the values of $[\eta]$ at 100 Hz were used to estimate $[\eta]$. The former method will overestimate $[\eta]$, and the latter method will underestimate it. Although the polymer solutions are not directly comparable, the fact that similar values of $\langle h^2 \rangle^{1/2}$ are obtained in both cases justifies the approximation.

The data in Table I indicate that in the concentration range studied by Okamoto *et al.*, one progresses from a situation of low molecular overlap

TABLE I

Average Polymer Coil Dimensions of PAA and PMAA under Various Conditions

Polymer series [reference]	P	α	C_p (g/dl)	C_s (N)	$\eta_0 - \eta_s$ (cP)	$[\eta]$ (dl/g)	$\langle h^2 \rangle^{1/2}$ (Å)	$V^{1/3}$ (Å)	$\langle h^2 \rangle^{1/2}/V^{1/3}$
(i) PAA [64]	7450	0	0.27	—	0.3	1.1	618	692	0.89
		1	0.27		1.0	2.34	870	692	1.25
		0	0.58	—	0.8	1.1	618	538	1.15
		1	0.58		1.7	2.34	870	538	1.61
(ii) PAA [72]	550	0	0.16	0.005	0.03	0.23	154	345	0.45
	1300				0.11	0.86	318	460	0.69
	3400				0.44	3.44	695	633	1.10
(iii) PAA [72]	260	1	0.16	0.005	0.08	0.63	168	269	0.63
	550				0.34	2.66	348	345	1.01
	1300				1.56	12.2	770	460	1.67
	3400				7.67	60.0	1800	633	2.84
(iv) PAA [72]	3300	0	0.14	0.21	0.10	0.88	435	655	0.66
		0.2			0.50	4.46	751		1.15
		0.4			0.95	8.48	930		1.42
		0.6			1.15	10.3	1010		1.54
		1.0			1.33	11.9	1080		1.65
(v) PAA [72]	1300	1	0.16	0.001	2.43	19.0	872	460	1.94
				0.002	2.17	17.0	860		1.87
				0.005	1.56	12.2	770		1.68
				0.01	1.22	9.54	710		1.54
				0.1	0.40	3.12	488		1.06
(vi) PMA [73]	4380	0	0.15	0.005	0.12	1.0	235	745	0.32
		0.17			0.46	3.8	368		0.49
		0.32			2.9	24	680		0.91
		0.43			4.9	41	862		1.16
		0.84			5.7	48	907		1.22

to high overlap with increasing degree of polymerization, since in series (iii), $\langle h^2 \rangle^{1/2}/V^{1/3}$ varies from less than 1 to more than 2. The resultant increase in molecular entanglements could account for the buildup of relaxation times with increasing P, as seen in Fig. 3.31. Similar increases in molecular overlap characterize the spectral variations with α and with C_s. $\langle h^2 \rangle^{1/2}/V^{1/3} = 1$ probably does not define the exact point of impingement; however, the point at which molecular overlap becomes significant undoubtedly lies in the range of polymer concentrations studied, because the variations in $\langle h^2 \rangle^{1/2}/V^{1/3}$ about unity parallel too closely the buildup of the excess long-time spectra to be mere coincidence.

The assignment of the short-time relaxation mechanism to conformational or internal relaxation is reasonable. However, the considerations discussed above indicate that its interpretation in terms of Zimm or Rouse models is only valid qualitatively. The study points to the need for a more sophisticated theoretical treatment of the viscoelastic properties of ion-containing polymer solutions.

3. Polyelectrolytes in Dilute Solution

(a) *Experimental Studies*

In truly dilute solution, relaxation effects due to polymer molecules are intramolecular. With dissolved polyelectrolytes, these could arise from conformational changes of the polymer itself or by rearrangement of the counterion atmosphere around the macroion. One would expect the relaxation rate of unbound counterions to be considerably greater than the rate of conformational relaxation because of the vast difference in size between the macroion and the small ion. Also, since counterion relaxation involves dipolar changes, it should be dielectrically active. Thus studies of the high-frequency dielectric response of polyelectrolyte solutions have been employed to investigate the dynamics of counterion motion.

Dielectric dispersion in dilute polyelectrolyte solutions has been studied as a function of a number of variables, including molecular weight, degree of neutralization, counterion type, and concentration of polymer and added salt. It will be seen that two dispersion regions exist for high-molecular-weight flexible polyelectrolytes—one at low (kilohertz) frequencies and the other at high (megahertz) frequencies. The lower dispersion is highly molecular weight dependent and involves fluctuations in the bound ion distribution over the whole macromolecule. The higher one is independent of chain length and involves local redistribution of ions in the counterion atmosphere. Various theories have been developed to account for the observed dielectric behavior. The presentation of results and theory follows largely the chronological order of the development of this field. Studies of counterion dynamics using other probes, such as ultrasound absorption and NMR, will also be discussed in this section.

Mandel and Jenard [76] measured the capacitance of various polyelectrolyte solutions at 23°C in the frequency range 10–700 kHz. The polymers used in the investigation were partially and fully neutralized poly(methacrylic acid) and poly(vinyl amine); typical concentrations were of the order of 0.05% by weight. The values of ε', the real part of the complex dielectric constant ε^* $(= \varepsilon' - i\varepsilon'')$, were determined at various frequencies from the measurements of capacitance. The decrease in ε' with frequency indicated the existence of a dispersion region in these polyelectrolyte solutions in the 1 to 100 kHz range. Since conductance measurements did not display any significant frequency dependence, because of the high specific conductivities of the polyelectrolyte solutions, the values of ε'' in this frequency range could not be determined. However, values of ε_0', ε_∞' and f_c, the critical frequency, were determined for each system. ε_0' and ε_∞' were estimated by direct extrapolation, and since ε'' could not be determined, f_c was estimated from the inflection points of the dispersion curves.

For any given degree of polymerization P and degree of ionization α, the dispersion curves were found to superimpose with respect to concentration C_p. From this it was concluded that f_c is independent of concentration. As will be seen, however, later studies have demonstrated that f_c is somewhat concentration dependent. An example of a reduced dispersion curve for a series of PMAA solutions ($P = 710$, $\alpha = 1$) is shown in Fig. 3.32. The reduced curves were found to fit the dispersion relation proposed by Cole and Cole [77]

$$\frac{\varepsilon' - \varepsilon'_\infty}{\varepsilon'_0 - \varepsilon'_\infty} = \frac{1}{2}\left[1 - \frac{\sinh \beta x}{\cosh \beta x + \cos \beta\pi/2}\right] \tag{9}$$

where $x = \ln(f/f_c)$, and β is a quantitative measure of the breadth of the distribution of relaxation times ($0 \leq \beta \leq 1$). For salt-free solutions of high-molecular-weight PMAA ($P = 5200$, $\alpha = 1$), no inflection point was observed within the experimental frequency range. For this series f_c was obtained by comparison with the best-fitting curve corresponding to Eq. (9), assuming that β is the same as that for the lower-molecular-weight material.

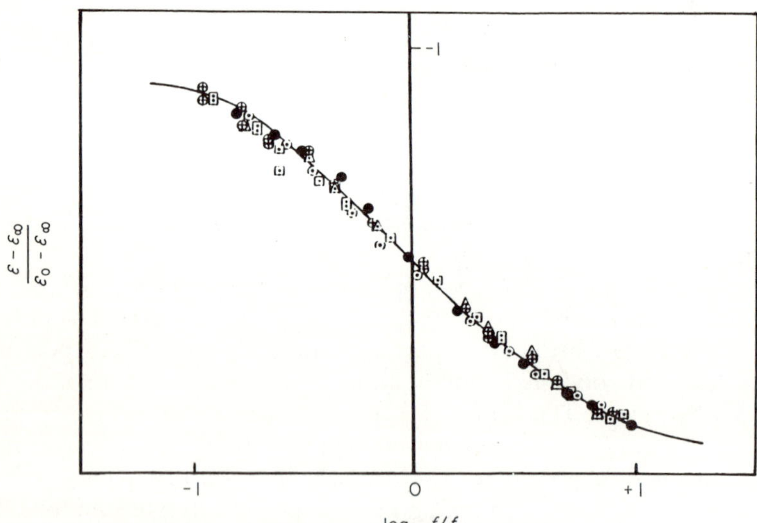

FIG. 3.32. Reduced curve of dielectric constant versus frequency for PMAA solutions ($P = 710$, $\alpha = 1$). Polymer concentrations (monomoles per liter) as follows: (○) 1.29×10^{-3}; (●) 2.50×10^{-3}; (⊕) 5.00×10^{-3}; (△) 7.50×10^{-3}; (□) 10.0×10^{-3}; $\log f_c = 5.0$ [76a].

The magnitude of the polarization effect, as evidenced by the values of ε_0', or alternatvely by $\Delta\varepsilon_1'$ ($=\varepsilon_0' - \varepsilon_\infty'$), is considerable; furthermore, the values of ε_0' were found to increase with increasing P. The relaxation time of the dispersion τ_1, corresponding to the inverse of f_c, was also found to increase with molecular weight. Finally, both $\Delta\varepsilon_1'$ and τ_1 were observed to increase with increasing α, although the dependence of τ_1 on α is rather weak.

The addition of excess univalent electrolyte was found to diminish the values of ε_0' and τ_1, while with the replacement of K^+ by Ba^{2+} at constant ionic strength, both ε_0' and τ_1 were observed to pass through maxima. It is worthwhile to note the extremely high values of dielectric constant obtained in the latter case; for example, values of ε_0' as high as 900 were measured for a PMAA solution ($P = 5200$, $\alpha = 1$) with 50% replacement of K^+ by Ba^{2+}.

The source of the dispersion in the 1- to 100-kHz region was attributed to polarization effects arising from the distribution of bound ions within the polyion coil. Although this view still holds, the specific model proposed at that time has since been modified; in view of this fact, the results of Mandel and Jenard will be discussed later in this section in terms of the revised theoretical approach.

These authors found that all the values of ε_∞' were significantly higher than the value corresponding to water, suggesting the existence of a second dispersion region at frequencies higher than 1 MHz. This was also indicated by the work of Dintzis *et al.* [78], whose measurements on poly(4-vinyl-1-butylpyridinium bromide) solutions extended from 1 kHz to 4 MHz. Dielectric dispersion in this high-frequency region was subsequently studied by Sachs *et al.* [79] in a comprehensive investigation covering the frequency range 400 Hz to 30 MHz. The polyelectrolytes used in the study were poly(vinyl sulfonates), poly(styrene sulfonates), and poly(sodium acrylates). As in the study by Mandel and Jenard, polymer concentrations were typically in the range of 0.05% by weight.

Conductance measurements on these polyelectrolyte solutions in the frequency range 400 Hz to 100 kHz indicated a small but measurable increase in equivalent conductance with increasing frequency, which had not been observed by Mandel and Jenard. This enabled the determination of ε'' for these frequencies. $\Delta\varepsilon''$ ($=\varepsilon'' - \varepsilon''_{H_2O}$) was observed to pass through a maximum which shifted to lower frequency (i.e., greater τ_1) with increasing molecular weight, but was insensitive to changes in polymer concentration or variations in the hydrated radius of the univalent counterion. Since capacitance measurements were not made in this frequency region and the variation in conductance was small, f_c could not be determined accurately; however, the results tended to confirm the conclusions of Mandel and Jenard [76] regarding the nature of the low-frequency dispersion.

Both conductance and capacitance measurements were made in the high-frequency region (0.6–30 MHz), and from these measurements, values of ε' and ε'' as a function of frequency were determined. Curves of $\Delta\varepsilon_2'$ versus frequency at different degrees of polymerization were found to superimpose, as seen in Fig. 3.33 for PSSA–Na, indicating the lack of a molecular-weight dependence in the high-frequency dispersion region. At the upper end of this frequency region, the values of ε' finally approach those of pure water, indicating that no significant dispersion exists at higher frequencies.

Cole and Cole [77] circular arc plots ($\Delta\varepsilon''$ versus $\Delta\varepsilon'$) were prepared for the various solutions of polyelectrolytes studied. An example of these, corresponding to the $\Delta\varepsilon'$ values in Fig. 3.33, is shown in Fig. 3.34. The right and left intercepts of the circular arc correspond to ε_0' and ε_∞', respectively. The fact that the center of the circle is depressed below the real axis indicates the existence of a distribution of relaxation times in this dispersion region; a quantitative measure of the breadth of the distribution is given by the parameter β (= 0.83), as indicated in Fig. 3.34. Neither β nor τ_2, the average relaxation time, were found to be affected by variations in P; evidence for this lies in the fact that data for the whole range of molecular weights studied superimpose in the construction of the Cole and Cole plot. The deviations from the arc at low frequencies indicate the onset of the low-frequency dispersion mechanism discussed previously.

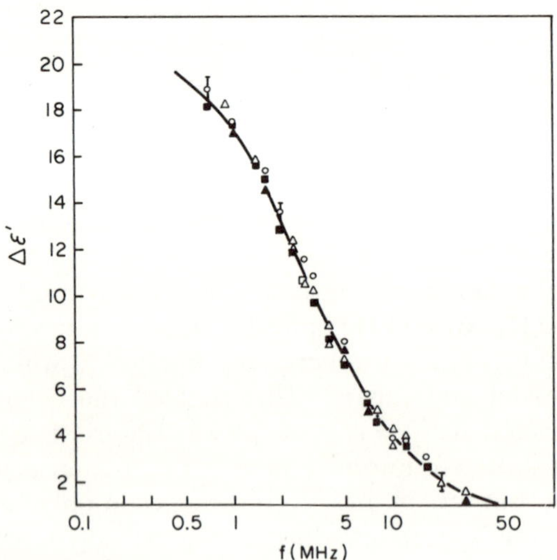

FIG. 3.33. Dielectric increment as a function of frequency for solutions of PSSA–Na with different values of P: (○) 875; (▲) 4380; (■) 10,100; (△) 64,300; $C_p = 2.5 \times 10^{-3}$ monomol/liter [79].

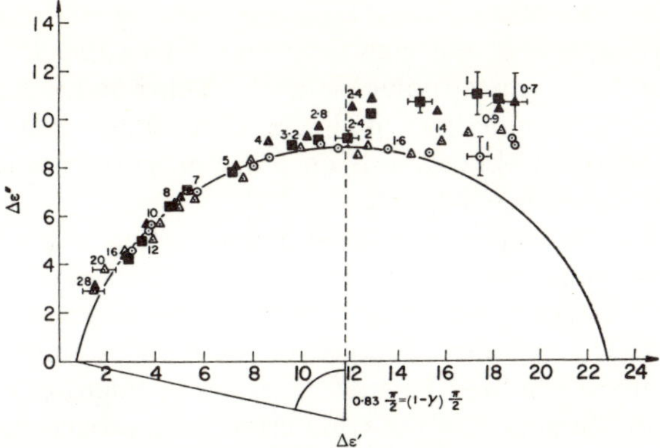

FIG. 3.34. Cole and Cole plot for solutions of PSSA–Na symbols and C_p as in Fig. 3.33. Numbers on experimental points indicate frequency in megahertz [79].

The concentration dependence of the dielectric dispersion was tested in solutions of PVSA–Na. The analysis of Cole and Cole plots in the high-frequency range indicated that while β remained unchanged, both τ_2 and $\Delta\varepsilon_2'/C_p$ decreased with increasing polymer concentration.

The variation of the high-frequency dispersion with the type of univalent counterion was also tested in the PVSA series. At constant polymer concentration, β was found to be independent of the counterion; however, $\Delta\varepsilon_2'/C_p$ and τ_2 were observed to decrease in the order

$$(nC_4H_9)_4N > (CH_3)_4N \sim Li > Na > K \sim Cs$$

which follows the approximate order of decreasing hydrated counterion radius.

Finally, the dependence of the dispersion on the degree of ionization in the PAA series was investigated. Two distinct types of Cole and Cole plots were obtained. At low ionization ($\alpha = 0.25$), the distribution of relaxation times was distinctly broader than at higher levels of ionization ($\alpha = 0.35$, 0.50, and 0.75), although the critical frequencies were the same in every case. The parameter β for $\alpha \geq 0.35$ was found to be invariant and nearly identical with the value of β observed in the completely ionized PSSA and PVSA series. $\Delta\varepsilon_2'/C_p$ increased while $\Delta\varepsilon_2'/\alpha C_p$ decreased with increasing α through the whole range studied.

Sachs *et al.* [79] concluded that the high-frequency dispersion represents the relaxation of the cylindrical counterion atmosphere surrounding the polyion. In other words, the dispersion is attributed to diffusion of the unbound but associated counterion fraction rather than the bound fraction whose contribution is seen at lower frequencies. This conclusion is supported by a number of experimental considerations: First of all, in contrast to the low-frequency dispersion, the high-frequency relaxation was found to be independent of the degree of polymerization. This implies that the polarizability of the ionic double layer is independent of the total length of the macroion, in line with observations on other properties which depend on the counterion atmosphere [49–51].

Assuming that the average relaxation time is directly related to the length of the diffusion path, the decrease in τ_2 with increasing C_p is interpreted qualitatively as corresponding to the decrease in the thickness of the counterion atmosphere. τ_2 was also observed to decrease in order of decreasing hydrated counterion radius. This effect could be interpreted as reflecting the variation in the ability of the counterion to approach the polyion, although it could simply represent the change in counterion mobility with size.

The reason that a distribution of relaxation times is observed is attributed to the fact that the charge density in the double layer varies with the radial distance from the polyion backbone. The lack of variation in β with the type of counterion or the degree of ionization (above $\alpha = 0.3$) or even with the polymer type indicates that the nature of the polyelectrolyte is nonspecific in determining the relative distribution of charge in the counterion atmosphere. The unusual behavior at low degrees of ionization is indicative of the conformational change from a tightly coiled structure to a loose open coil that occurs at about $\alpha = 0.3$ in PAA. The broadness of the distribution below $\alpha = 0.3$ seems to imply that the counterion atmosphere is different from the cylindrical atmosphere which surrounds the polyion segments at higher degrees of ionization.

The insensitivity of the high-frequency dispersion to changes in the degree of neutralization above $\alpha = 0.3$, in contrast to the strong dependence on α at low frequencies, is explained by the fact that the effective number of counterions (the unbound fraction) does not increase beyond $\alpha = 0.3$, whereas the number of bound ions, on which the low-frequency dispersion depends, does [50]. It should be noted, however, that although τ_2 did not vary, $\Delta\varepsilon_2'/C_p$ did increase somewhat with α in the PAA system, indicating that this mechanism might not be completely insensitive to ion binding.

In another study, Allen *et al.* [80] investigated the dielectric behavior of aqueous solutions of sodium carboxymethyl cellulose—NaCMC—in the frequency range 2–50 MHz. Their work indicated that the dielectric dispersion in this system is very similar to that observed by Sachs *et al.* [79] in the same frequency range. As in the latter study, the relaxation time in the NaCMC system was found to decrease with increasing polymer concentration and to be essentially independent of variations in molecular weight. In addition, the relaxation time was found to decrease with increasing concentration of added NaCl and also to decrease with increasing temperature (in the range 5–45°C). From the temperature dependence of f_c, the activation energy of the relaxation process was determined to be 4.8 kcal/mol.

These results confirm that the dispersion mechanism operative in the megahertz region is due to the relaxation of the counterion atmosphere. Allen *et al.* interpreted their results according to the theory of O'Konski [81], which describes the dielectric response due to the diffuse (i.e., associated but unbound) counterion fraction in a suspension of elongated polyions. The polyion conductivity is assumed to be the sum of the bulk conductivity within the polyion domain and the surface conductivity. Relaxation effects are attributed to the difference in conductivity between the polyion domains and the solvent phase, which arises mainly from the difference in the mobile ion concentration in the polyion atmosphere and that of the surrounding medium. The theory predicts that, for highly elongated polyions, the critical frequency varies directly with the mobile counterion concentration and the intrinsic ionic conductance. The relative variation in τ_2 with temperature as well as with the concentration of polymer and of added salt was shown by Allen *et al.* to be in close agreement with the theory of O'Konski, although the absolute values of τ_2 were smaller by a factor of about 2 than those predicted by the theory.

Minikata and Imai [82] investigated the dilute-solution behavior of ionized PAA over a wide frequency range, 300 Hz to 6 MHz, covering both dispersion regions. Their results on the PAA–$(nC_4H_9)_4N$ system ($P = 1220$, $\alpha = 1$) as a function of polymer concentration are shown in Fig. 3.35. Both dispersion regions are clearly indicated in these curves. The dielectric data (ε' versus frequency) were fitted to a two-component form of the Cole–Cole dispersion formula [Eq. (9)] to obtain the contribution of each dispersion (low and high frequency) to the total dielectric increment. The best-fitting values of the Cole–Cole parameter β, which gives a measure of the breadth of the distribution, were found to be about 0.65 for both the low- and the high-frequency components, indicating a slightly broader distribution than that observed by Sachs *et al.* [79] for the PAA–Na system.

Minikata and Imai found that τ_2, the high-frequency relaxation time, obtained from the inflection point of the Cole–Cole plot, decreased markedly with increasing polymer concentration, in line with the observations of Sachs *et al.* [79] and Allen *et al.* [80]. The values of τ_1 also decreased with increasing C_p, although the variation was slight. Mandel and Jenard [76], over a narrower concentration range, had found no significant variation in τ_1 with C_p. Both reduced increments $\Delta\varepsilon_1'/C_p$ and $\Delta\varepsilon_2'/C_p$ decreased nonlinearly with increasing concentration, as seen previously.

With increasing degree of neutralization, both low- and high-frequency increments increased, at first rapidly at low α and then more gradually as α approached one; however, no appreciable change in either τ_1 or τ_2 was

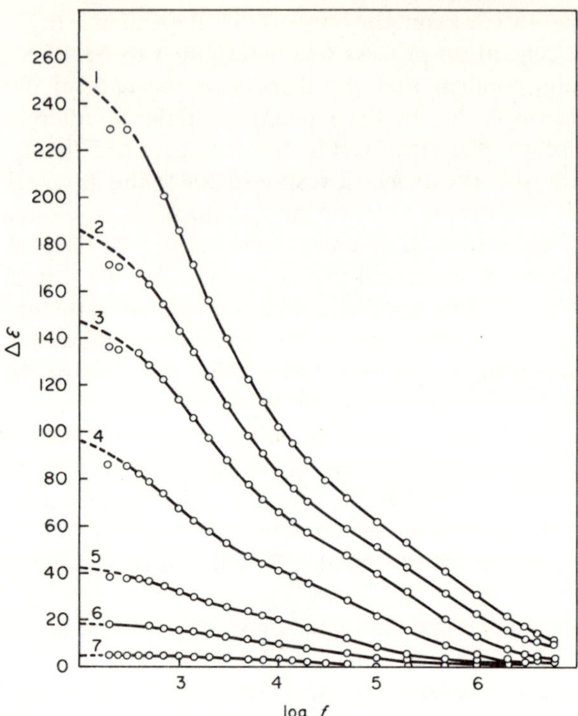

FIG. 3.35. Dielectric dispersion of PAA–n-Bu_4N as a function of polyion concentration. Curve (1) $C_p = 5 \times 10^{-3}$ monomol/liter; (2) 2.5×10^{-3}; (3) 1.25×10^{-3}; (4) 5×10^{-4}; (5) 2×10^{-4}; (6) 1×10^{-4}; (7) 2×10^{-5} [82a].

observed for variations in the degree of neutralization. Counterion effects similar to those observed by previous workers were also seen in the present study; i.e., substitution of Na for $(nC_4H_9)_4N$ was found to decrease both $\Delta\varepsilon'$ and τ, in both dispersion regions, and replacement of monovalent counterion by equinormal divalent ion led to a maximum in $\Delta\varepsilon_1'$ at an intermediate counterion ratio.

The origin of the high-frequency dispersion was attributed to the motion of mobile counterions as described by the theory of O'Konski [81], although quantitative agreement between theoretical and observed values of τ_2 was lacking, as was also the case in the study of Allen *et al.* [80]. The low-frequency dispersion was considered to be due to the displacement of condensed counterions in the longitudinal direction along the polyion. A theory for the dielectric increment due to ion fluctuations was presented by Minakata *et al.* [83]. Their expression for the low-frequency dielectric increment $\Delta\varepsilon_1'$ can be written as

$$\Delta\varepsilon_1' = \pi e^2 N_{av} C_p \alpha L^2 K' f(m)/9 \times 10^3 kT \tag{10}$$

where C_p is the polymer concentration in monomoles per liter, α is the degree of neutralization, L is the polyion length, and K' is an internal field correction factor usually set equal to one. The expression $f(m)$ is a function of the mean charge of the bound counterions and is calculated by a matrix method which takes into account the interaction between site-bound ions and their nearest neighbors. The theory predicts that the dielectric increment is directly related to the "looseness" of the counterion binding. The theory was used to explain the maximum in $\Delta\varepsilon_1'$ which occurs at intermediate monovalent–divalent ion ratios.

Van der Touw and Mandel [84] have recently proposed a new model to account for both the low- and high-frequency dispersions in solutions of flexible polyelectrolytes. Polymer flexibility is incorporated by considering the molecule to be a collection of rigid rodlike segments (of length b) along which bound counterions can move. An energy barrier greater than kT is assumed to impede the diffusion of ions between the segments. Fluctuations in the bound counterion distribution can occur locally, along the segments, or over the whole molecule. The theory predicts the existence of two separated dispersion regions if both the exchange between bound and free ions and the rotation of the whole molecule are relatively slow in comparison to the local bound counterion density fluctuations.

The dielectric increment due to the perturbation of the bound counterion distribution over the whole chain is given by [84]

$$\Delta\varepsilon_1' = \pi z^2 e^2 N_{av} C_p \alpha \gamma K' \langle R^2 \rangle / 9 \times 10^3 kT \tag{11}$$

where ze is the counterion charge, γ the degree of counterion association, and $\langle R^2 \rangle$ the mean-square end-to-end displacement of the polyion chain. All the other symbols have the same meaning as in Eq. (10). Both Eqs. (10) and (11) are very similar in form to the expression for $\Delta\varepsilon_1'$ derived earlier by Mandel and Jenard [76]. It should be noted, however, that in the theory of Minakata *et al.* [83], $\Delta\varepsilon_1'$ is considered to depend not on γ but on $f(m)$, which depends inversely on the strength of the counterion association. The high-frequency dielectric increment $\Delta\varepsilon_2'$ in the theory of van der Touw and Mandel is given by the same expression as above [Eq. (11)] with b^2 replacing $\langle R^2 \rangle$.

The relaxation time for the high-frequency dispersion is related both to b, the length of the subunit, and to u, the counterion mobility. The expression

$$\tau_2 = b^2/\pi^2 ukT \tag{12}$$

obtained for unidimensional diffusion of small ions was considered by van der Touw and Mandel to be most consistent with their model. As far as the low-frequency dispersion is concerned, the authors point out that the interpretation of the average relaxation time is more difficult, since the dispersion may derive either from the electric polarization due to the rotation of the whole macromolecule or from the axial diffusion of bound counterions from one subunit to another or from both. The average relaxation time τ_1 for the low-frequency dispersion could therefore contain contributions from both processes.

For the rotational relaxation process, van der Touw and Mandel employed an expression for the rotational relaxation time τ_R of a cylindrical particle about its short axis. For a cylindrical polyion of length L and diameter d at infinite dilution

$$\tau_R = \pi\eta_s L^3/6kT[\ln(2L/d) + K] \tag{13}$$

where K is a constant of the order of -0.5.

No theoretical expression was derived for the relaxation time τ_Q of the axial diffusion process in the flexible polyion, although for polyions whose conformation can be considered to approximate a rigid rod of length L, a single dispersion mechanism with a relaxation time

$$\tau' = L^2/\pi^2 ukT \tag{14}$$

is predicted, in anology with Eq. (12). Since the flexible polyion model requires a barrier to intersegmental diffusion, τ_Q obviously must be greater than τ'. It is assumed that the relaxation times are related as follows

$$1/\tau_1 = 1/\tau_R + 1/\tau_Q \tag{15}$$

Van der Touw and Mandel [84a,b] obtained results for the dilute-solution dielectric behavior of several synthetic polyelectrolyte series over a wide frequency range, i.e., 5 kHz to 100 MHz. The polymers studied were PMAA–0.4Na, PAA–0.4Na as a function of chain length ($C_p = 5.2 \times 10^{-3}$ monomol/liter), and PSSA–Na ($P = 500$) as a function of polymer concentration. The results for the PAA–0.4Na series are illustrated in Fig. 3.36. For each polyelectrolyte system studied, the results showed clearly the existence of two separated dispersion regions, as had been shown previously by Minakata and Imai [82] for the PAA–$(nC_4H_9)_4N$ system, and a similar method of analysis was used to obtain the relaxation times of each dispersion. The mean relaxation time τ_1 and the static electric permittivity ε_0' of the low-frequency dispersion were both found to be molecular weight dependent, in line with previous observations [76a,b]. For the high-frequency dispersion, both τ_2 and $\Delta\varepsilon_2'$ were found to be independent of molecular weight, also as observed by earlier workers [79, 80].

Van der Touw and Mandel showed that the reciprocal values of the specific increments $C_p/\Delta\varepsilon_i'$ and the relaxation times $1/\tau_i$ are all linearly dependent on the macromolecular concentration, and extrapolations to infinite dilution could be used to obtain intrinsic values of the dielectric

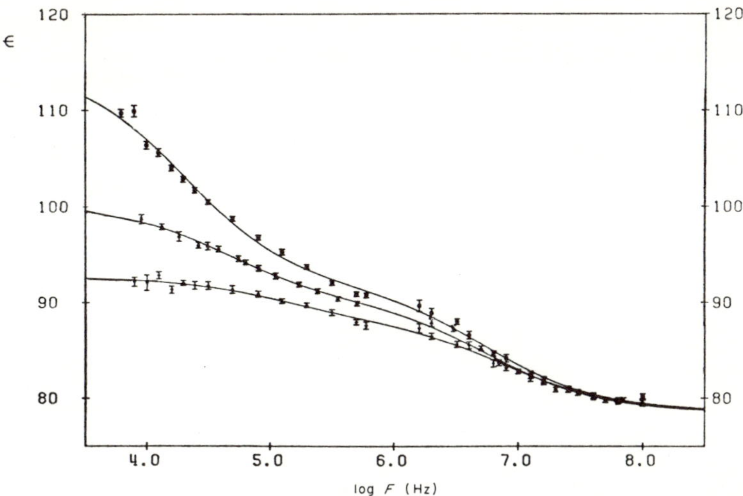

FIG. 3.36. Dielectric dispersion of PAA–0.4Na as a function of chain length. $C_p = 5.2 \times 10^{-3}$ monomol/liter. $P = 640$ (+), 1530 (▲), 7500 (●) [84b].

parameters. An example of this treatment for the PSSA–Na series is shown in Fig. 3.37 for both the low- and high-frequency dielectric dispersions. The authors also reanalyzed the results of several earlier workers [76a,b, 79, 82a,b] in the same fashion and showed that the extrapolation procedure applied in these cases as well. Using Eq. (11), they were able to calculate values for $\langle R^2 \rangle$ and b^2 from the values of the intrinsic increments, assuming values for γ in the range of 0.6 to 0.8. For example, for PSSA–Na ($P = 500, \alpha = 1$), $\langle R^2 \rangle^{1/2} = 1040$ Å and $b = 430$ Å were obtained, assuming $\gamma = 0.7$. Values of similar magnitude were obtained for the other high-molecular-weight polymers studied. Also for PSSA–Na, the values $(\tau_1)_0 = 7 \times 10^{-6}$ sec and $(\tau_2)_0 = 0.9 \times 10^{-7}$ sec were found from the data illustrated in Fig. 3.37. Using the above-calculated value of b (= 430 Å) in Eq. (12), and assuming the mobility of sodium ion to be that in infinitely dilute solution, they obtained $(\tau_2)_{pred} = 1.5 \times 10^{-7}$ sec, in reasonable agreement with the observed value.

Values of τ_R were obtained from Eq. (13) and the previously determined values of $\langle R^2 \rangle^{1/2}$, assuming a range of values of d. For the PSSA–Na series, using $\langle R^2 \rangle^{1/2} = 1040$ Å and 2 Å $< d <$ 10 Å, Eq. (13) yields a value for τ_R between 1.5×10^{-5} sec and 4.5×10^{-5} sec, as compared with the observed value of $\tau_1 = 7 \times 10^{-6}$ sec. They point out that the discrepancy may be due in part to the unknown contribution of the axial diffusion mechanism to the low-frequency dispersion. It should also be pointed out, how-

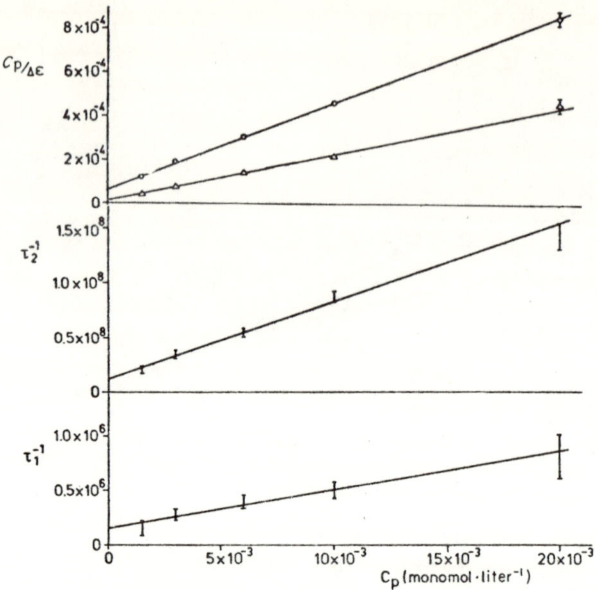

FIG. 3.37. Variation in reduced dielectric increments and relaxation times with polymer concentration for PSSA–Na. (△) $C_p/(\Delta\varepsilon_1 + \Delta\varepsilon_2)$; (○) $C_p/\Delta\varepsilon_2$ [84b].

ever, that the agreement between τ_1 and τ_R would not be as good as shown above if a smaller degree of polyion asphericity were assumed. (Van der Touw and Mandel used $100 < L/d < 500$.) For $L/d = 10$, for example, the value of τ_R is approximately twice the value at $L/d = 100$.

The low-frequency relaxation mechanism proposed by van der Touw and Mandel is really two mechanisms. The rotational relaxation mechanism is very similar to that proposed by Okamoto *et al.* [72] to account for the excess relaxation in PAA–Na solutions, and in fact both authors use the same expression for τ_R—Eq. (13). The axial diffusion mechanism is similar to that proposed by Minakata *et al.* [83], with the exception of the type of dependence on counterion association, as noted earlier.

The proposed high-frequency mechanism differs from that of previous workers [79, 81] in that the dispersion depends on the length of the polymer segment rather than on the thickness of the ionic atmosphere surrounding the polyion. Differentiating between thse two possibilities would probably be difficult, because for dilute polyelectrolyte solutions of low salt content, the values of b calculated for an equivalent chain model, such as that proposed by Alexandrowicz [85], are of the same order of magnitude as the average radius of the counterion atmosphere, and changes in the segment length with such parameters as polymer concentration and added salt content are paralleled by changes in the counterion atmosphere radius. (See Chapter V for a more complete discussion of polyelectrolyte models.)

The model of van der Touw and Mandel was tested by Muller [86] for the polycondensate of L-lysine and 1,3-benzene-disulfonyl chloride—PLL—whose structural unit is

$$[—SO_2—\phi—SO_2—NH—CH(COOH)—(CH_2)_4—NH—]_n$$

This polymer undergoes a conformational transition upon ionization, as evidenced by potentiometric and viscosimetric data [87]; the variation in viscosity is analogous to that seen in PMAA solutions [65]. The dielectric data (ε' versus frequency) were fitted to a two-component form of Eq. (9) (for two separated dispersions), using $\beta = 0.8$ to represent the breadth of the distribution of relaxation times. The variation in $\Delta\varepsilon_1'$ and $\Delta\varepsilon_2'$ with α is illustrated in Fig. 3.38. At low degrees of neutralization, only a single broad dispersion was observed; however, above $\alpha = 0.2$, two distinct dielectric dispersions were seen. Above $\alpha = 0.2$, neither τ_1 nor τ_2 was found to be very dependent on α, the values being approximately 3×10^{-6} sec for τ_1 and 3×10^{-8} sec for τ_2 ($C_p = 2 \times 10^{-3}$ monomol/liter).

With decreasing polymer concentration, Muller [86a] found that the relaxation times τ_1 and τ_2 and the reduced increments $\Delta\varepsilon_1'/C_p$ and $\Delta\varepsilon_2'/C_p$ all decreased with increasing C_p. The dependence of $\Delta\varepsilon_1'/C_p$ on C_p and α was found to parallel the change in η_{sp}/C with C_p and α, and was related to the associated change in overall polymer dimensions. The variation in the high-frequency dispersion with C_p and α was attributed to changes in the thickness of the ionic atmosphere.

Muller *et al.* [88] also studied the dielectric properties of poly-L-glutamic acid—PGA—in salt-free aqueous solutions over the frequency range 2.5

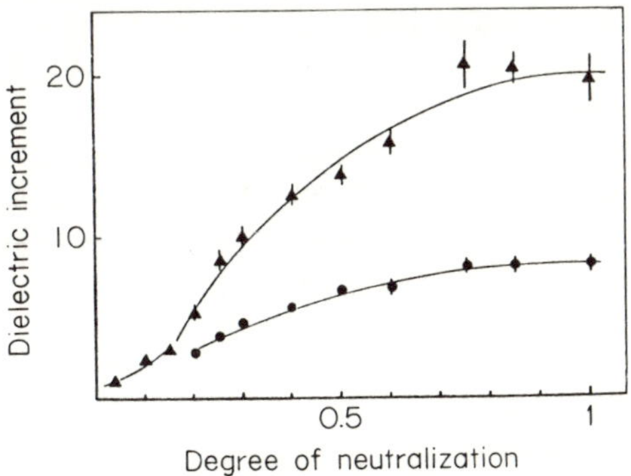

FIG. 3.38. Dielectric increment as a function of the degree of neutralization for PLL–Na. $C_p = 2 \times 10^{-3}$ monomol/liter. Triangle represents total increment ($\Delta\varepsilon_1 + \Delta\varepsilon_2$); circle represents $\Delta\varepsilon_2$ only [86a].

kHz to 100 MHz. As in the other polymer solutions studied, two dielectric dispersions are seen, the low-frequency parameters being molecular weight dependent, the high-frequency ones not. The variation in the low-frequency dispersion with the degree of ionization was found to reflect the conformational changes associated with the helix-coil transition that occurs in the region $\alpha = 0.3$–0.5. The experimental findings for the PLL and PGA systems are in general agreement with those found by the previous workers on other polyelectrolyte systems, and confirm the conclusions regarding the experimental nature of the two dispersion regions.

Relaxation effects due to counterions can also be studied by ultrasonic absorption in polyelectrolyte solutions. Several investigations of this nature have been reported. In one, Atkinson *et al.* [89] measured ultrasonic absorption in aqueous solutions of PAA and CMC salts in the frequency range 1–190 MHz. Although the polymer concentrations ($\sim 5\%$ by weight) employed in this study were much higher than in the dielectric studies, the results are discussed in this section because of the association of the ultrasound absorption process with counterion dynamics.

Atkinson *et al.* observed that the transmission of ultrasound through the polyelectrolyte solutions was characterized by excess absorption. With sodium neutralized polymers, they found that the process associated with the excess absorption could be fitted to a distribution of relaxation times similar to the Cole and Cole distribution [Eq. (9)] for dielectric dispersion.

The ultrasonic absorption in the megahertz range was found to be independent of variations in polymer molecular weight. This fact led to the conclusion that the high-frequency process does not involve the entire polymer chain but only repetitive segments of it; this conclusion was also found applicable to the dielectric dispersion process in the same frequency range. In contrast to the behavior of dielectric dispersion, however, the relaxation frequencies due to ultrasonic absorption were not affected by variations in polymer concentration or by the addition of excess univalent electrolyte. Also, with increasing degrees of neutralization of PAA, the ultrasonic absorption was observed to shift to higher frequencies as shown in Fig. 3.39, whereas f_c did not change with α in the dielectric studies. Finally, with the substitution of Ca^{2+} or Mg^{2+} for Na^+, a second absorption band appeared in the lower range of frequency studied, and the Cole and Cole analysis could no longer be applied. Similar findings were reported by Baumgartner *et al.* [90] for ultrasound absorption in the polyethyleneimine system.

These observations indicate that the ultrasonic absorption in the high-frequency region is sensitive to ion-binding phenomena. Tondre and Zana [91], in fact, used ultrasonic absorption as a probe for the study of site binding of counterions in polyelectrolyte solutions. They point out that the existence of excess absorption in these cases implied that a positive volume change ΔV is associated with the relaxation process whose equilibrium is

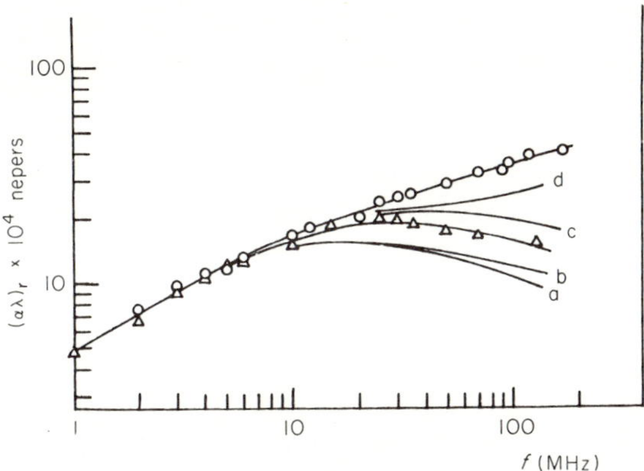

FIG. 3.39. Ultrasound energy dissipated per wavelength, $(\alpha\lambda)_r$, for PAA solutions (C_p = 0.5 monomol/liter) at various degrees of neutralization with NaOH: (a) 0.1, (b) 0.0, (c) 0.4, (d) 0.6; (○) 1.0, (△) 0.2 [89].

perturbed by the transmission of the sound wave. Since the association process presumably involves partial release of the water of hydration of the unassociated ions, ultrasound absorption is expected to be particularly sensitive to ion-binding phenomena since ΔV will likely vary a great deal with the type of ions involved. The assignment of the high-frequency absorption process to ion binding allows one to account for the dependence on the degree of ionization (which increases the number of bound ions), for the specific counterion effects, and for the lack of sensitivity to excess univalent electrolyte or concentration changes. It should be pointed out that the absorption observed in polyelectrolytes is similar in many aspects to absorption in solutions of low-molecular-weight associating electrolytes, where the association process involves the release of bound water [92].

Motion of counterions within the diffuse or unbound layer, on the other hand, will not lead to significant volume changes if the hydration shells of the counterions do not overlap with that of the polyion. The latter process, however, involving large fluctuations in charge separation should be highly sensitive to polarizability phenomena, as seen by the dielectric behavior at high frequencies. Variations in the dimensions of the unbound counterion layer and the diffuse counterion concentration both lead to changes in the polarizability, as evidenced by the dependence on polymer concentration as well as that of added electrolyte.

It is worth noting that the low-frequency dielectric dispersion occurs in the same frequency region as the mechanical dispersion involving conformational relaxation [72]. Although both are strongly molecular weight dependent, it is not clear what mechanistic relationship exists between these two processes.

TABLE II
Dilute-Solution Relaxation Phenomena

Relaxation frequency (Hz)	Ultrasonic (U) or dielectric (D)	Proposed mechanism	Material	P	C_p (monomol/liter)	Ref.
3×10^2–4×10^4	U	Rotational relaxation of elongated polyion	PAA–Na ($C_s = 0.005M$)	260–3400	0.02	[72]
3×10^3–4×10^5	U	Conformational relaxation	PAA–Na ($C_s = 0.005M$)	260–3400	0.02	[72]
7×10^2–1.5×10^3	D	Longitudinal polarization of bound counterions	PAA–$(nC_4H_9)_4N$	1200	2×10^{-5}–0.005	[82]
7×10^3–10^5	D	Longitudinal polarization of bound counterions[a]	PMAA–K	710–4.1×10^4	0.005	[76a,b]
10^4	D	Longitudinal polarization of bound counterions[a]	PVA–Cl	$<2 \times 10^3$	0.005	[76a,b]
2×10^4–1.2×10^5	D	Longitudinal polarization *or* rotational relaxation	PAA–0.4Na	640–7500	0.005	[84a,b]
2.5×10^4–1.3×10^5	D	Longitudinal polarization *or* rotational relaxation	PSSA–Na	500	0.0015–0.02	[84a,b]
5×10^4	D	Longitudinal polarization *or* rotational relaxation	PLL–Na	—	0.002	[86a,b]

1.8×10^4–7×10^5	D	Short-range polarization of counterion atmosphere around polyion	PAA–$(nC_4H_9)_4N$	1200	2×10^{-5}–0.005	[82]
6×10^5–4.6×10^6	D	Short-range polarization of counterion atmosphere around polyion	PVSA–Na	500	0.0003–0.004	[79]
2.5×10^6	D	Short-range polarization of counterion atmosphere around polyion	PSSA–Na	875–6.4×10^4	0.0025	[79]
5.3×10^6	D	Short-range polarization of counterion atmosphere around polyion	PAA–0.5Na	830	0.0075	[79]
3.3×10^6–2.3×10^7	D	Ion fluctuations over subunit	PSSA–Na	500	0.0015–0.02	[84a,b]
6×10^6	D	Ion fluctuations over subunit	PAA–0.4Na	640–7500	0.005	[84a,b]
6×10^6	D	Ion fluctuations over subunit	PLL–Na	—	0.002	[86a,b]
6×10^6–2.2×10^7	D	Short-range polarization of counterion atmosphere	CMC–0.6Na	340–400	0.0032–0.013	[80]
5×10^7	U	Counterion association/dissociation	PAA–Na	350–3500	0.5	[89]

[a] Reinterpreted as longitudinal polarization *or* rotational relaxation [84].

In Table II the characteristic frequencies for the various relaxational processes, which occur in dilute solutions of polyelectrolytes, are summarized. Two separated regions of frequency dependence are seen—the low-frequency (kilohertz) range, involving relaxations over the whole polyion, and the high frequency (megahertz) range, involving relaxational processes over polymer segments.

Finally, one additional counterion relaxation process should be mentioned here. Nuclear magnetic relaxation of ^{23}Na in polyelectrolyte solutions has been studied by van der Klink *et al.* [93], who used the cylindrical cell model for polyelectrolytes to obtain an expression for the nuclear spin-lattice relaxation of counterion nuclei possessing a quadrupole moment (such as ^{23}Na). The expression showed that, above $\alpha \simeq 0.3$, the relaxation rate should be a quadratic function of α, while in the concentration range 0.05–1 *N*, the concentration dependence is only slight. Experimental results for the ^{23}Na NMR in PAA–Na solutions were shown to be in reasonable agreement with the theory. The authors point out that this technique should be a very sensitive means of studying the polyion atmosphere, since NMR rates of the co- and counterions in polyelectrolytes depend directly on details of the ion distribution.

It has been shown that there is considerable range of agreement on the experimental nature of the two dielectric dispersion regions in aqueous polyelectrolytes. By contrast, no such unanimity exists in regard to the theoretical interpretation. While some authors ascribe the low-frequency mechanism to rotational relaxation and/or axial diffusion of bound counterions [84], others interpret it as due to counterion fluctuations alone [83]. In addition, the dependence on counterion association is conceptually different in the two treatments.

The high-frequency mechanism is ascribed either to local fluctuations in bound counterion density [84a,b], or to relaxation of the unbound counterions in the counterion atmosphere [79], or to conductivity differences between the polyion coils and the surrounding medium [80, 83]. Again the theories differ in their dependence on counterion association or binding. Furthermore, the ultrasonic high-frequency mechanism is shown to be directly related to the process of counterion association [91]. Since it is doubtful that all of these mechanisms are correct, it is to be hoped that future work will lead to a theory which attracts a broad range of support and is capable of combining the high-frequency ultrasonic and dielectric aspects, which must be related, as well as being consistent with the features of conformational relaxation at lower frequencies.

(*b*) *Theoretical Approach: Infinite Dilution*

In Section C2, several of the papers that were discussed dealt with the viscoelastic properties of polyelectrolyte solutions in terms of dilute-solution viscoelastic theory. Okamoto *et al.* [72, 73] used the dilute-solution theory of Zimm as a basis for the interpretation of their results. However, as pointed out in Section C2, intermolecular contributions to the viscoelastic properties were found to be significant, and thus the work was included in the previous section. It was also shown that the Zimm and Rouse theories, developed for nonionic dilute solutions, were not quantitatively accurate in describing the intramolecular portion of the relaxation. The study pointed out the need for the development of viscoelastic theory to deal with ion-containing polymers. In this section, a theoretical approach to dilute-solution viscoelasticity is described. This theory predicts for a simple ionic polymer model some of the deviations in viscoelastic parameters which result from ionization. It is, as a first approach, oversimplified and awaits a more thorough development. It also awaits further experimental results in truly dilute solution, where the exclusively intramolecular component of the viscoelastic response can be obtained. It should be emphasized that this theory deals only with conformational relaxation, and no attempt is made to include the effects of counterion dynamics.

Several models have been developed to represent the viscoelastic behavior of nonionic polymers in solution. The approach to the study of ion-containing polymers has been to modify one of these models. The most commonly employed representation of a linear, flexible macromolecule is the bead–spring assembly; its application to the viscoelasticity of nonionic polymers is well known [70, 74, 94–98]. In the following treatment, the additional factor of charge–charge interaction is incorporated in the model.

The charged bead–spring model is identical to that described by Rouse [70] with the additional feature that the beads are considered to contain electrostatic point charges of equal magnitude [99]. Because of this consideration, the forces acting on the beads contain Coulombic components, and these additional terms modify the analysis of the viscoelastic properties of this system. However, the same basic approach that is used for the analysis of uncharged models is used with the charged systems.

Owing to the complexity of the problem, only the two simplest members of the charged bead–spring series have been treated. These arc referred to as the dumbbell and the three-bead/two-spring—3B2S—assembly for simplicity. In this section only a brief summary of the theory will be presented, essentially emphasizing the differences between this development and that corresponding to the uncharged bead–spring model. A more complete account has been presented elsewhere [55, 99]. First, the analysis of the viscoelastic functions of the ionic dumbbell model will be summarized insofar as it differs from existing nonionic theory. Then a summary of calculations on this model will be given. Finally, the analysis and results of the charged 3B2S system will be presented.

In the bead–spring model, all of the frictional effect is considered to be localized in the beads, and the viscous drag force on each bead is proportional to its relative velocity in the solvent. Since the beads are small enough for the inertial force (mass times acceleration) to be considered negligible, the viscous drag force may be equated to the force which causes the bead to move. This force is the sum of the (random) Brownian motion force, the Hookean force exerted by the spring which joins the beads, and the Coulombic force due to the charges on the beads. This may be written symbolically as

$$F_{\text{visc}} = F_{\text{Brownian}} + F_{\text{Hookean}} + F_{\text{Coulombic}} \tag{16}$$

$F_{\text{Coulombic}}$ assumes the form $-(q^2/\varepsilon R^2)\hat{R}$, where q is the magnitude of charge on each bead, ε is the dielectric constant of the solvent, R is the interbead separation, and $\hat{R}$ is the unit vector along the line joining the two beads. The other force expressions take their usual forms.

All of the force expressions can be cast in terms of ψ, the time-dependent distribution function of the coordinates of the two beads. The time and spatial derivatives of ψ are related by the equation of continuity, which can be combined with Eq. (16) to give the diffusion equation, from whose solution the viscoelastic properties of interest can be calculated.

The complete solution of the diffusion equation is not known, even for the uncharged dumbbell. However, for simple shear flow an approximate solution of the equation can be obtained by substituting a power series for the distribution function, i.e.,

$$\psi = \psi_0 \sum_{n=0}^{\infty} (\kappa^n \Lambda_n), \qquad \Lambda_0 \equiv 1 \tag{17}$$

where κ is the rate of shear. This procedure results in a recursion relation for the Λ_n, in which Λ_{n+1} can be determined from Λ_n.

For low shear rates, the area of interest in most dynamic studies, only the first two terms of the power series expansion need be considered in order to determine the complex viscosity. All higher-order terms will vanish in the limit $\kappa \to 0$. The first term, ψ_0, is the distribution function which applies for the condition of no flow. For the uncharged dumbbell, it represents the Gaussian distribution of end-to-end separation

$$(\psi_0)_{\text{nonionic}} = \exp[-(H/2kT)R^2] \tag{18}$$

where H is the Hookean force constant of the spring. For the ionic dumbbell, ψ_0 is the same function multiplied by a Coulombic factor

$$(\psi_0)_{\text{ionic}} = (\psi_0)_{\text{nonionic}} \cdot \exp[-(q^2/\varepsilon kT)/R] \tag{19}$$

For the nonionic dumbbell, a closed-form solution of Λ_1 has been obtained. For the ionic dumbbell, Λ_1 cannot be expressed in closed form because of the additional Coulombic terms which appear in the differential equation; however, the solution can be obtained by a boundary-value method. $(\Lambda_1)_{\text{ionic}}$ is considered to be asymptotic with $(\Lambda_1)_{\text{nonionic}}$ at large values of end-to-end separation where Coulombic interaction is negligibly small. As the bead separation approaches zero, the Coulombic repulsion increases without limit, and the distribution function must vanish. With these two boundary conditions, an iterative solution of $(\Lambda_1)_{\text{ionic}}$ can be obtained.

From the approximate solution $\psi = \psi_0 + \kappa\Lambda_1$, the complex viscosity $[\eta^*]$ is evaluated in the usual fashion. $[\eta^*]$ may be expressed as a function of the reduced frequency $\omega\lambda_0$ for different sets of values of the parameters H, q, and ε, defined previously. These parameters are obtained by comparison of the ionic dumbbell model with the real system to which it most closely corresponds—the series of linear dicarboxylic acid salts [100]. For sufficiently long diacid chains, the distribution of end-to-end distances can be considered Gaussian, and the chain is assumed to have viscoelastic properties equivalent to those of an elastic dumbbell—two beads joined by a Hookean spring whose constant is inversely proportional to the mean-square separation of the ends. Each bead is considered to have one electronic charge, and the dielectric constant is taken to be that of the pure solvent, for simplicity. Linear zwitterions [101] can also be considered, with certain mathematical reservations since the Coulombic force is attractive.

From the calculations performed, it is found that both the intrinsic viscosity and the relaxation time of a suspension of like-charged dumbbells are greater than those corresponding to the uncharged material. It is further found that the shapes of the curves of complex viscosity as a function of reduced frequency are essentially unchanged, and that the curves can be superimposed by a simple multiplicative shift in the relaxation time. The magnitude of the increase in intrinsic viscosity and relaxation time is found to be a function of the dimensionless parameter $q^2H^{1/2}/\varepsilon(kT)^{3/2}$. The ratios of the intrinsic viscosity and relaxation time of ionized and unionized forms of linear dicarboxylic acids and zwitterion are presented in Fig. 3.40 as a function of this parameter. Four individual cases are shown: (a) a C_{10} diacid in water; (b) a C_{30} diacid in water; (c) a C_{30} diacid in ethanol; and (d) a C_{30} zwitterion in water. The examples demonstrate the effect of various changes in chain length, dielectric constant, and sign of charge on the viscoelastic functions. The examples chosen are based on hypothetical polymers and are mainly intended to illustrate the type and the relative magnitude

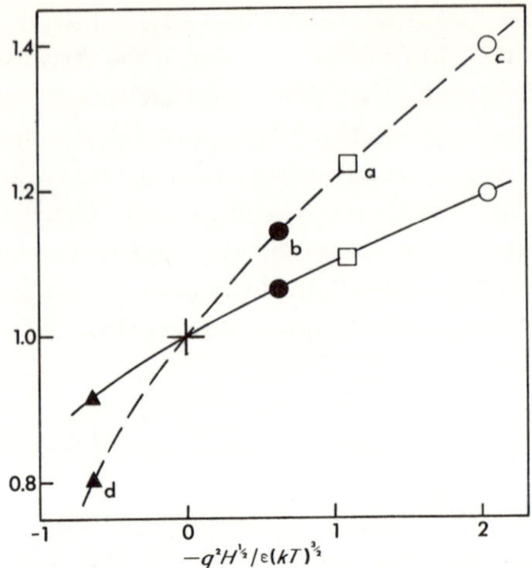

FIG. 3.40. $[\eta]/[\eta]_0$ (dashed line) and λ/λ_0 (solid line) plotted against the dimensionless parameter $-q^2H^{1/2}/\varepsilon(kT)^{3/2}$ for the ionic dumbbell [99].

of the effect that one should expect ions to exert on dilute-solution viscoelasticity.

The analysis of the charged 3B2S system proceeds in basically the same way as that of the ionic dumbbell. One additional simplification is necessary in order to solve the differential equation of the distribution function. In the dumbbell analysis, the result that the relaxation time of the ionic dumbbell is obtained as a simple multiplicative shift from that of the uncharged dumbbell could be obtained equivalently by combining the Coulombic and Hookean forces, which act along the same line, into a "pseudo-Hookean" force, with an associated constant which can be predicted from Fig. 3.40. In the 3B2S system, the 1–2 and 2–3 Coulombic forces are combined in the same way with the respective 1–2 and 2–3 Hookean forces as "pseudo-Hookean" forces.

The calculations were performed for the linear triacids which can be considered the dimers of examples (a), (b), and (c) above. Here, two relaxation times λ_1 and λ_2 are found, as in the nonionic case. Both of the relaxation times, as well as the intrinsic viscosity, are increased from those of the unionized material. In addition, the ratio of relaxation times, λ_1/λ_2, which has the exact value of 3 in the nonionic case, is increased on ionization. The importance of these calculations is that the distribution of relaxation times is broadened in the 3B2S case, and that the broadening of the distribution of relaxation times would likely occur in more complex, multiply charged bead–spring models, as well.

In the above treatment, no account was taken for the concentration of small ions. Nakajima and Wada [102] extended the study of the ionized 3B2S system to determine the viscoelastic behavior as a function of small-ion concentration as well as the charge on the beads. The effect of shielding by small ions was taken into account by the Debye–Hückel approximation, which introduces an effective shielding radius, κ^{-1}, where κ^2 is proportional to c_s, the added salt concentration (see Chapter V).

Fujimori *et al.* [103] continued the investigation of the charged bead–spring model with a study of the computer simulation of Brownian motion. The method provided a means of calculating the dynamic viscosity and relaxation times of the polyelectrolyte model, quantities that depend on the distribution function of the coordinates of the beads, for which an analytical solution cannot be found. For bead–spring systems up to $N = 9$, polymer dimensions and shape, relaxation times, and the components of the complex viscosity $\eta^*(\omega)$ were determined for various values of the bead charge q and the Debye–Hückel parameter κ.

They found that for any fixed value of N and κ, $\overline{h^2}$ increases with increasing q and that the rms length of the principal axis L_1 increases at a faster rate than that of the other two axes L_2 and L_3, indicating an increasing degree of asphericity with increasing charge. For any given value of q and N, increasing κ (i.e., increasing c_s) reduced $\overline{h^2}$ and the axial lengths, and also reduced the degree of anisotropy.

Relaxation times of vectorial parameters, such as the end-to-end vector $\mathbf{h}$ or the first principal axis $\mathbf{L}_1$ were interpreted as rotational relaxation times, while relaxation times of scalar quantities, such as $\overline{h^2}$ and $\overline{L_1{}^2}$, were interpreted as conformational relaxation times. They found that values of τ_{rot} and τ_{conf} were comparable for the nonionic model, but for increasing values of q or decreasing κ, the values of τ_{rot} increased much more rapidly than the values of τ_{conf}.

The real and imaginary components η' and η'' of the complex viscosity were then computed by Fourier transform analysis. They found that the low-frequency limiting values of η' increased rapidly with increasing q at constant N and κ or with increasing N at constant q and κ, the latter case reproducing qualitatively the pattern illustrated in Fig. 3.29. The curves of η'' versus ω were split into two peaks, with the major peak, representing the rotational contribution, shifting rapidly to lower frequencies with increasing charge or bead number. The frequency range studied by Okamoto *et al.* [72, 73], however, did not permit the observation of this latter phenomenon.

REFERENCES

1. A. Eisenberg, *Adv. Polym. Sci.* **5,** 59 (1966).
2. H. von Rotger, *Glastech. Ber.* **19,** 192 (1941).
3. A. Eisenberg, M. King, and T. Yokoyama, *in* "Water-Soluble Polymers" (N. M. Bikales, ed.), p. 349. Plenum Press, New York, 1973.
4. J. R. Van Wazer, "Phosphorus and Its Compounds," Vol. I. Wiley (Interscience), New York, 1958.
5. A. V. Tobolsky, "Properties and Structure of Polymers." Wiley, New York, 1960.
6. C. F. Callis, J. R. Van Wazer, and J. S. Metcalf, *J. Am. Chem. Soc.* **77,** 1471 (1955).
7. U. P. Strauss and E. H. Smith, *J. Am. Chem. Soc.* **75,** 6168 (1953).
8. J. D. Ferry, W. C. Child, Jr.. R. Zand, D. M. Stern, M. L. Williams, and R. F. Landel, *J. Colloid Sci.* **12,** 53 (1957).
9. A. Eisenberg and T. Sasada, *J. Polym. Sci. Part C* **16,** 3473 (1968).
10. J. D. Ferry, "Viscoelastic Properties of Polymers," 2nd ed., Chapter XI. Wiley, New York, 1970.
11. A. V. Tobolsky and K. Murakami, *J. Polym. Sci.* **40,** 443 (1959).
12. H. Sobue and K. Murakami, *J. Polym. Sci. Part A* **1,** 2039 (1963).
13. A. Eisenberg, S. Saito, and T. Sasada, *J. Macromol. Sci. Phys.* **A1,** 121 (1967).
14. A. Miyake, *J. Polym. Sci.* **22,** 560 (1956).
15. A. V. Tobolsky, *J. Chem. Phys.* **37,** 1575 (1962).
16. A. V. Tobolsky and R. B. Taylor, *J. Phys. Chem.* **67,** 2439 (1963).

17a. A. Eisenberg and L. A. Teter, *J. Polym. Sci. Part B* **7,** 471 (1969).

17b. A. Eisenberg and L. A. Teter, *J. Am. Chem. Soc.* **87,** 2103 (1965).

18. W. M. van Beck, T. J. Odijk, F. van der Touw, and M. Mandel, *J. Polym. Sci. Polym. Phys.* **14,** 773 (1976).
19. A. Eisenberg and S. Saito, *J. Macromol. Sci. Chem.* **A2,** 799 (1968).
20. E. Bamann and M. Meisenheimer, *Ber.* **71,** 1711 (1938).
21. W. W. Butcher and F. H. Westheimer, *J. Am. Chem. Soc.* **77,** 2420 (1955).
22. A. Eisenberg, M. King, and M. Navratil, *Macromolecules* **6,** 734 (1973).
23. I. L. Hopkins and R. W. Hamming, *J. Appl. Phys.* **28,** 906 (1957).
24. A. M. Beuche, *J. Chem. Phys.* **21,** 614 (1953).
25. J. P. Berry and W. F. Watson, *J. Polym. Sci.* **18,** 201 (1955).
26. H. Yu, *J. Polym. Sci. Part B* **2,** 631 (1964).
27. A. V. Tobolsky, *J. Polym. Sci. Part B* **2,** 636 (1964).
28. W. Eitel (ed.), "Silicate Science," Vols. I to V. Academic Press, New York, 1964–1966.
29. J. A. Prins (ed.), "Physics of Non-crystalline Solids." North-Holland Publ., Amsterdam, 1965.
30. N. H. Ray, *in* "Ionic Polymers" (L. Holliday, ed.), p. 363. Applied Science Publ., London, 1975.
31. J. W. Marx and J. M. Sivertsen, *J. Appl. Phys.* **24,** 81 (1953).
32. K. E. Forry, *J. Am. Ceram. Soc.* **40,** 90 (1957).
33. R. Ryder and G. E. Rindone, *J. Am. Ceram. Soc.* **44,** 532 (1961).
34. J. V. Fitzgerald and K. M. Laing, *J. Soc. Glass Tech.* **1,** 14 (1952).
35. A. Eisenberg and K. Takahashi, *J. Non-Cryst. Solids* **3,** 279 (1970).
36. R. D. Lundberg, F. E. Bailey, and R. W. Collard, *J. Polym. Sci. Part A-1* **4,** 1563 (1966).
37. K. J. Liu, *Macromolecules* **1,** 308 (1968).
38. R. C. Little, *J. Appl. Polym. Sci.* **15,** 3117 (1971).
39. K. J. Liu and J. E. Anderson, *Macromolecules* **2,** 235 (1969).
40. K. J. Liu and R. Ullman, *J. Chem. Phys.* **48,** 1158 (1968).

41. J. Moacanin and E. F. Cuddihy, *J. Polym. Sci. Part C* **14,** 313 (1966).
42a. M. J. Hannon and K. F. Wissbrun, *J. Polym. Sci. Polym. Phys.* **13,** 113 (1975).
42b. K. F. Wissbrun, *Makromol. Chem.* **118,** 211 (1968).
43. D. B. James and R. E. Wetton, *J. Chem. Soc. Faraday Trans.* (to be published).
44. M. M. Labes, *J. Polym. Sci. Part C* **17,** 95 (1967).
45. T. Sulzberg and R. J. Cotter, *Macromolecules* **2,** 146 (1969).
46. A. Rembaum, *J. Polym. Sci. Part C* **29,** 157 (1970).
47. J. Kumanotani, R. Miyatake, H. Mitsui, and S. Saito, paper presented to Polymer Chem. Soc., Japan, Kyoto (1972).
48. T. Sulzberg and R. J. Cotter, *Macromolecules* **1,** 554 (1968).
49. S. A. Rice and M. Nagasawa, "Polyelectrolyte Solutions." Academic Press, New York, 1961.
50. A. Katchalsky, Z. Alexandrowicz, and O. Kedem, *in* "Chemical Physics of Ionic Solutions" (B. E. Conway and R. G. Barradas, eds.), p. 295. Wiley, New York, 1966.
51. F. Oosawa, "Polyelectrolytes." Dekker, New York, 1971.
52. W. E. Fitzgerald and L. E. Nielsen, *Proc. Roy. Soc.* (*London*) **A282,** 137 (1964).
53. J. E. Fields and L. E. Nielsen, *J. Appl. Polym. Sci.* **12,** 1041 (1968).
54. L. E. Nielsen, *Polym. Eng. Sci.* **9,** 356 (1969).
55. M. King, Ph.D. Thesis, McGill Univ. (1973).
56. A. Eisenberg and T. Sasada, *in* "Physics of Non-crystalline Solids" (J. A. Prins, ed.), p. 99. North-Holland Publ., Amsterdam, 1965.
57. A. Eisenberg, H. Matsuura, and T. Yokoyama, *J. Polym. Sci. Part A-2* **9,** 2131 (1971).
58. M. Takayanagi, *Mem. Fac. Eng. Kyushu Univ.* **23,** 41 (1963).
59. L. L. Chapoy and A. V. Tobolsky, *Chem. Scripta* **2,** 44 (1972).
60. F. R. Schwarzl and L. C. E. Struik, *Adv. Mol. Relaxat. Proc.* **1,** 201 (1968).
61. A. Silberberg and P. F. Mijnlieff, *J. Polym. Sci. Part A-2* **8,** 1089 (1970).
62. A. Silberberg, J. Eliassaf, and A. Katchalsky, *J. Polym. Sci.* **23,** 259 (1957).
63. N. Nishida, *J. Polym. Sci. Part A-2* **4,** 845 (1966).
64. A. Konno and M. Kaneko, *Makromol. Chem.* **138,** 189 (1970).
65. V. Crescenzi, *Adv. Polym. Sci.* **5,** 358 (1968).
66. I. Noda, T. Tsuge, and M. Nagasawa, *J. Phys. Chem.* **74,** 710 (1970).
67. H. Eisenberg and D. Woodside, *J. Chem. Phys.* **36,** 1844 (1962).
68. M. Sakai, I. Noda, and M. Nagasawa, *J. Polym. Sci. Part A-2* **10,** 1047 (1972).
69. See Ferry [10], Chapter III.
70. P. E. Rouse, Jr., *J. Chem. Phys.* **21,** 1272 (1953).
71. W. W. Graessley, *J. Chem. Phys.* **43,** 2696 (1965).
72. H. Okamoto, H. Nakajima, and Y. Wada, *J. Polym. Sci. Polym. Phys.* **12,** 1035 (1974).
73. H. Okamoto and Y. Wada, *J. Polym. Sci. Polym. Phys.* **12,** 2413 (1974).
74. B. H. Zimm, *J. Chem. Phys.* **24,** 269 (1956).
75. P. J. Flory, "Principles of Polymer Chemistry," Chapter XIV. Cornell Univ. Press, Ithaca, New York, 1953.
76a. M. Mandel and A. Jenard, *Trans. Faraday Soc.* **59,** 2158 (1963).
76b. M. Mandel and A. Jenard, *Trans. Faraday Soc.* **59,** 2170 (1963).
77. K. S. Cole and R. H. Cole, *J. Chem. Phys.* **9,** 341 (1941).
78. H. M. Dintzis, J. L. Oncley, and R. M. Fuoss, *Proc. Nat. Acad. Sci. U.S.* **40,** 62 (1954).
79. S. B. Sachs, A. Raziel, H. Eisenberg, and A. Katchalsky, *Trans. Faraday Soc.* **65,** 77 (1969).
80. D. J. Allen, S. M. Neale, and P. J. T. Tait, *J. Polym. Sci. Part A-2* **10,** 433 (1972).
81. C. T. O'Konski, *J. Phys. Chem.* **64,** 605 (1960).
82a. A. Minakata and N. Imai, *Biopolymers* **11,** 329 (1972).
82b. A. Minakata, *Biopolymers* **11,** 1567 (1972).

83. A. Minakata, N. Imai, and F. Oosawa, *Biopolymers* **11,** 347 (1972).
84a. F. van der Touw and M. Mandel, *Biophys. Chem.* **2,** 218 (1974).
84b. F. van der Touw and M. Mandel, *Biophys. Chem.* **2,** 231 (1974).
85. Z. Alexandrowicz, *J. Chem. Phys.* **47,** 4377 (1967).
86a. G. Muller, *J. Polym. Sci. Polym. Lett.* **12,** 319 (1974).
86b. G. Muller, *Polymer* **15,** 585 (1974).
87. J. C. Fenyo, *Eur. Polym. J.* **10,** 233 (1974).
88. G. Muller, F. van der Touw, S. Zwolle, and M. Mandel, *Biophys. Chem.* **2,** 242 (1974).
89. G. Atkinson, E. Baumgartner, and R. Fernandez-Prini, *J. Am. Chem. Soc.* **93,** 6436 (1971).
90. E. Baumgartner, G. Atkinson, and M. Emara, *J. Am. Chem. Soc.* **95,** 5881 (1973).
91. C. Tondre and R. Zana, *J. Phys. Chem.* **75,** 3367 (1971); *Biophys. Chem.* **1,** 367 (1974).
92. J. Struehr and E. Yeager, *in* "Physical Acoustics," (W. P. Mason, ed.), Vol. II, Part A, Chapter VI. Academic Press, New York, 1965.
93. J. J. van der Klink, L. H. Zuiderweg, and J. C. Leyte, *J. Chem. Phys.* **60,** 2391 (1974).
94. F. Bueche, *J. Chem. Phys.* **22,** 603 (1954).
95. N. Tschoegl, *J. Chem. Phys.* **39,** 149 (1963).
96. R. E. DeWames, W. F. Hall, and M. C. Shen, *J. Chem. Phys.* **46,** 2782 (1967).
97. A. V. Tobolsky and D. B. DuPre, *Adv. Polym. Sci.* **6,** 103 (1969).
98. R. B. Bird, H. R. Warner, Jr., and D. C. Evans, *Adv. Polym. Sci.* **8,** 1 (1971).
99. M. King and A. Eisenberg, *J. Chem. Phys.* **57,** 482 (1972).
100. E. P. Otocka, M. Y. Hellman, and L. L. Blyer, *J. Appl. Phys.* **40,** 4221 (1969).
101. V. Jaacks and N. Mathes, *Makromol. Chem.* **131,** 295 (1970).
102. H. Nakajima and Y. Wada, *Repts. Progr. Polym. Phys. Japan* **17,** 51 (1974).
103. S. Fujimori, H. Nakajima, Y. Wada, and M. Doi, *J. Polym. Sci. Polym. Phys.* **13,** 2135 (1975).

Chapter IV

Viscoelastic Properties of Copolymers

In Chapter III, it was shown that ionic forces can effect considerable changes in the mechanical behavior of homopolymers, especially when the concentration of ions is large, as it is with solid-state polyelectrolytes. It was seen, for example, that ionization of poly(acrylic acid) produced a twofold increase in modulus and a notable increase in the glass-transition temperature. Accompanying these changes, unfortunately, were increased brittleness and general intractability [1].

It is worth recalling that high levels of ionization in inorganic polymers, e.g., window glass materials, provide a wide range of useful properties. In these materials the presence of bond interchange at high temperatures ensures tractability and ease of manufacture. This, coupled with the fact that these materials, which are in the highest oxidation state, make for thermal stability, in contrast to organic polymers which decompose and oxidize at high temperatures. Also, with organic polymers at high levels of ionization and in the absence of plasticizers, the effects of ionic forces are too severe to yield useful mechanical properties. Nevertheless, at lower concentrations of ions, worthwhile modifications in the properties of organic polymers can be achieved.

Polymers containing low concentrations of ions can be prepared by copolymerization of a nonionic monomer with an ionizable or ionic monomer; alternatively, many nonionic polymers can be modified by appropriate chemical techniques to yield partly ionic materials. Both approaches can be considered to result in the same end product—copolymers whose major component is nonionic and whose minor component is ionizable or ionic. Considerable attention has been devoted to the development of systems of this nature.

In this chapter, the viscoelastic behavior of ionic copolymer systems is discussed. It will be seen that, just as in homopolymer systems, the influence of ions on the mechanical behavior of copolymers occurs in large part through the changes in structure which result from ionization. Structural modifications can occur through the aggregation of ionic material, such as the multiplet or cluster formation discussed in Chapter II. This can also be accompanied by the breakdown of existing structure in the nonionic material, such as the drastic modification of crystalline morphology in ethylene ionomers.

For the sake of convenience, the work on ion-containing copolymers is presented in three sections. The first concerns itself with ionomers based on noncrystalline materials of relatively high T_g such as the styrene-based ionomers. The second is devoted to ion-containing elastomers, for example, ionomers based on butadiene. The third section deals with partially crystalline polymers, independent of T_g. It should be stressed that this arrangement is primarily one of pedagogical convenience and does not necessarily follow the historical development of the field. It does, however, follow the order of increasing structural complexity of the base polymer, and thus should enable the reader to appreciate most easily the effect of ion-related structural modifications on the viscoelastic properties of ionomer systems.

The mechanical properties that can be anticipated in these copolymer systems are fairly similar to those of the parent polymers, since usually only a small fraction of ionic material is incorporated in them. Thus, conventional techniques for determining viscoelastic parameters can be used. However, the interpretation of results is expected to be somewhat more complicated.

This chapter also includes a description of the viscoelastic properties of polyelectrolyte complexes. Two types of complexes can be considered—those formed from mixtures of homopolymeric bases and acids and those formed from mixtures of copolymers. The interactions in the former case can be expected to be much stronger, and it has been shown that special techniques are required for the preparation of these materials. Although the equilibrium properties of polyelectrolyte complexes have received considerable attention, their viscoelastic properties and structure have not been as well explored as those of simple homopolymer systems.

A. NONCRYSTALLINE COPOLYMERS OF HIGH T_g

1. Styrene-Based Ionomers

In Chapter II, the structural aspects of styrene-(methacrylic acid) ionomers were discussed. From a range of physical measurements, it is apparent that there are two distinct structural forms for copolymers of low and high ion content. Below about 6 mole % ionic material, the physical evidence indicates the existence of multiplet aggregates of only a few ion pairs; there is no evidence of more extensive aggregation at low ion contents. Above 6 mole % ions, the appearance of new small-angle X-ray scattering sites along with other changes to be discussed below provides evidence for a different type of ion aggregation, i.e., clustering or microphase separation. It will be seen that these variations in structure are highly significant in determining the viscoelastic behavior of this ionomer system.

The works of Fitzgerald and Nielsen [1] and Erdi and Morawetz [2] represent the pioneering investigations into the viscoelastic properties of styrene-based ionomers. In addition to the studies on poly(acrylic acid) salts discussed in Chapter III, Fitzgerald and Nielsen investigated the thermomechanical properties of some S–MAA ionomers. They determined the glass-transition temperatures of neutralized and unneutralized copolymers by dynamic mechanical means. They found that the T_g of the unneutralized copolymer increased as the acid content increased, and that the conversion of the acid to the sodium salt further increased the value of T_g. However, no increase in the shear modulus at room temperature was observed in either the neutralized or the unneutralized copolymer up to 40% acid or salt content. By contrast, it should be recalled from Chapter III that the room temperature modulus of PAA increased by a factor of about 2 upon neutralization and that the modulus of the polyacid itself was larger by about 1.5 than that of polystyrene.

Two important differences were noted in the mechanical behavior of the acid and salt forms of the copolymers. The stress relaxation rates of pure polystyrene and the copolymer S–0.075MAA were found to be nearly identical when compared relative to their T_g values, while the relaxation rate of the sodium salt S–0.075MAA–Na was found to be significantly smaller. Also, T_g damping peaks of the ionomers were observed to broaden appreciably with increasing ion content, while no such effect was observed in the acid copolymers. Thus, the results of the study by Fitzgerald and Nielsen indicate that ionic bonding has a very important effect on the solid-state mechanical properties of polymers, while the effect of hydrogen bonding under the same conditions is minor.

Erdi and Morawetz [2] performed creep and creep recovery experiments on various salts of the plasticized copolymer S–0.023MAA–0.25DOP. Isochronal (100 sec) compliances at 60°C were obtained for the Li and Ba salts of the copolymer and compared with those of the unneutralized material and a styrene homopolymer of similar molecular weight. They found that the inclusion in polystyrene of 2.3 mole % —COOH, —COOLi, and —$COOBa_{1/2}$, respectively, decreased the isochronal compliance at 60°C in the ratio 1.0/0.58/0.035/0.012; alternatively, the temperature at which the equivalent compliance is attained is increased by 5°, 35°, and 55°, respectively. They also found that the irrecoverable creep decreases in the order of —COOH > —COOLi > $COOBa_{1/2}$ when compared at the temperatures corresponding to the same compliance. At the same time, however, the activation energy for viscous flow was found to decrease with increasing ionic content, especially for the barium salt. The authors hypothesized that a decrease in the size of the flow unit was responsible for this trend.

Ide and co-workers [3, 4] studied the ionic cross-linking of a range of S–MAA copolymers, neutralized with Na, Ca, and Zn, although mainly with the latter. Copolymers containing up to 10 mole % MAA were employed in the study; the degrees of neutralization as determined by IR spectroscopy varied from 0 to 50%. They found from dynamic mechanical measurements that the position of T_g shifted to higher temperatures with increasing degrees of neutralization. There was little change in the absolute value of the sub-T_g storage modulus; however, the height of the rubbery plateau in G' versus temperature plots increased. Melt index values of the ion-containing copolymers were strongly dependent on the degree of neutralization, but were also related to the MAA content and to the intrinsic viscosity of the base polymer. The authors reported that the ionically cross-linked copolymers were readily moldable by all conventional thermoplastic techniques.

The mechanical properties of the Zn and Na salts of an S–MAA copolymer were also investigated by Tamura *et al.* [5]. The copolymer was produced by emulsion polymerization, with a monomer mixture S:MAA = 85:15 by weight. Equivalent ion contents of ~8% Zn and 2% Na were used. The authors observed that the stress relaxation behavior of these ionomers is similar to that seen in some partially crystalline polymers. Since different sets of WLF [6] parameters were observed above and below T_g for both the acid and salt forms of the copolymers, they proposed the existence of two relaxation mechanisms in each case. The viscoelastic response below T_g was attributed to normal amorphous relaxation, in which the hydrogen or ionic bonding is stable. They assumed that T_g corresponds to the transition from carboxyl dimers to monomers, and that above T_g, relaxations due to the micro-Brownian motion of side chains contribute to the overall behavior. In the light of current understanding, these mechanistic assignments seem speculative.

They observed a lower T_g and faster relaxation in Na salts, as compared to the unionized polymer and attributed this finding to a lower bonding force in Na^+COO^- dimers. This is in direct contrast with the observations to be presented later in this section and may have resulted from the presence of water. The Zn salts exhibited higher T_g values and slower rates of relaxation, as anticipated. No evidence for the breakdown of time–temperature superposition was presented. Also, the modulus-temperature curves of these ionomers show indications of an inflection point or partially developed plateau above T_g, in the region $E = 10^8 - 10^9$ dyn/cm^2. This phenomenon, which results on ionization, was not investigated further by Tamura *et al.*; however, it will be discussed at length in the work to be presented later in this section.

Otocka also briefly investigated some S–MAA ionomers as part of a larger study [7], but only with respect to modulus determinations below T_g. No major effects were observed.

The effects of a range of molecular parameters on the viscoelasticity of S–MAA ionomers were described in a series of papers by Eisenberg and Navratil [8–11]. Variations in viscoelastic behavior with ion concentration and the type of counterion or plasticizer were related to the structural changes that result from ionic interaction. The rheological aspects of these studies involved extensive stress relaxation measurements in the region of T_g. For each material, in a series of constant temperature runs, the Young's modulus E was determined as a function of time. From these measurements, the preparation of a master curve of reduced modulus versus reduced time

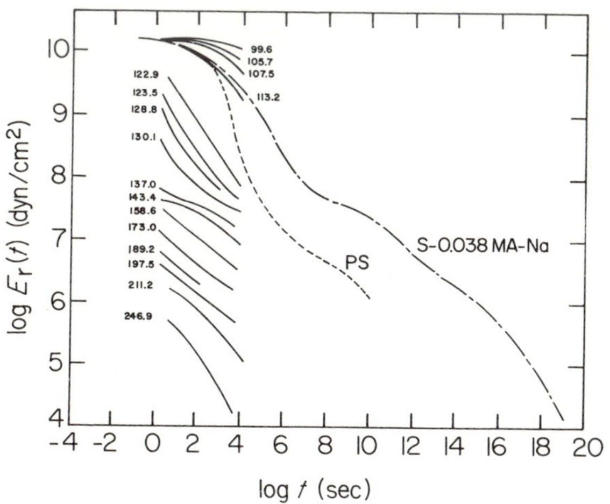

FIG. 4.1. Stress relaxation curves and the master curve for S–0.038MAA–Na, $M_n = 4.0 \times 10^5$. The master curve for polystyrene $M_n = 2.3 \times 10^5$ is shown as the dashed line. Temperature (°C) is given on each curve [8].

was attempted by first applying the small vertical reduction parameter [6]

$$E_r(t) = E(t)\rho_0 T_0/\rho T \tag{1}$$

and then following the usual procedure of shifting successive curves of $\log E_r$ versus $\log t$ horizontally with respect to a reference curve to produce the fit with the maximum overlap.

An example of the stress relaxation data obtained for S–MAA ionomers of low ion concentration (<6 mole %) is presented in Fig. 4.1. The specific composition of the sample, in correspondence with the terminology described in Chapter I, is S–0.038MAA–Na. In Fig. 4.1, the individual curves of $\log E_r$ versus $\log t$ are shown in the central portion of the figure, while the master curve, obtained by shifting the individual curves relative to T_g, is shown as the solid line. It can be seen that good time–temperature superposition of the data is obtained, with overlap observable over several orders of magnitude of modulus or time. The shift factors are of the WLF type, although the WLF parameters are different from those of pure polystyrene. The master curve for polystyrene, included for the sake of comparison, shows a much narrower transition region than the one for the copolymer. The existence of two inflection points (at $\log E_r \sim 7.5$ and 6.5) in the master curve of the copolymer is noteworthy. The upper inflection point indicates the existence of some kind of cross-linking effect due to the ions. The lower inflection point coincides with the usual modulus associated with entanglements in pure polystyrene and similar polymers.

In Fig. 4.2, the stress relaxation data for a sample of higher ion concentration—S–0.077MAA–Na—are shown. Attempts to form a master curve by shifting the individual modulus-time curves in the usual manner were unsuccessful. If the overlap was maximized in the short-time region, then pronounced deviations were observed at long times. This is evident from the solid lines shown in Fig. 4.2. It was found that pseudo master curves of modulus versus time could be obtained by overlapping data from narrow time segments (two orders of magnitude). However, none of the curves

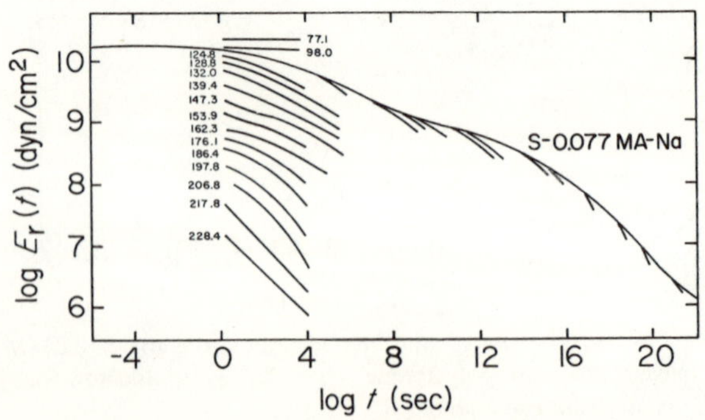

FIG. 4.2. Original stress relaxation curves and attempted master curve for S–0.077MAA–Na, $M_n = 1.4 \times 10^5$; temperature (°C) is given on each curve [8].

obtained can be considered true master curves since they do not describe the viscoelastic response over the entire region of time and temperature and since data from different time segments yield curves of substantially different shape. It is interesting to note that, even in the cases where time–temperature superposition breaks down, the pseudo shift factors are still of apparent WLF type. However, both WLF parameters appear to be dependent on the time segment employed in the construction of the pseudo master curve.

The critical ion concentration for the breakdown of time–temperature superposition in the S–MAA system was found to be about 6 mole %. This point coincides with the ion concentration at which the onset of microphase separation is observed, as discussed in Chapter II. It is apparent that the breakdown of time–temperature superposition at this composition is due to the onset of a second relaxation mechanism associated with the change in structure. The nature of this second mechanism will be discussed in more detail below.

To illustrate the variation in viscoelastic behavior with ion concentration and other parameters, the next few figures depict groups of master curves without the original stress relaxation data. Where no superposition of data was obtained, the pseudo master curves obtained from the short-time segments are presented. Where necessary, the curves are normalized to obtain overlap in the glassy region in order to account for experimental errors in the absolute value of modulus (up to 20%) and in T_g ($\pm 2°$).

In Fig. 4.3, four different master curves are plotted. They correspond to the polystyrene sample ($M_n = 2.3 \times 10^5$, $M_w = 3.3 \times 10^5$) shown previously in Fig. 4.1, and three copolymers: S–0.038MAA, $M_n = 4.0 \times 10^5$; S–0.037MAA–Na, $M_n = 0.5 \times 10^5$; and S–0.038MAA–Na, $M_n = 4.0 \times 10^5$. The latter curve is also taken from Fig. 4.1. It can be seen that the curves for pure polystyrene and the unneutralized copolymer are similar in shape, with the height of the inflection point ($\log E_r \sim 6.5$) unaffected by the presence of the carboxylic groups. The effects of hydrogen bonding, which is observed in the transition region, seems to diminish with increasing temperature. The difference in the flow region for these two samples is probably due to the difference in their molecular weights only.

The curves in Fig. 4.3 corresponding to the two ionized samples indicate a much broader transition region than that observed for the nonionic samples. The two ionomer curves are identical until the entanglement region ($\log E_r \sim 6.5$) is reached; below this point, the effect of molecular weight is apparent. The identity of the curves in the transition region indicates that the inflection point at $\log E_r \sim 7.5$ is independent of molecular weight.

The family of master curves for high-molecular-weight copolymers containing different amounts of the ionic comonomer was shown together with that for the polystyrene sample in Fig. 2.4. Two features are noteworthy: The broadening of the transition and flow regions can be observed with increasing ion content, and two inflection points are visible on each master

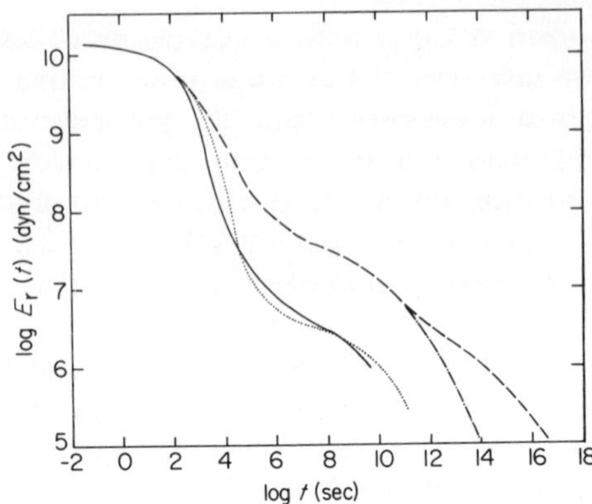

FIG. 4.3. Stress relaxation master curves for four different samples: polystyrene (———), high-molecular-weight S–0.038MAA (····), low-molecular-weight S–0.037MAA–Na (—·—·), high-molecular-weight S–0.038MAA–Na (— — —). All curves are drawn with T_g as reference [9].

curve, except for those corresponding to PS and the copolymer S–0.006-MAA–Na.

The height of the upper inflection point is found to be a linear function of the ion content and can be expressed by the relationship

$$\log E_r = 30c + 6.5 \tag{2}$$

where c is the mole fraction of neutralized carboxylic groups, and the intercept 6.5 is the logarithm of the rubberlike modulus of pure polystyrene. The lower inflection point remains unchanged at $\log E_r = \sim 6.5$ (dyn/cm^2). It is also seen that the upper inflection points appear in a relatively constant time region, corresponding to $\sim 10^7$ sec relative to the T_g of each polymer. It is significant that for $c > 0.06$ the observed values of the height of the upper inflection point are much greater than those calculated from the theory of ideal rubber elasticity with the assumption that each ion pair is incorporated in a crosslink.

The distributions of relaxation times were calculated for the polymers in which time–temperature superposition was valid [9]. The curves showed a considerable broadening with increasing ion concentration. It was found that both the slope of the linear portion of the distribution of relaxation times and the half-width ($\Delta \log \tau$, as defined in Chapter III, Section A1) of the relaxation spectrum became fairly linear functions of the ion concentration for copolymers containing more than 1 mole % salt.

The slope of the linear portion of the distribution of relaxation times has been correlated with the dimensionality of the intersegmental forces, i.e., with the degree of interchain interaction, using one-dimensional and three-dimensional oscillators as the two extremes of the mechanical model [12]. This approach predicts an increase in the magnitude of the slope (or correspondingly a decrease in $\Delta \log \tau$) with increasing three-dimensional character of the polymer interaction. For some ionic polymers, such as the polyphosphate system discussed in Chapter III, the results are in agreement with this prediction. However, in the styrene ionomers the trend is exactly the opposite, that is to decreasing steepness with increasing ion content. The reason for this is not clear; it may be due to the superposition of other effects such as cross-linking or copolymerization.

As mentioned before in connection with Fig. 4.1, the shift factors of the master curves from Fig. 4.4, as well as from others not shown, were all found to be representable by the WLF equation. C_1 and C_2, the WLF parameters, are plotted in Fig. 4.4 as a function of the ion content of S–MAA copolymers [13]. For ion contents greater than 6 mole %, the WLF parameters correspond to the pseudo shift factors employed in the treatment of data for samples in which time–temperature superposition was not applicable. The sudden pronounced increase in both C_1 and C_2 at this critical concentration is thus very noteworthy, since it indicates that two distinct types of viscoelastic behavior exist above and below this point.

An attempt was made to correlate the rate of relaxation with the ion concentration. The time required for the modulus to reach a given value at T_g can be readily determined from the master curve plotted with T_g as the reference temperature. For instance, in the case of polystyrene (Fig. 4.3), the value $E_r = 3 \times 10^6$ dyn/cm^2 is reached at $\sim(5 \times 10^7)$ sec in a hypothetical stress relaxation run performed at the T_g of PS (102°C). For the copolymer S–0.038MAA–Na, the same modulus is obtained after $\sim 10^{12}$ sec if measured at its T_g (112°C). By plotting the master curves at their respective T_g's, the effect of ions on T_g is eliminated, and the only observed effect is the slowing down or retardation of the primary diffusional relaxation mechanism under the influence of the ions. In the above example, the time required to reach the chosen value of modulus (3×10^6 dyn/cm^2) is greater by a factor of 2×10^4 for S–0.038MAA–Na than for PS.

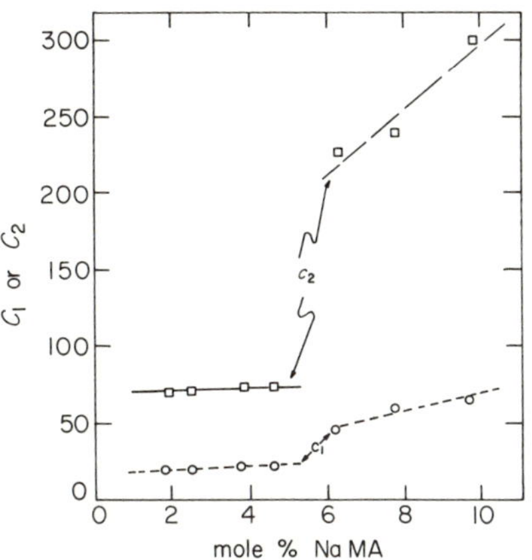

FIG. 4.4. The WLF constants C_1 and C_2 for high-molecular-weight S–MAA ionomers. Above 6% NaMA, C_1 and C_2 were obtained from pseudo master curves for 2–100-sec data [13].

One can expect the retardation factor to be a function of both the chosen modulus and the ion concentration. If the relative displacement of a given modulus value along the time axis of the master curve (drawn with respect to T_g) is plotted against the ion concentration, a measure of the retardation of the primary relaxation mechanism due to the presence of ions is obtained. These retardation factors for different values of modulus are shown in Fig. 4.5; the concentration of ions is expressed both in mole % and as an average number of styrene units between successive sodium methacrylate groups. The points corresponding to the highest ion concentration (6.2 mole %) were evaluated from the pseudo master curve obtained from short-time relaxation moduli.

It was found that, for low-molecular-weight copolymers containing more than 6 mole % NaMA, time-temperature superposition is reestablished at a temperature of about $T_g + 60°$; in the copolymer S–0.079MAA–Na, $M_n = 0.7 \times 10^5$, for example, this corresponds to moduli less than 10^8 dyn/cm^2. Phenomena of this type were not observed in ionomers of high molecular weight.

Finally, the 10-sec moduli obtained in stress relaxation experiments are plotted as a function of temperature in Fig. 4.6 for samples of different ion concentration. The increase of T_g with increasing ion concentration as well as the enhanced rubberlike plateaus can be clearly seen in this figure. The molecular-weight effect is also quite apparent here. (Compare the curves for 3.7 and 3.8 mole % NaMA).

The slopes m of the $\log E$ versus temperature plots at $\log E = 8.5$ are shown in Fig. 4.7, plotted as $-\log(-m)$ versus mole % NaMA. This curve

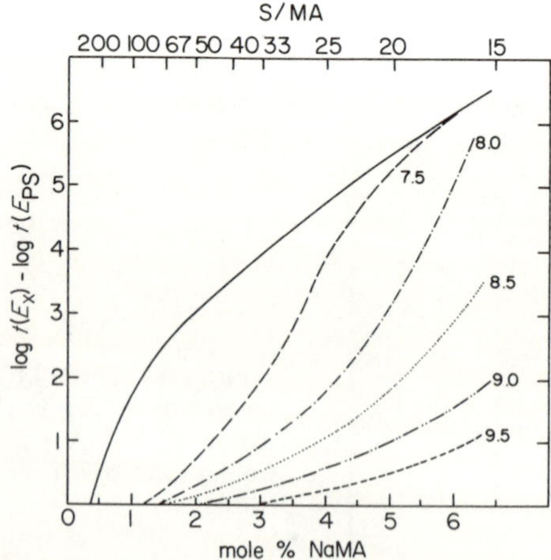

FIG. 4.5. Retardation factor of primary stress relaxation process as a function of ion content or S/MAA ratio. The values of $\log E_r$ shown on each curve indicate the modulus values for which the retardation factor is computed [9].

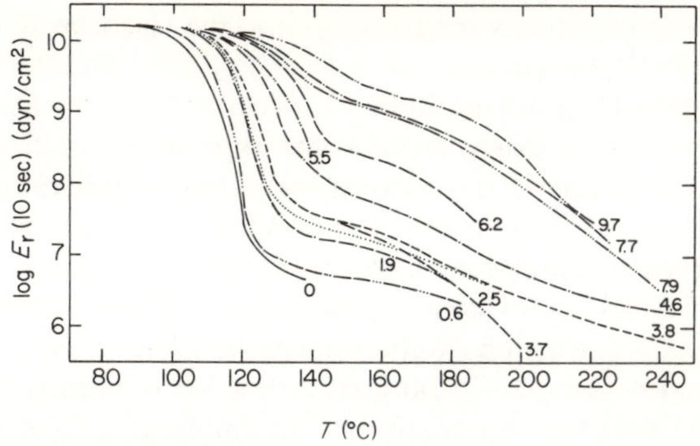

FIG. 4.6. 10-sec modulus versus temperature for samples of different NaMA content. All curves are for high-molecular-weight polymers, except those for 3.7, 5.5, and 7.9% NaMA. Mole % NaMA is indicated on each curve [9].

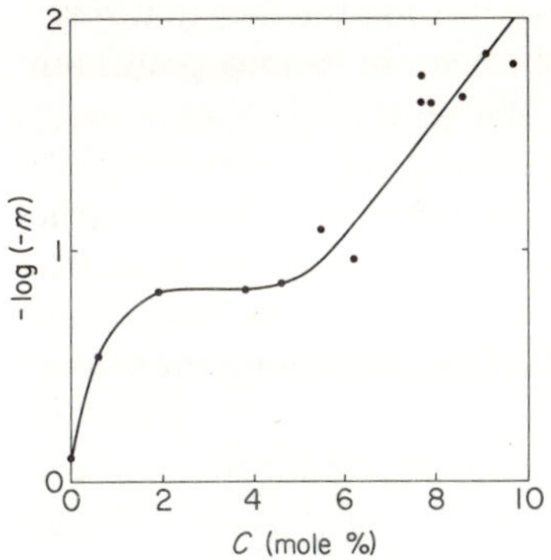

FIG. 4.7. Plot of $-\log(-m)$ versus NaMA content of S–MAA ionomers, m is the slope of the $\log E$ versus $\log t$ curves at $\log E = 8$, computed from the data shown in Fig. 2.4.

bears a very striking resemblance to a result of a melt rheology study, shown in Fig. 4.17, and discussed much more fully in a subsequent section. As a matter of fact, Fig. 4.7 was drawn only after the study of the melt rheology was completed and an interpretation of Fig. 4.17 had been suggested. As will be seen below, the shape of the curve is strongly related to the types of ion aggregation in various concentration regions. It should be noted that the shape of this curve is in no way related to the change in T_g, which is linear in this concentration region.

The rheological evidence presented so far, along with the other structural considerations (X-ray, water uptake) discussed in Chapter II, support the hypothesis that in S–MAA ionomers containing less than 6 mole % of ionic material, the ions form simple multiplets, which act as transient cross-links and slow down the primary relaxation process; however, for ion contents greater than 6 mole %, cluster formation, which bears many of the characteristics of microphase separation, is encountered. In the subsequent analysis, the data will be interpreted with this idea in mind, and the most important consequences of it will be discussed, especially with regard to the relationship between the polymer morphology and the viscoelastic properties above the critical ion concentration.

Figure 4.6 and Fig. 2.4 in Chapter II provide the most clear-cut rheological justification for the existence of some type of ionic aggregates in these materials. The fact that the secondary inflection point in the relaxation modulus correlates with the ion concentration cannot be interpreted as anything but some type of temporary cross-linking or microphase separation. This becomes particularly evident on inspection of the modulus-temperature curves in Fig. 4.6, which bear all of the hallmarks of a system of polymers of increasing cross-link density, at least at low ion concentrations. At high ion concentrations these curves strongly resemble an incompatible block copolymer or polyblend system [14].

Several experimental facts make it clear that the morphology of the styrene ionomers is of two distinctly different forms above and below the critical ion concentration of ~6 mole %. These include the fact that time–temperature superposition is obeyed below but not above that point, and the fact that a sharp if not discontinuous change in the WLF parameters occurs at that concentration. Furthermore, the results of the water uptake experiments presented in Chapter II also indicate the existence of a structural transition at ~6 mole % NaMA. The X-ray diffraction results described in that chapter reinforce the concept of microphase separation above 6 mole %, since new scattering centers appear in ionomers containing more than 6 mole % CsMA.

The fact that the WLF equation is generally applicable below 6 mole % NaMA suggests that the relaxation mechanism in this region is quite similar to that of normal polymers. However, it is also clear that the presence of ions slows down the diffusional mechanism, the magnitude of the retardation increasing with ion concentration and size of the moving segment. This is interpreted as arising from an ion-hopping process which results from the finite lifetimes of ionic cross-links (multiplets). The ionic multiplets dissociate at a finite rate to temporarily free one or more ion pairs; once dissociated, the ion pairs can reform in the same multiplet or they can diffuse to form new multiplets, in either case maintaining the same overall cross-link density. This process would occur whether the sample is under stress or not, but it is likely, from energetic considerations, that the fraction of dissociated multiplets at any given time is low. However, if the sample were deformed, the ion-hopping process would lead to a relaxation of stress.

The relaxation processes occurring in the clustered material (above 6 mole % NaMA) are undoubtedly much more complex than in the simpler structures below 6 mole %. The fact that time–temperature superposition does not hold in the clustered material is a likely indication that multiple relaxation mechanisms (with different temperature dependences) are operative here.

Accordingly, the relaxation data obtained from S–MAA ionomers containing more than 6 mole % NaMA were analyzed by the method described in Chapter III, Section A2. Briefly, this method assumes that two concurrent relaxation mechanisms, each with its own temperature dependence, contribute separately to the total compliance. Thus the stress relaxation results were first converted to compliance curves as described previously. At any given temperature, the compliance at short-time values was assumed to be due to the contribution of one mechanism only. The contribution of the second mechanism, which does not become apparent until longer-time values, was obtained by subtracting the excess compliance at long times from the envelope of short-time compliances (the pseudo master curve). The resulting subtracted compliance curves were then reshifted into a new master curve with a new set of shift factors.

The temperature dependences of the shift factors corresponding to the two "master curves" were analyzed in order to elucidate the nature of the two relaxation mechanisms. It was found that the shift factors corresponding to the short-time mechanism were fitted well by a WLF-type equation. The WLF parameters obtained from these short-time shift factors are plotted in Fig. 4.4. The values of C_1 and C_2 are, of course, very different from those at low ion concentrations, implying considerable diminution in the values of the glassy free volume and free-volume expansion coefficient [6]. This is not unreasonable, however, in view of the constricted structure required by clustering.

The apparent activation energy at T_g, $(\Delta H_a)_{T_g}$, associated with the short-time mechanism can be evaluated from the slope of a plot of $\log a_T$ versus $1/T$ at the glass-transition temperature. It can also be evaluated from the WLF parameters and Eq. (3) of Chapter III. Application of either method to S–MAA ionomers above the critical concentration gives values of $(\Delta H_a)_{T_g}$ of the order of 160 (± 15) kcal/mol. These values are quite close to those observed in polystyrene and other amorphous nonionic polymers near T_g for the primary (diffusional) mechanism that characterizes the glass transition. The activation energy for the secondary mechanism, also obtained from a plot of $\log a_T$ versus $1/T$, yields a value of ~ 40 kcal/mol, with apparent Arrhenius-type temperature dependence.

In clustered (microphase-separated) ionomers, it is reasonable to expect two major relaxation mechanisms corresponding to the yielding of each phase. In view of the magnitude of the activation energy and the WLF-type temperature dependence of the short-time mechanism, it is probable that this relaxation process is due to the yielding of the largely nonionic phase. Thus, it is likely that the long-time mechanism is related to the presence of the ion-rich regions and corresponds to some kind of yielding process in the clusters. At this time, it seems premature to speculate about the details of this mechanism. It is noteworthy that the compliance due to the second mechanism yields a slope of 1 on a log J versus log t plot, suggesting that a pure viscosity is involved [14]. This bears a strong similarity to simple bond interchange.

In low-molecular-weight copolymers above the critical ion concentration, it was found that time–temperature superposition is reestablished for temperatures above $\sim(T_g + 60°)$. This probably indicates that cluster lifetimes at that temperature are very much shorter than the relaxation times due to the retarded polymer motions, in which case the clusters would not contribute appreciably to the relaxation. The reestablishment of time–temperature superposition is not observed in high-molecular-weight ionomers, suggesting that, even if the residence time of any one ion pair in a cluster were very short, a chain of very high molecular weight has so many ion pairs that the effect of clusters would not be dissipated until much higher temperatures. It is also conceivable, although not probable, that $T_g + 60°$ represents the temperature above which clusters cease to exist. In the higher-molecular-weight ionomers this cluster decomposition temperature would presumably be much higher. Alternately, the difference in cluster decomposition temperatures could possibly be a solubility effect of the same type as seen with the polyether–salt complexes discussed in Chapter III, Section B1.

The results of a dynamic mechanical study [10] on S–MAA ionomers reinforce the views discussed above on structure in this ionomer system. The dynamic shear modulus and the loss tangent were determined as a function of temperature for resonant frequencies of the order of 0.1 Hz, using a torsion pendulum. The temperature dependence of the storage modulus for copolymers containing from 0 to 10% NaMA is shown in Fig. 4.8. The difference in the absolute heights of the glassy moduli is due to the fact that $E \sim 3G'$ in the glassy region. As in Fig. 4.6, the increase in the height of the inflection point with increasing ion content is notable. The loss tangent data for samples below the critical ion concentration are shown in Fig. 4.9. It can be seen that, for samples containing between 1 and 6 mole % salt, a second peak, above T_g, is visible. Although the glass-transition peak increases with increasing ion concentration, the position of the upper peak ($\sim$150°C) is relatively insensitive to ion concentration. For samples above the critical ion concentration, a continuous increase in tan δ above $\sim$150°C is observed, as illustrated in Fig. 4.10. Here it can be seen that the T_g peak moves to higher temperatures and broadens with increasing ion content, Figure 2.16 shows the positions of the maxima in the tan δ versus temperature curves as a function of the ion concentration. For samples containing more than 6 mole % NaMA, for which no second peak could be found, the temperature at which tan δ equals 0.5 was chosen arbitrarily to illustrate the shift in corresponding behavior with increasing ion concentration.

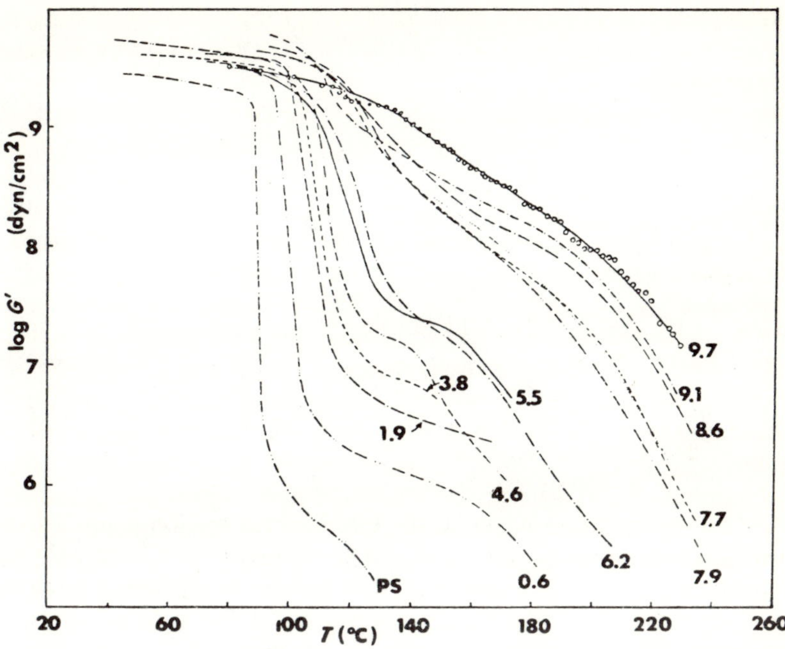

FIG. 4.8. Shear storage modulus versus temperature for S–MAA ionomers. Mole percent NaMA indicated on each curve [10].

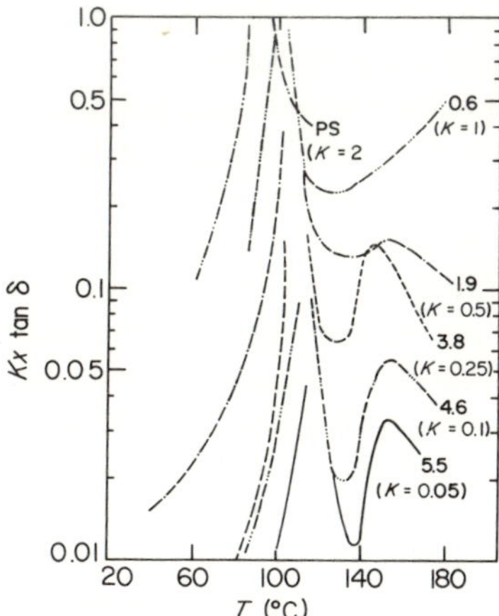

FIG. 4.9. Loss tangent versus temperature for S–MAA ionomers below the critical ion concentration. Mole percent NaMA indicated on each curve. K represents a multiplicative vertical shift by the amounts indicated [10].

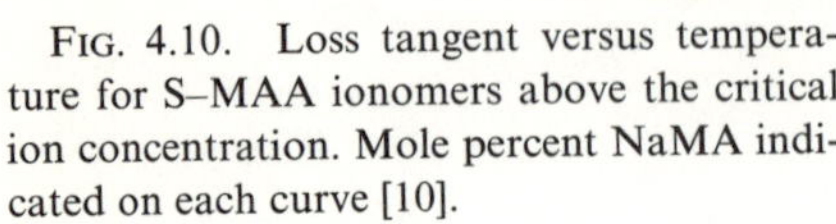
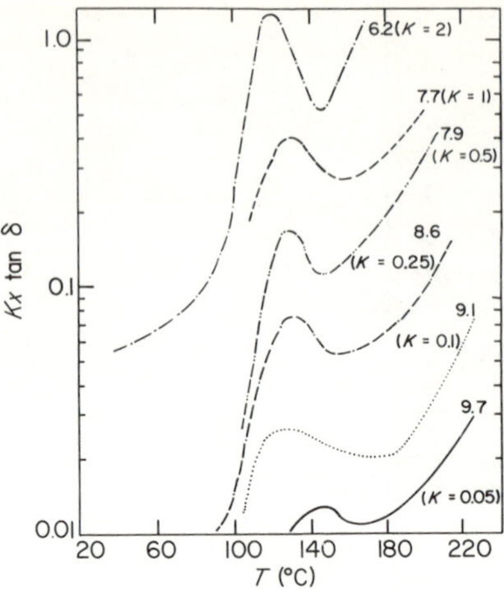

FIG. 4.10. Loss tangent versus temperature for S–MAA ionomers above the critical ion concentration. Mole percent NaMA indicated on each curve [10].

The high-temperature peaks in the tan δ curves for ionomers below the critical ion concentration can likely be attributed to a dissipative process in the ion multiplets. The nature of this process is not clear; however, it cannot represent the complete decomposition of the ion multiplets, since the primary relaxation is still slowed down above this temperature. For ionomers above the critical ion concentration, only a continuously increasing tan δ without a maximum is found above T_g. This fact suggests that although clusters do soften with increasing temperature the transition is very broad, and likely indicates that the structure of the clusters changes continuously over a wide temperature range.

The effect of different counterions on the viscoelastic properties of S–MAA ionomers was also tested by Eisenberg and Navratil [11]. Cesium-neutralized copolymers were prepared by a method analogous to that used for sodium salts. For ionomers containing 3.7 and 3.8% CsMA, time–temperature superposition was also found to be applicable. The master curves prepared for these salts are compared with those of the corresponding sodium salts in Fig. 4.11. It can be seen that the rate of relaxation is considerably faster for the cesium salts; however, the shapes of the curves, with respect to the molecular-weight effect and the height of the upper plateau, are similar. These changes in viscoelastic behavior can be rationalized if it is assumed that the cesium and sodium salts have identical or at least very similar structures, which in this case means that the ions are all incorporated into multiplets which form time-dependent cross-links. Thus, any variation in the rate of relaxation in this concentration region should be related to the ease with which ion multiplets dissociate. This is clearly a function of the interion distance in the multiplets—the larger the interion distance, the

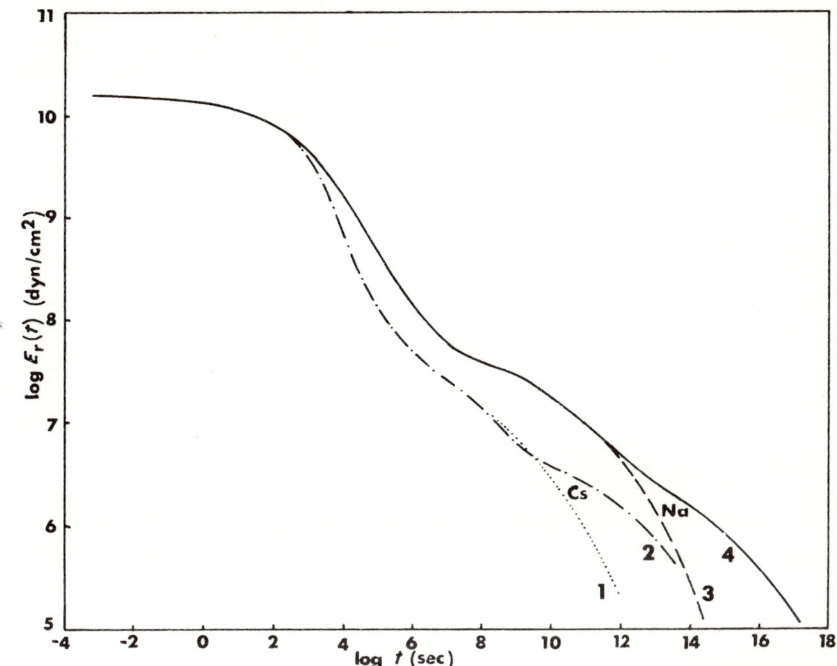

FIG. 4.11. Stress relaxation master curves for Na and Cs salts of S–MAA copolymers. Curves 1 and 3: mole % MAA = 3.7, $M_n = 0.5 \times 10^5$; curves 2 and 4: mole % MAA = 3.8, $M_n = 4.0 \times 10^5$ [11].

lower the energy required to remove an ion pair. Thus, it is not surprising that multiplets containing Cs should form less stable cross-links than those containing Na, whose ionic radius is much smaller, and that this should manifest itself in the less well-developed rubberlike plateaus seen in the stress relaxation curves.

The effect of Ba substitution in S–MAA ionomers is more surprising. Instead of the decreased rates of relaxation that would be predicted from the above reasoning (the charge to radius ratio for Ba is greater than that for Na), the stress relaxation curves show increased rates of relaxation here too, as shown in Fig. 4.12. It can be seen that the height of the rubberlike plateau due to ionic cross-links is considerably lower for a given Ba salt than for the equivalent Na (or Cs) salt; however, the sharpness of the maximum in the distribution of relaxation times indicates that the Ba plateau is much better developed than the Na and Cs plateaus [11]. Thus, it appears that, while the effective cross-link density in the Ba salts is lower (as evidenced by the height of the rubberlike plateau), the cross-links are more stable than those of the equivalent Na or Cs salts. This finding suggests

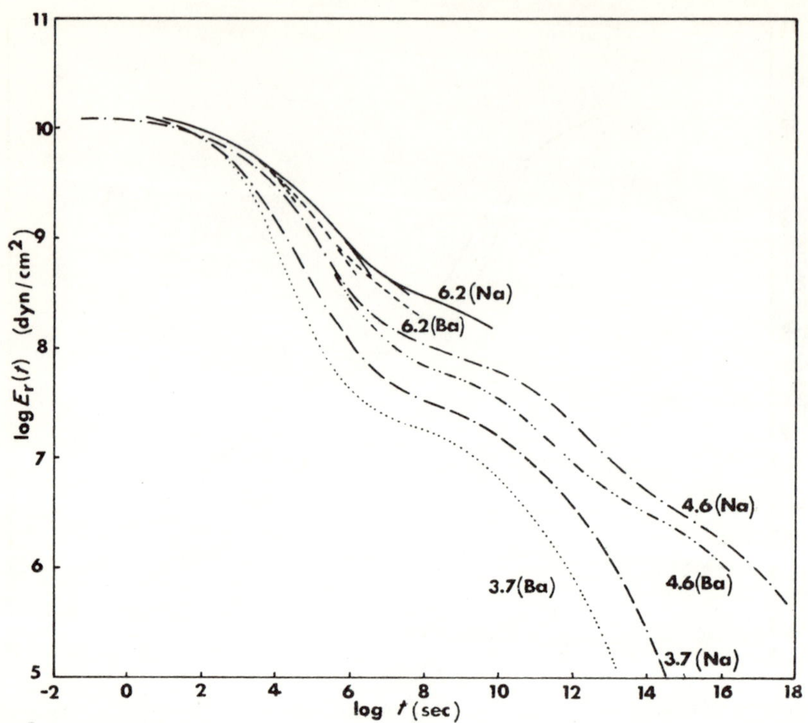

FIG. 4.12. Stress relaxation master curves for Na and Ba salts of S–MAA copolymers. The mole % MAA and the counterion are indicated on each curve [11].

that some of the carboxyl groups are not incorporated into multiplets. This may very well be due to the presence of residual water or unneutralized carboxyl groups.

Two experiments were performed in which the viscoelastic properties of partially neutralized S–MAA ionomers were examined [11]. In one, the degree of ionization was varied at constant comonomer composition; the master curves obtained for these materials are shown in Fig. 4.13. It is apparent from the disappearance of the upper inflection point that ions cease to act as cross-links if the degree of neutralization is below ~50% for this carboxyl concentration. Simple calculations based on rubber elasticity theory indicate that the apparent number of cross-links (which determines the height of the rubberlike plateau) is less than the number that would occur if each $—COO^-Na^+$ group participated in a cross-link and each —COOH group did not. For example, for the 60% neutralized copolymer, the apparent cross-link density is only about 25% of that of the fully neutralized material. In addition to the decrease in the apparent number of cross-links, the average

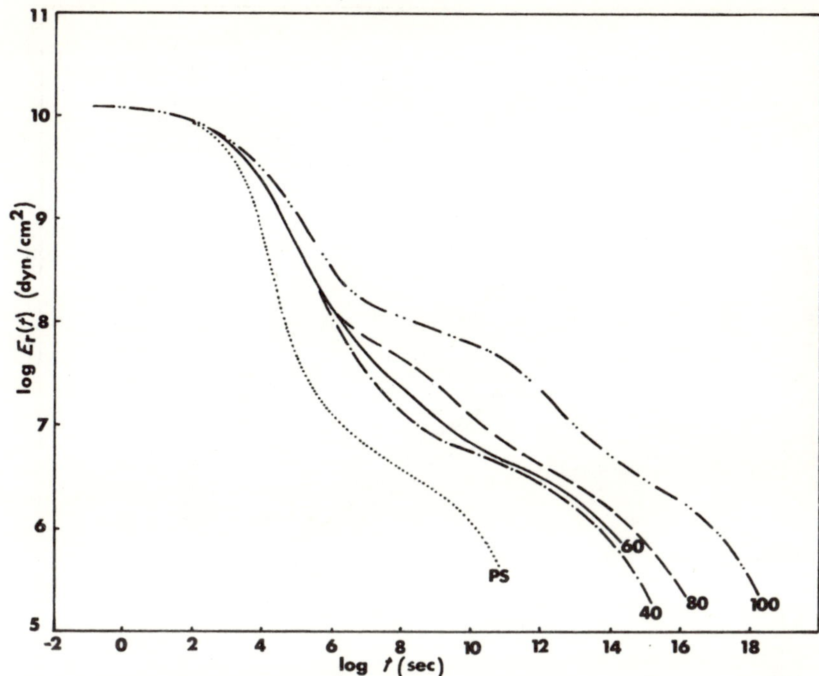

FIG. 4.13. Stress relaxation master curves for the partially neutralized copolymer system S–0.046MAA–Na. The degrees of neutralization (%) are indicated on each curve. The master curve for polystyrene is also shown [11].

lifetime of those remaining seems to be reduced significantly, as manifested by the poorly developed inflection points (plateaus) for lower degrees of neutralization.

In the second experiment, stress relaxation data were obtained for S–MAA ionomers in which the concentration of ions was held constant while the carboxyl group content was varied. Four different ionomers of high molecular weight, each containing 3.8 mole % NaMA and varying amounts of unneutralized MAA, were tested; their modulus-temperature relationships are shown as curves 1 to 4 in Fig. 4.14. The most notable feature about this graph is that, although the curves are divergent in the upper transition region, they all coalesce into one single curve at high temperature (> 180°C), with the exception of curve 5 to be discussed later.

In the treatment of stress relaxation data from these four materials (not shown here), time–temperature superposition was observed to fail only in the ionomer S–0.086MAA–0.44Na, as it did in the corresponding fully neutralized material S–0.086MAA–Na. The overall rate of relaxation of

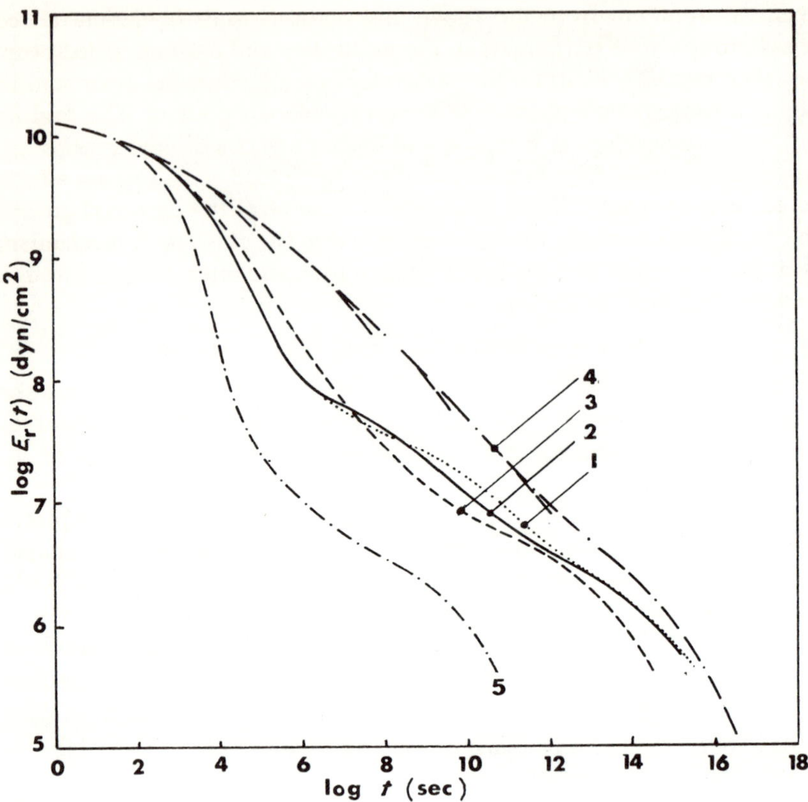

FIG. 4.14. Stress relaxation master curves for five samples of similar ion content: (1) S–0.038MAA–Na, (2) S–0.046MAA–0.80Na, (3) S–0.062MAA–0.60Na, (4) S–0.086MAA–0.44Na, (5) PS (56 mole %), S–0.086MAA–Na (44 mole %) [11].

the partially neutralized copolymer is much greater than that of the fully neutralized ionomer, and there is no indication of any high modulus plateau in the pseudo master curve. Thus, this result is consistent with the results obtained for the partially neutralized S–0.046MAA copolymers (Fig. 4.13) in which no ionic plateau was observed until 50% neutralization. Diminished ionic plateaus were observed in the master curves of the other partially neutralized S–MAA copolymers.

It was determined that the observed effects are not due to inhomogeneous neutralization. If this were the case, then the mechanical behavior of a partially neutralized S–MAA copolymer should be similar to a mixture of polystyrene and the fully neutralized copolymer. However, this was not true, as can be seen from Fig. 4.14, by comparing curve 4 for S–0.086MAA–0.44Na with curve 5, obtained for a mixture of PS (56 mole %) and S–0.086-MAA–Na (44 mole %).

Thus it is apparent from the above observations that the unneutralized carboxyl groups must participate in the multiplets and clusters. If this were not so, then partially neutralized copolymers would show behavior similar to fully neutralized copolymers of lower total carboxyl content. This finding supports the suggestion of Bonotto and Bonner that carboxyl groups are incorporated in clusters [15]. In view of the decreased cross-linking effects observed, it is obvious that the presence of unneutralized carboxyl groups in ionic multiplets weakens the interaction within the clusters. A mechanism whereby the —COOH groups could weaken the ionic interaction is through an exchange reaction of the type

$$—\mathrm{COOH} + —\mathrm{COO^-Na^+} = —\mathrm{COO^-Na^+} + —\mathrm{COOH}$$

If this exchange occurred at a sufficiently rapid rate, it would increase the mobility of any given $—\mathrm{COO^-Na^+}$ group because of the much weaker interaction of the unneutralized carboxyl groups and the consequent increase in mobility.

The experiments with plasticized styrene ionomers [11] show that plasticizers change not only the glass-transition temperatures but also the shapes of the stress relaxation curves. For example, dioctyl phthalate (DOP), which is known to be a good plasticizer for polystyrene, was used to plasticize the ionomer S–0.025MAA–Na. In addition to the expected decrease in T_g, a marked broadening of the stress relaxation master curve (or pseudo master curve) was observed. Furthermore, at higher degrees of plasticization, the breakdown of time–temperature superposition was found to occur, while the unplasticized ionomer is well within the concentration region of superposition. However, the broadening of the transition region by DOP cannot be regarded as specific for ion-containing polymers, since the same type of broadening of the transition region was observed in PS plasticized by DOP or DEP. Similar observations were made previously by Chapoy and Tobolsky [16].

By contrast, the use of dimethyl sulfoxide (DMSO) as a plasticizer for styrene ionomers results in a narrowing of the glass-transition region. This type of specific plasticizer effect on the broadeness of the transition region was also seen in the PAA system, as illustrated in Fig. 3.16 of Chapter III. It suggests that different plasticizers associate preferentially with either the ionic or the nonionic regions of the polymer matrix.

Rafikov *et al.* [17], as part of a broader study, investigated the thermomechanical behavior of a random S–MAA copolymer (20 mole % MAA) partially neutralized (30, 60, and 90%) with Ca and Zn. They found only slight differences in the deformation-temperature curves for this range of samples, the rubbery plateau in all cases extending from ~150°–180°C. By contrast, a block copolymer of styrene and methacrylic acid (30 wt % MAA), neutralized to 90% with Zn, exhibited a very well-developed rubbery plateau, extending from ~180°–380°C. They interpreted the results in terms of the model proposed by Bonotto and Bonner [15].

Gaertner *et al.* [18] investigated the modulus and T_g of S–MAA and S–AA copolymers containing up to 20 mole % COOH. They found that T_g was a linear function of both the COOH content and the degree of neutralization with Na or with Mg; however, the tensile strength changed very little with the changes in copolymer composition.

In a subsequent publication, Schade and Gaertner [19] investigated the flow properties of the same materials. They found that with Na neutralization, the intrinsic viscosity in DMF increased with increasing degree of neutralization, whereas it decreased with Mg neutralization for both copolymers. As might be expected, the melt index I_m (at 190°C) decreased with increasing degree of neutralization for both Na and Mg ionomers (8.8 mole % MAA), the rate of decrease of I_m being sharper for Mg than for Na. In comparing the 5.0 and 8.8 mole % MAA samples, the rate of decrease of I_m was higher for the former.

Iwakura and Fujimura [20] investigated by rheogoniometry the melt rheology of blends of polystyrene with styrene ionomers. Only a small portion of the study was devoted to the unblended ionomers. The most interesting result is embodied in Fig. 4.15, which shows the zero-shear viscosity η_0 and the creep compliance J at various temperatures for an ionomer containing 15 wt % MAA (~18 mole %) neutralized to various degrees with Na and Zn, the degree of neutralization in each case being reported as weight percent metal. The most interesting detail in Fig. 4.15 is the sigmoidal-shaped curve of log η_0 versus wt % Zn. The upswing at ~0.5 wt % Zn is probably associated with the onset of clustering, although the

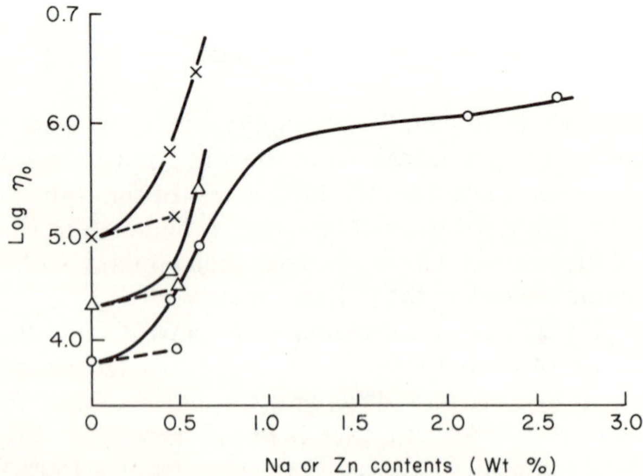

FIG. 4.15. Log η_0 versus ion content for S–0.18MAA partially neutralized with Na or Zn. (×) 220°C, (△) 240°C, (○) 260°C, (——) Zn, (---) Na [20].

authors did not discuss it in these terms. Most of the study dealt with the melt rheology of ionomer blends, and although many interesting observations were made, structural correlations are impossible without considerable additional information.

In a study of the melt rheology of styrene ionomers, Shohamy and Eisenberg [21] determined G', the dynamic modulus, and η', the viscosity, for various temperatures in the frequency range 10^{-2}–40 rad/sec. The range of modulus values was 10^4–10^7 dyn/cm^2, with a corresponding viscosity range. In an attempt to eliminate any possible effects due to variations in molecular weight or polymer microstructure and thereby isolate the specific effect of ions, aliquots of each acid sample were neutralized or esterified, and the rheological studies were performed on all three.

It was found that within the range investigated, frequency–temperature superposition was applicable. Furthermore, it was observed that by a judicious choice of reference temperatures, the master curves of G' versus ω for the Na salt, the methyl ester, and the unmodified acid could all be made to superimpose. An example for the S–0.015MAA system is shown in Fig. 4.16, the reference temperatures being 164°C for the ester, 172°C for the acid, and 202°C for the salt. This behavior suggests that at the selected reference temperatures, the dynamics of the polymer chains in all three systems are identical, the increase in temperature having served to loosen the interchain forces to the same degree. The differences in temperature (ΔT = 8°C for the acid,

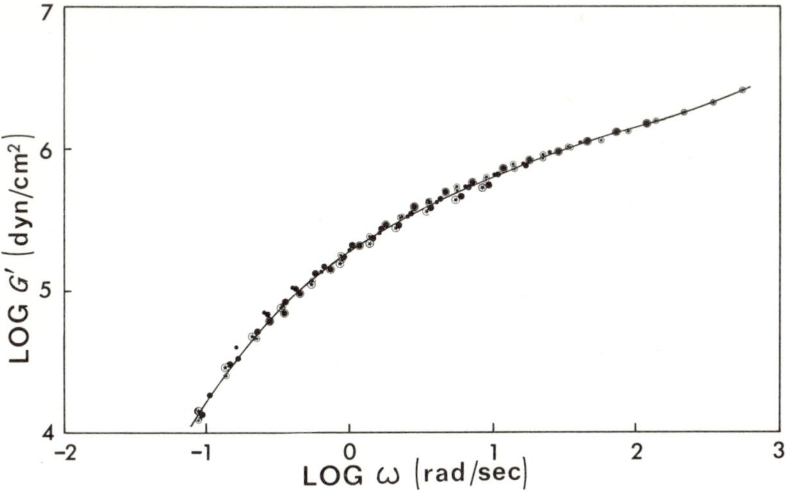

Fig. 4.16. Master curve of G' (dyn/cm^2) versus ω (rad/sec), for the copolymer S–0.015 MAA, its Na salt, and its methyl ester. Reference temperatures are 164°C for the ester [the original temperatures were 150°C (⊙) and 160° (◉)]; 172° for the acid [the original temperatures were 170°C (⊗) and 180°C (⊗)]; and 202°C for the salt [the original temperatures were 195° (·), 205° (•), and 210° (●)] [21].

38°C for the salt) are directly related to the strength of the interactions involved in hydrogen bonding (11 kcal/mol) and ionic multiplet formation (23 kcal/mol).

In addition to the method described above, another method was proposed for the determination of ΔT, based on the temperature difference at which G' reaches a particular value, i.e., 2×10^5 dyn/cm^2 at $\omega = 1$ rad/sec. Using either of these methods, the ΔT values between ester and salt pairs were determined as a function of concentration; they are displayed in Fig. 4.17. It is significant that molecular weight has no effect on the value of ΔT, since a low-molecular-weight ($M_n = 0.5 \times 10^5$) and a high-molecular-weight ($M_n = 4 \times 10^5$) sample at $C = 3.8$ mole % gave identical values of ΔT.

The shape of the ΔT versus C plot is exceedingly similar to the shape of the $-\log(-m)$ versus C plot shown in Fig. 4.7 and undoubtedly reflects the same underlying structural effects. It is suggested that between 0 and 1.5 mole % NaMA, multiplets are formed with increasing efficiency as the concentration increases, the inefficiency being due to entropic considerations, i.e., distances between ions too large to be overcome by electrostatic interactions. Between 2–4 mole %, the multiplets seem to be identical in structure, as reflected by the constancy of the ΔT value. This conclusion is based on the assumption that the energy required to remove an ion pair from a multiplet depends on the size of the multiplet. In the region above 6 mole %, the increase of ΔT with C suggests that the size of the aggregate is increasing with concentration. The region correlates with the existence of a SAXS peak, as well as with high water uptake, etc., all of which are considered to be related to the presence of clusters, as discussed in Chapter II.

2. Perfluorinated Ionics

This section is devoted to the description of an exploratory study of the structure and viscoelastic properties of Nafion, a polymer developed by

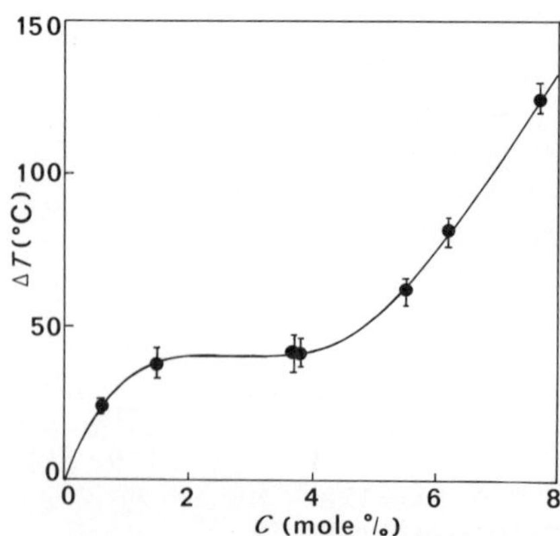

Fig. 4.17. Temperature shift ΔT for the ester–salt pairs, determined at $G' = 2 \times 10^5$ dyn/cm^2 and $\omega = 1$ rad/sec, as a function of ion concentration c [21].

du Pont for use as separators in electrochemical applications [22–24]. The polymer is based on a perfluoroethylene backbone with a side chain of the structure (I), where x is small. Equivalent weights are in the range of 1000 to 2000. The acid polymer, which is itself ionic, is readily neutralized by immersing the polymer sheet in an aqueous solution of an appropriate base. A number of physical properties relevant to the proposed commercial applications have been summarized [25]. In the subsequent discussion, the acid form will be referred to as Nafion–H and the salts as Nafion–Na, etc.

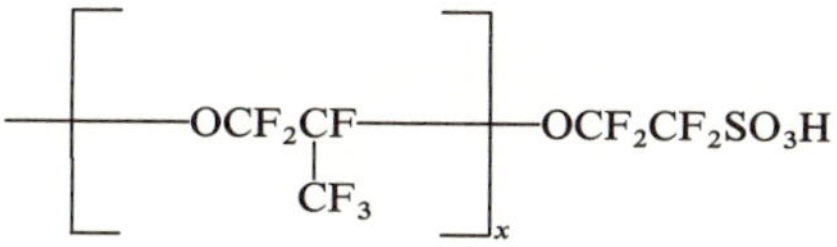

$$-\left[-OCF_2\underset{\substack{|\\CF_3}}{CF}-\right]_x-OCF_2CF_2SO_3H$$

(I)

A study by Yeo and Eisenberg [26] examined the mechanical and dielectric properties of a "Nafion" system, of equivalent weight 1365. As discussed in Chapter II, the presence of a SAXS halo in the cesium salt, along with the mechanical evidence to be presented here, suggests that the ions in this material are clustered.

The drying behavior and thermal stability of the acid and salt forms of this Nafion were examined first. It was observed that Nafion–H showed considerably less thermal stability, based on weight loss on heating under vacuum, than did the salt forms. This appears to be a common phenomenon in sulfonated systems; in sulfonated polysulfones, for example, increased thermal stability of the neutralized material is also observed [27].

The water sorption behavior of the Nafion system was also studied. The diffusion constant for H_2O in Nafion–H at 28°C was found to be 2.6×10^{-6} cm^2/sec. This value of D_{H_2O} is much higher than for most other polymers, and in fact is comparable to D_{H_2O} in sulfonated ion-exchange networks. The activation energy for diffusion was determined to be 4.8 kcal/mol, which is close to the value for the self-diffusion of water (4.41 kcal/mol) [28]. This high permeability to water in Nafion–H is suggestive of a very loose structure. This idea is supported by the fact that the density of the acid form is appreciably lower than that of the nonionic precursor, in which the SO_3H group is replaced by SO_3F. It should be noted that in carboxyl-containing polystyrene, the effect of ionization on the permeability to water is just the opposite, e.g., D_{H_2O} in S–0.097MAA–Na is about 1/500 the value in pure PS.

The stress relaxation behavior of the Nafion system exhibits some unusual features: In dry Nafion–H, time–temperature superposition of stress relaxation data seems to be valid, at least over the time scale in which measurements were made (3.5 to 4 decades of time). Figure 4.18 shows the individual stress

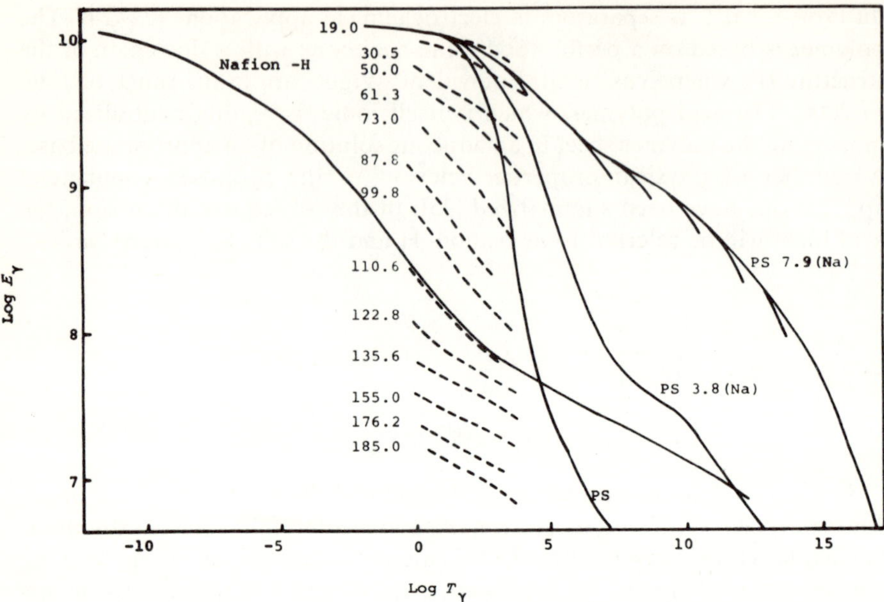

FIG. 4.18. Stress relaxation curves and master curve for Nafion–H, as well as master curves for polystyrene (PS) and two ionomers (S–0.038MAA–Na and S–0.079MAA–Na). $T_{ref} = T_g$ [26].

relaxation curves and the master curve for Nafion–H drawn with respect to T_g, as determined by DSC. Master curves for polystyrene and two styrene ionomers are included for comparison. In the extreme broadness of the transition and the unusual shape, the Nafion–H curve resembles more that of the ion-clustered styrene ionomer than the others. It is worth noting, however, one remarkable difference—the unusually low value of the Young's modulus in the glass-transition region ($T_g = 104°C$).

The addition of small amounts of water to Nafion–H leads to a breakdown in time–temperature superposition. This was also found to be the case for dry Nafion salts, as illustrated in Fig. 4.19, which shows stress relaxation data for Nafion–K. In the latter instance, superposition is reestablished above 180°C, in the terminal region of relaxation. Similar behavior at high temperatures was observed for low-molecular-weight ($\sim 5 \times 10^4$) styrene ionomers [9], such as the example shown in Fig. 4.18. The shape of the curve of Nafion–K in the terminal region would suggest that the molecular weight is low in this case too, but there were no independent means to verify this. As in the acid, the Nafion salts exhibit an unusually low value of modulus in the glass-transition region. In each case, the value of E (10 sec) at T_g is

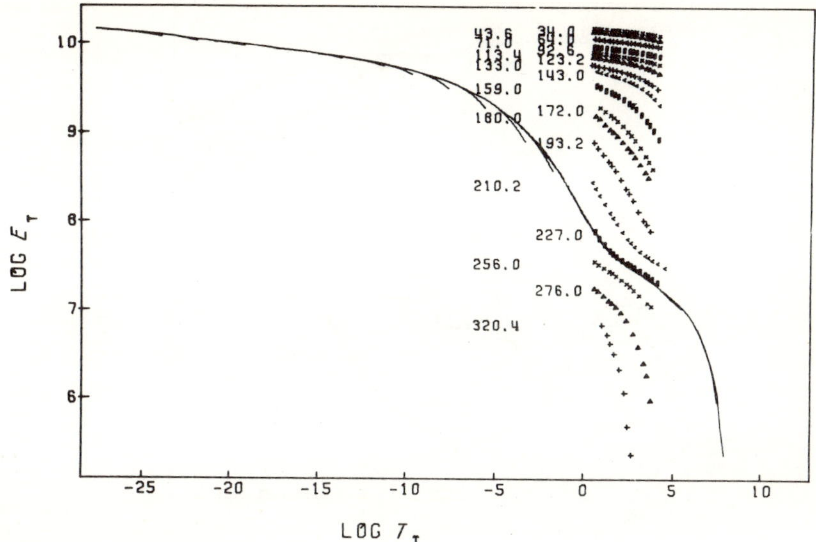

Fig. 4.19. Stress relaxation curves and pseudo master curve for Nafion–K. $T_{ref} = T_e$. [26].

$\sim 10^8$ dyn/cm^2, lower by about two orders of magnitude than the value for styrene or its ionomers.

As in the styrene ionomer system, the Nafions revealed no trace of crystallinity in X-ray scattering studies; in the Cs salt, however, a SAXS halo at $2\theta = \sim 1.7°$ was observed, similar to the ones seen in ion-clustered S–MAA–Cs samples. In view of this fact, the similarity of the stress-relaxation data suggests very strongly the existence of common structural features, i.e., an ion-clustered structure, with regions of high and low ion content.

The dynamic mechanical behavior of Nafion at ~ 1 Hz was investigated by means of a free vibration torsion pendulum. Figure 4.20 shows a plot of tan δ_M and G' versus temperature for Nafion–H and its Cs salt. Three dispersion regions, labeled α, β, and γ, are evident in each case. Similar behavior was observed for Li, Na, and K salts. The α peak corresponds in temperature with the T_g as measured by DSC. The β region is present in each case as a shoulder on the α peak, with tan δ values above background of 0.03–0.06. The γ dispersion occurs at the same temperature ($\sim 100°C$) as that of pure poly(tetrafluoroethylene) and has the same activation energy, 13 kcal/mol [29].

An investigation of the mechanical α relaxation in the acid revealed that water ($\sim 3H_2O/SO_3H$) has only a minor effect on the magnitude or position of the peak. This fact suggests that the α peak is due to the glass transition

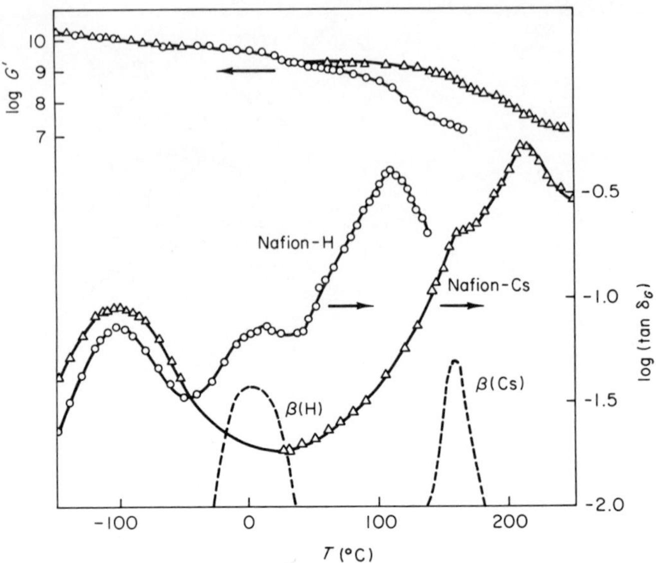

FIG. 4.20. Storage modulus and mechanical loss tangent versus temperature for Nafion–H and Nafion–Cs at ~1 Hz. Dashed lines represent values for β peak above background. [26].

of the nonionic phase, since water would not be expected to interact with the PTFE backbone, but would interact strongly with the ionic regions. The position of the γ peak, which is associated with PTFE as well, is also unchanged in the acid upon water absorption. The β peak in Nafion-H, on the other hand, is depressed considerably by the presence of small amounts of water, as shown in Fig. 4.21. The same is true of the β peak in the salts, where the decrease in temperature with increasing water content is perhaps even more drastic (+160° to −50°C with the addition of 1.4 H_2O/SO_3Li, for example). This remarkable sensitivity to water makes it reasonable to associate the β relaxation with the polar regions of the polymer. The effect of water on the mechanical β dispersion will be discussed further in connection with the dielectric results.

The β relaxation region in the Nafions was also studied dielectrically, in the frequency range 100 Hz–10 kHz.

The results at 100 Hz for Nafion–H are shown in Fig. 4.22. A major peak with a minor peak adjacent to it appear in the β dispersion region of the acid, while the γ relaxation was not observed. In each case, the β dispersion shifts to lower temperature as the water content increases. In Nafion–H, for water contents less than 1.7 H_2O/SO_3H, the minor peak appears at temperatures lower than the major peak, while above 1.7 H_2O/SO_3H, the reverse occurs. The salts exhibited somewhat similar behavior in the same region.

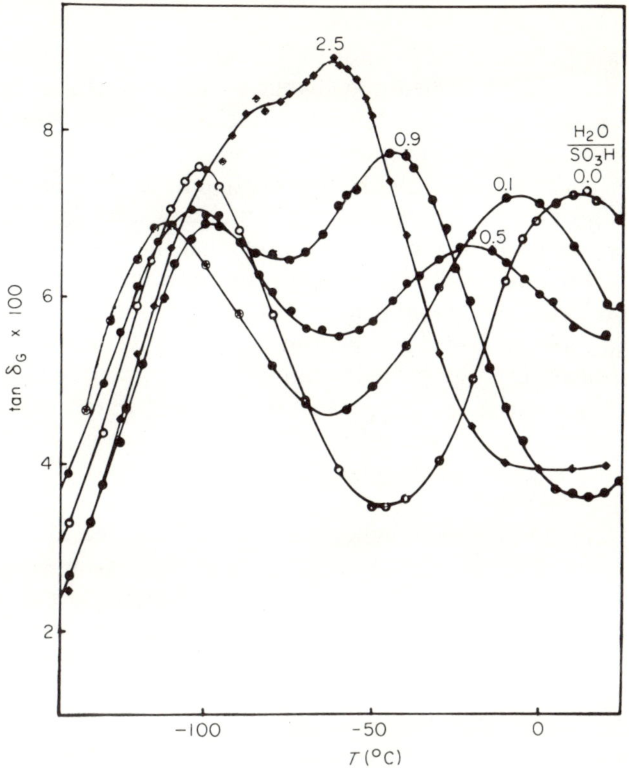

FIG. 4.21. Mechanical loss tangent versus temperature for Nafion–H with varying water content at ~1 Hz [26].

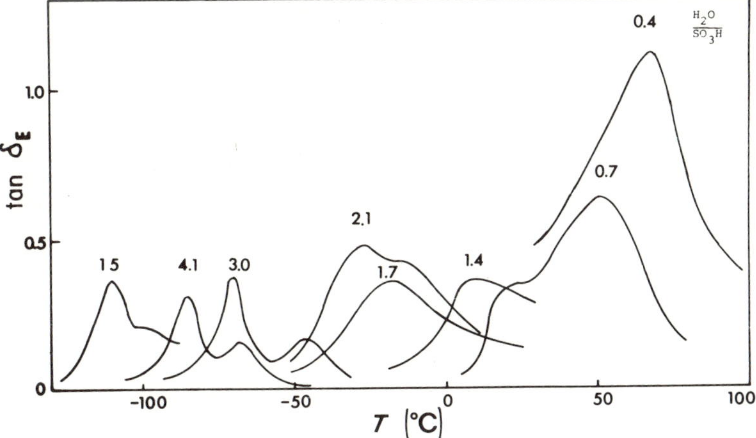

FIG. 4.22. Dielectric loss tangent versus temperature for Nafion–H with varying water content at 100 Hz [26].

The major component of the β region of Nafion–H has an activation energy of ~26 kcal, the minor 16. The rapid change of the β transition region with water content makes it clear that this relaxation is associated with the polar regions. The variation in peak position with water content for the mechanical β transitions is illustrated in Fig. 4.23. It is worth noting that the β relaxation is much more prominent dielectrically than mechanically. This fact, along with the water sensitivity, makes it conceivable that the major β peak represents the T_g of the polar regions, although the constancy of the activation energy argues against this assignment.

Perhaps the most interesting conclusion of this investigation concerns the supermolecular structure of the Nafions. It is suggested that the ions in these materials are clustered, i.e., present in large aggregates, possibly also containing some fluorocarbon material. This suggestion is based on both the rheological properties of the material and the existence of a low-angle X-ray scattering halo.

The glass transition is much higher than would be expected on the basis of rheological data alone. The discrepancy suggests an unusual packing effect, an idea which is supported by the high value of the diffusion coefficient for water and the low activation energy, as well as by the decrease of density on ionization.

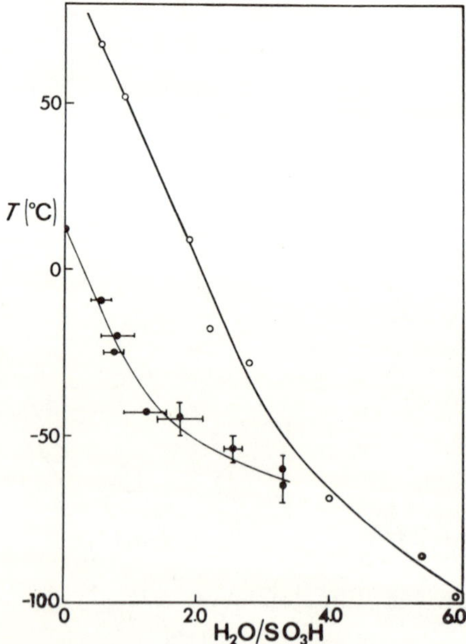

FIG. 4.23. Mechanical β peak position (~1 Hz) (●) and dielectric major peak position (100 Hz) (○) versus water content for Nafion–H [26].

The failure of time–temperature superposition in these materials suggests the existence of two relaxation mechanisms, one due to chain diffusion (as in normal organic polymers) and the other due to the presence of ionic clusters (as also found in the styrene ionomers, among others). The secondary mechanism probably involves the hopping of ion-terminated chains from one cluster to another. The clusters appear to become mobile above 180°C, as suggested by reestablishment of time–temperature superposition above that temperature in the salts.

The β dispersion in dynamic studies is probably also related to the ionic regions. The peak position is strongly affected by the presence of water and may reflect the glass transition of those regions. Dielectrically, a double peak is observed which is also highly sensitive to water. The γ peak is of the same origin as in PTFE.

The diffusion coefficient for water is extremely high—higher than in any other nonionic or partly ionic polymer for which data are available, and comparable to that of ion-exchange resins, although the polymer has a very different structure. The activation energy is close to that for self-diffusion in pure water.

It is evident that the Nafions resemble other organic ionomers in a wide range of properties, notably in the presence of ion clustering and the resultant effect on the rheology of the materials. By contrast, the dramatic decrease in the density upon ionization and the accompanying increase in the permeability to water are novel features, not encountered in other ionomers. The reasons for this difference need to be elucidated. Finally, the dynamic mechanical studies suggest that, at least in the presence of water, the glass transition of the ionic regions may occur at a lower temperature than that of the matrix. This has not been encountered in other ionics either.

B. RUBBER-BASED IONOMERS

Another common and useful class of polymers whose properties can be modified by the introduction of ionic forces are the rubber-forming polymers such as polybutadiene. Ionic character is commonly introduced to these materials by the incorporation of carboxylic or amine-containing end groups or by copolymerization with corresponding monomers and subsequent neutralization.

The structural studies performed on ionomers of butadiene indicate that the ions exist in aggregates, although the state of aggregation, whether multiplets or larger clusters, is still open to question. In any case, it can be expected that the predominant effect of ion aggregation on the viscoelastic properties of these ionomers will be time-dependent cross-linking. The studies made to date make this very clear.

The original purpose in introducing carboxylic groups into rubbers was to modify the rubbery characteristics through the polar nature of the carboxyl group. The properties of a number of carboxyl-modified rubbers were reviewed by Brown in 1957 [30]. Vulcanization of such copolymers can take place by conventional means, such as with sulfur or organic peroxide cures. However, the changes in elastomeric character obtained with the small percentages of acid groups normally incorporated in rubbers are small. Generally speaking, the main effects observed are an increased T_g and a decreased crystallization tendency.

Cross-linking through the carboxyl group can be achieved by a number of chemical reactions [30], but the method of greatest interest is salt formation, especially with the use of divalent metal oxides. Frequently, a combination of cures—sulfur and ZnO, for example—is used. In general, metal oxide cures on carboxyl-containing rubbers, alone or in combination with sulfur or peroxide, result in elastomers with enhanced strength properties—increased moduli and tensile strengths, as well as ultimate elongations which exceed those of equivalent covalently cross-linked rubbers. The cross-linking effect is observed with both monovalent and divalent cations, although more strongly with the latter [31]. Ionomeric butadiene elastomers have also been prepared by the direct cross-linking of ester copolymers of butadiene with basic oxides or hydroxides [31].

Quantitative observations on the transient nature of ionic cross-links in carboxylic rubbers have been made by a number of authors. Cooper [32] studied the mechanical properties of various divalent salts of butadiene-(acrylic acid) and butadiene-(methacrylic acid) copolymers. The salts of low-molecular-weight copolymers (M_v between 3×10^3 and 20×10^3, —COOH content between 2 and 7 mole %), as indicated in Fig. 4.24, were found to be elastic when subjected to a rapid impact, but plastic under

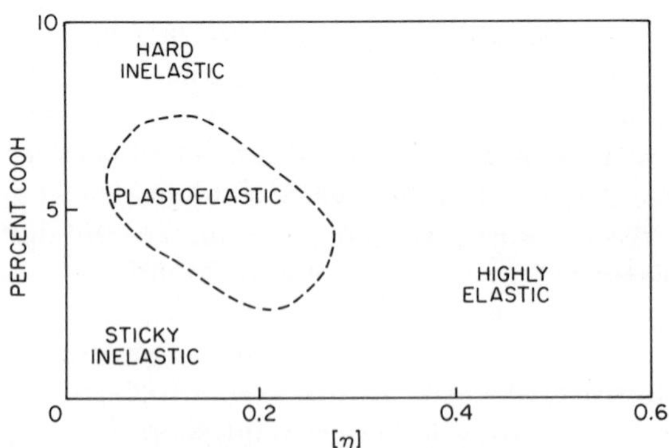

FIG. 4.24. Properties of B–MAA–Zn copolymers as a function of intrinsic viscosity and ion content [32].

slowly applied stress. In this respect, the properties of these materials are very similar to the familiar silicone "bouncing putty." The effect was found to be most pronounced for Zn and Cd salts, and less so for salts of the alkaline earth series. In accord with the observations of earlier workers, Cooper found that the amount of metal oxide required is not critical as long as it is somewhat in excess of the available —COOH content.

The rebound resilience, a measure of the elasticity under rapid deformation, was found to be sharply temperature dependent, passing, in the case of Zn salts, through a maximum at ~20°C. The temperature dependence of the plasticity of a number of these salts was measured and used to determine the activation energies for viscous flow. The activation energy was found to increase with increasing acid content of the copolymer and to be dependent on the nature of the cation.

The neutralized copolymers of low-molecular-weight (~10^4) were found to be generally soluble in organic solvents such as benzene. This is in contrast with the behavior of the same copolymers cross-linked covalently. The latter materials were observed to be insoluble, elastic products with no unusual properties. In dilute ionomer solutions, intermolecular bonding was not found to occur to any great degree. The intrinsic viscosities of the neutralized copolymers were found to be very close to those of the corresponding unneutralized materials. Osmotic pressure measurements also indicated only slight degrees of aggregation in dilute solution. However, the state of aggregation was found to rise rapidly with increasing ionomer concentration, as indicated by viscosity measurements; ultimately, gelation occurred at ~10% polymer. This behavior appears to be very similar to that which occurs in the poly(methacrylic acid) solutions described in Chapter III, Section C2, and likely indicates a transfer from intra- to intermolecular bonding as the polymer concentration increases. The limits of solubility in butadiene ionomers depend on both the molecular weight and the carboxyl content. Cooper found that for B–MAA salts in benzene, insolubility results with 8 to 16 carboxyl units per chain, depending on the nature of the cation.

Butadiene ionomers of high molecular weight ($M_v > 2 \times 10^4$, —COOH content ~5 mole %), as shown in Fig. 4.24, were found to be insoluble, elastomeric materials of considerable strength. The results of mechanical measurements reported by Cooper [32] indicate that the rate of creep in butadiene ionomers is considerable; furthermore, the rate is greater for ionomers containing Zn or Pb than for those containing Mg or Ca. This seems to correlate with the fact that the strength properties of Zn and Pb vulcanizates are significantly better than those of corresponding alkaline earth vulcanizates, which in turn seems to suggest that ionic bond interchange is easier with the former than with the latter. This effect might conceivably be related to the *q*/*a* values, as discussed in Chapter II in connection with glass transitions.

Stress relaxation measurements, the results of which paralleled the findings from the creep data, indicated that the rate of relaxation increased considerably with increasing temperature. Interestingly, however, Cooper found that there was very little permanent set associated with the creep or stress relaxation. This latter observation was also made in connection with the metal-oxide-cured, carboxyl-modified natural rubber prepared by Cuneen *et al.* [33], and confirmed in subsequent work on a number of metal oxide vulcanizates of carboxyl-containing rubbers [34–36].

Cooper attributed the creep and relaxation phenomena to a chemical mechanism in which interchange reactions between ionic cross-links serve to relieve the stress by reformation of the network to a more relaxed state. This postulate would help to explain the high values of ultimate elongation, which correlate with high creep rates, since a partial relaxation of the polymer network would relieve local stresses and prevent premature failure and thus lead to high elongations and high tensile strengths. However, this postulate assumes that the copolymers are cross-linked exclusively through multiplets, i.e., associations between pairs of carboxyl groups from neighboring chains and any relaxation process involving exchange between simple ion multiplets would result in a high level of permanent set, since a reformed network should resist a return to its original state. The fact that there is little permanent set leads to the conclusion that the interaction of metal oxides with carboxylic rubbers is considerably more complicated than simple multiplet formation. It suggests, in fact, that if the bonding is entirely ionic, some of the aggregates must be very stable indeed. There is a possibility, in addition, that some heterogeneities are introduced into the system by incomplete dissolution of the metal oxide, and that these particles act as a very high-energy reinforcing filler. Both of these possibilities have been suggested by workers in the field (see below).

Halpin and Bueche [34] put forth a different postulate, based on a suggestion by Zakharov [35], to explain the relaxation behavior of carboxylic elastomers. They proposed that the high creep rates and high-strength properties were merely the reflection of a sparse network of chemical cross-links, and that the high initial values of modulus result from an increased T_g. In their study they compared the fracture properties of the sulfur and ZnO vulcanizates of a carboxylic terpolymer of butadiene and acrylonitrile, denoted by B–0.26AN–0.04MAA. They determined the elongation at break as a function of the temperature and the strain rate. The curves of ultimate strain versus time to break were found to be superimposable with temperature. For the sulfur vulcanates, the curves followed a normal pattern, with both the ultimate stress and the ultimate strain decreasing with increasing temperature; however, for the ZnO vulcanizate, although the loss of strength with increasing temperature was similar to the sulfur-cured polymer, the elongation at break was found to increase rapidly with increasing temperature.

By plotting the reduced stress at break versus the ultimate strain, "failure envelopes" which are unique to any given material are obtained. It can be seen from Fig. 4.25 that the envelopes for the two types of cross-linked materials diverge with increasing temperature. Halpin and Bueche were able to predict the shapes of the failure envelopes by applying their molecular fracture theory to the generalized creep response of these two materials. From this agreement, they concluded that the differences in the strength properties must be related to differences in the creep response of the two vulcanizates.

Halpin and Bueche also found the activation energies for the viscoelastic response, obtained from the temperature dependences of the respective shift factors, to be very similar (26 ± 2 kcal/mol) for the two types of vulcanizates. However, this could be merely coincidence, since similar values of the activation energy for plastic deformation of metal oxide vulcanizates were found by Cooper [32].

Halpin and Bueche attributed the differences in creep response to a sparse network of chemical cross-links and a higher T_g for the metal oxide vulcanizate, rather than to any significant chemical relaxation. However, as pointed out by Tobolsky *et al.* [36], the T_g values of both vulcanizates (sulfur and ZnO) do not differ significantly from each other or from that of

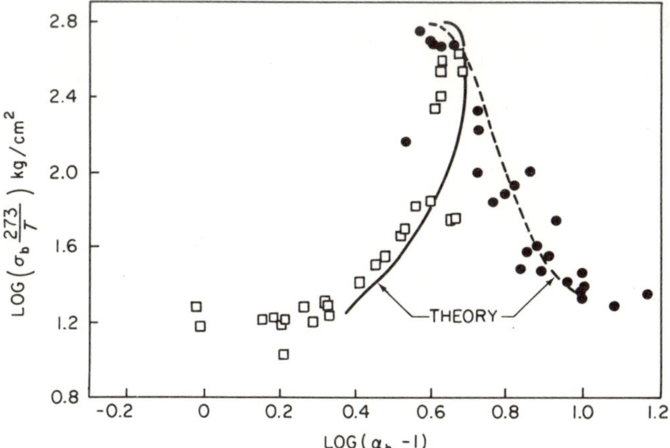

FIG. 4.25. Comparison of failure envelopes—reduced stress at break versus ultimate strain—for two B–AN–MAA vulcanizates. The theoretical curves are predicted from creep data; (□) sulfur cross-linkages, (●) metal carboxylic cross-linkages [34].

the uncured polymer (see Fig. 4.29). Hence, it seems unlikely that the high initial moduli of the metal oxide vulcanizates could result from a sparse network of chemical cross-links.

The chemical exchange mechanism was investigated further by Otocka and Eirich [37]. The ionomers used in their study were lithium salts (90% neutralized) of three B–MAA copolymers and three quaternized copolymers of butadiene and 2-methyl-5-vinylpyridine (MVP). The modulus temperature curves for these two sets of ionomers are shown in Figs. 2.3 and 4.26. In contrast with the curves for the parent (nonionic) copolymers, which display no particularly unusual features, the curves for the ionomers all exhibit pseudoequilibrium rubbery plateaus in the region of 10^8 dyn/cm^2. In this respect, these curves bear a great deal of similarity to the family of curves for the styrene ionomer system, shown in Fig. 4.6; in each case, the height of the rubbery plateau is an increasing function of the ion concentration.

The plateaus in the B–MQMVP system (MQMVP ≡ methyl-quaternized MVP) are particularly well defined. It is notable that the decrease in modulus beyond the plateau level occurs at roughly the same temperature (~20°C)

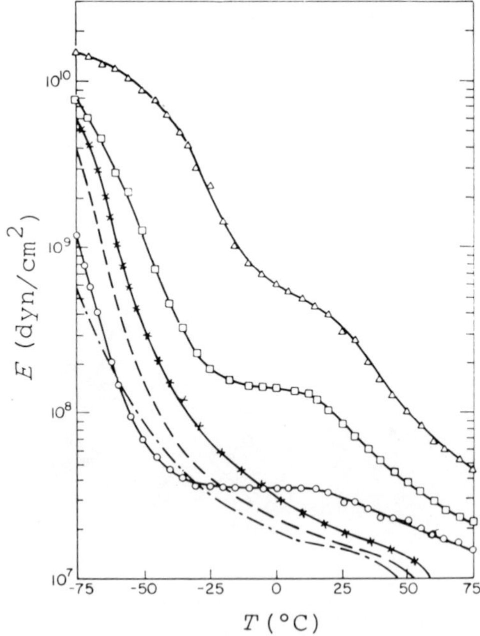

Fig. 4.26. Modulus–temperature curves for B–MVP copolymers and their methyl-quaternized salts. *Symbols:* 1.8% MVP: (·—·) base, (○) salt; 6.5% MVP; (— —) base, (□) salt; 11.8% MVP: (×) base, (△) salt. [37a].

in each case, despite the fact that the primary transitions of the three ionomers differ by more than 50°C. From this observation and the fact that irregularities in the DSC trace occur at ~20°C, Otocka and Eirich defined a new transition in the B–MQMVP system [37]. This transition is designated here as T_m, since it is possibly of the same nature as the transition observed in styrene ionomers above their T_g values. Although the plateaus are not as sharp in the B–MAA–Li system, it is probable that a similar transition exists in this system as well, since the relaxation spectra of both types of ionomers show maxima above T_g.

The stress relaxation results presented by Otocka and Eirich [37] provide additional evidence for a secondary transition in the B–MQMVP system. The stress relaxation curves for the B–MAA–Li copolymers were similar to those for styrene ionomers containing less than 6 mole % ions, and the shift factors in each case were of the WLF type. However, in the B–MQMVP system, the shift factors appear to give rise to two segments of WLF curves whose arcs are joined near T_m. This is illustrated in Fig. 4.27, in which log a_T is plotted versus $T - T_m$. In this respect, the B–MQMVP copolymers exhibit viscoelastic behavior very similar to certain phase-separated block copolymers in which the T_g values of the two phases are widely separated [38]. No evidence for the breakdown of time–temperature superposition in

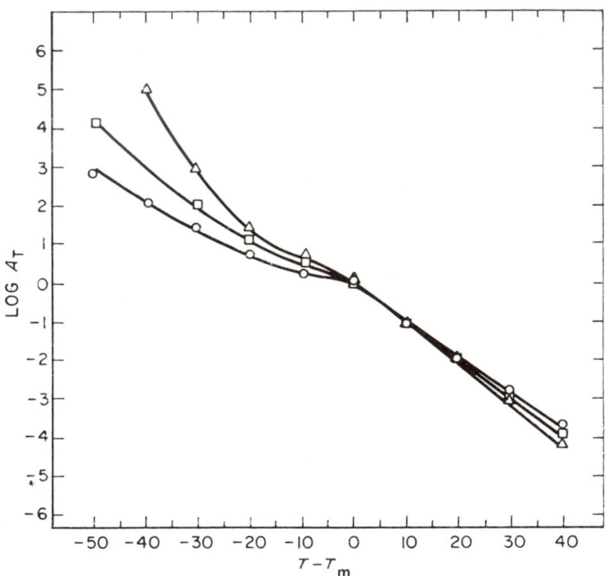

FIG. 4.27. Log a_T versus $(T - T_m)$ for B–MQMVP copolymers. Mole percent pyridinium: (○) 1.7, (□) 6.0, (△) 11.7 [37b].

these ionomers was presented; however, the time span for superposition was only two orders of magnitude.

Otocka and Eirich calculated the activation energies for viscous flow in the two ionomer systems they studied. In the case of the B–MAA–Li copolymers, the value of E_{visc} above T_m was determined to be ~ 18 kcal/mole—about the same value predicted by the WLF equation for that temperature. This fact may provide an explanation for the lack of a well-defined second mechanism above T_g in the carboxyl-containing system. By contrast, E_{visc} for the B–MQMVP copolymers was observed to rise sharply above T_m to values in the range of 40–50 kcal/mol. The mechanism that contributes to the relaxation above T_m was attributed by Otocka and Eirich to the migration of ion pairs from ion quadrupolar cross-links, essentially the same mechanism proposed by Cooper [32].

An investigation by Tobolsky *et al.* [36] employed ZnO and sulfur vulcanizates of the same B–AN–MAA terpolymer used in the study by Halpin and Bueche [34]. The modulus-temperature behavior of the two types of vulcanizates is shown in Figs. 4.28 and 4.29. It can be seen that while the

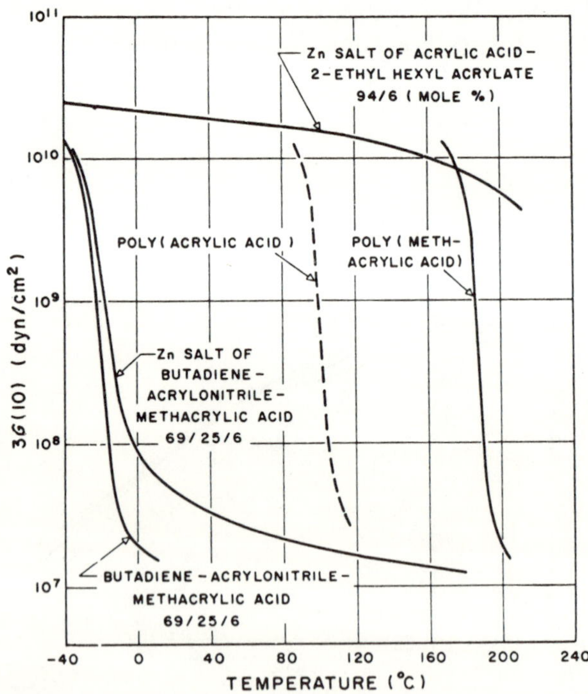

FIG. 4.28. Modulus–temperature curves of PAA, PMAA, a B–AN–MAA terpolymer (69:25:6) and their zinc salts [36].

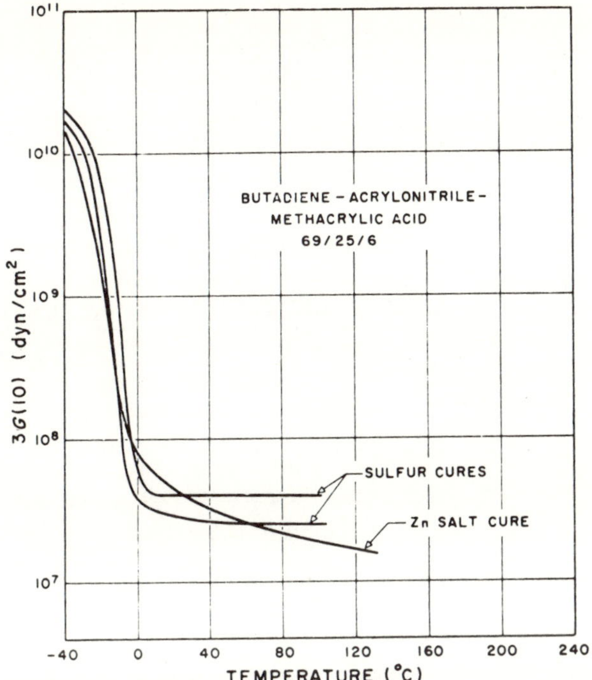

FIG. 4.29. Modulus–temperature curves of a B–AN–MAA (69:25:6) terpolymer with sulfur cures and a ZnO cure [36].

covalently cross-linked sulfur networks exhibit a true rubbery plateau, the modulus of the Zn salt decreases slowly with increasing temperature over the entire span above 0°C. A combination ZnO–sulfur vulcanizate (not shown in Fig. 4.29, for reasons of clarity) was also tested and found to exhibit an intermediate response, reaching a plateau around 80°C.

It was found to be possible to solvate selectively the ionic domains of the combination ZnO–sulfur vulcanizate and thereby plasticize the viscoelastic effect of the ionic clusters. This was accomplished by swelling the sample with a mixture of *p*-xylene and glacial acetic acid and subsequently deswelling by evaporation, which selectively removed the *p*-xylene. The effect of the high dielectric constant plasticizer is illustrated in Fig. 4.30, which shows that the pseudoequilibrium plateau is lowered by a factor of 3, while the primary transition is changed only slightly. This selective solvation is similar to that which occurs in styrene ionomers swollen with DMSO [11].

From stress relaxation measurements on the ZnO vulcanizate, it was found that the stress decay is continuous and practically uniform over the

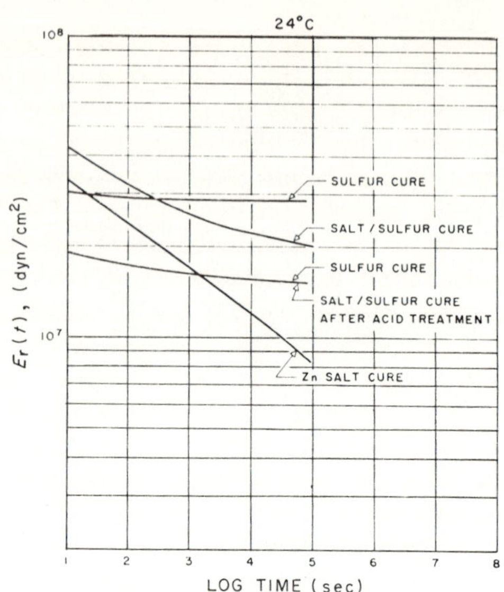

FIG. 4.30. Stress relaxation curves at 24°C for a B–AN–MAA terpolymer (69:25:6) with sulfur cures, ZnO cure, and a combination ZnO/sulfur cure [36].

entire temperature span between 0° and 150°. This contrasts with the behavior that would be predicted from either the chemical exchange theory or the sparse network theory. The latter theory would predict a decreasing relaxation rate as the modulus approached the plateau corresponding to the cross-link density. Hence, the stress relaxation curves would be concave upward, as they are for the covalently cross-linked materials in Fig. 30.

The former theory would predict an increasing rate of relaxation due to chemical flow, and hence downward concavity of the stress relaxation curves. Such behavior was observed in the butadiene ionomers studied by Cooper [32] and Otocka and Eirich [37], and was associated by the latter with chemical exchange above a second transition temperature. In the ZnO vulcanizates tested by Tobolsky *et al.* [36], the stress relaxation behavior shows no indication of chemical flow, although its possible existence above 150°C cannot be ruled out. Nevertheless, the fact that no true rubbery plateau is obtained for the B–AN–MAA–Zn ionomers indicates that relaxations involving the ionic bonds must occur over the entire temperature span of 0°–150°C. Tobolsky *et al.* proposed that the high strength of the carboxylic rubbers cured with metal oxides is due to the presence of ionic clusters, which act as both a reinforcing filler and as time-dependent cross-links.

In reviewing the evidence presented thus far, it is evident that the metal-oxide-cured, carboxyl-containing elastomers present us with an unusual combination of properties, i.e., high initial modulus, high elongation at break, and low permanent set. This combination of properties suggests that the bonding is very complex and consists of different types of species, i.e., multiplets and clusters or ionic fillers of high surface energy. The low permanent set can be explained by the presence of a small concentration of highly stable cross-links, probably small multiplets. The high elongation at break is probably due to the fact that those strained network elements, which would initiate rupture in permanently cross-linked systems, can relax by ionic bond interchange. Finally, the high initial modulus is probably due to the presence of clusters, which act as a reinforcing filler, but which can fall apart under lower strain values than the multiplets.

Dynamic mechanical evidence for a transition involving ion clusters in butadiene ionomers has been presented by Pineri and co-workers [39a]. In divalent metal salts (Mn, Cu, Zn) of a B–0.09MAA copolymer ($M = \sim 2 \times 10^5$), the high-intensity peak above T_g in the internal friction curves, as illustrated in Fig. 4.31, was attributed to a transition in the ionic domains, for which X-ray evidence had previously been presented. The unionized copolymer also showed an internal friction peak above T_g, at $T = 270°C$, suggesting that significant carboxyl group association occurs in the unionized material as well.

In a very recent study, Meyer and Pineri [39b] investigated the B–0.10S–0.05VP (VP = vinylpyridine) terpolymer system complexed with nickel chloride. This system was already discussed in Chapter II in connection with electron microscopy and Mössbauer studies. The techniques employed in this work included stress relaxation and loss tangent measurements in a

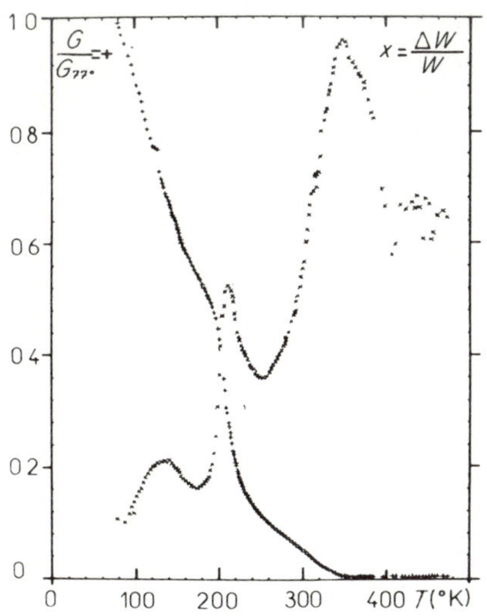

FIG. 4.31. Loss tangent and storage modulus of B–0.09MAA copolymer 90% neutralized manganese acetate [39a].

torsion pendulum. In the stress relaxation study, it was found that both a horizontal and a vertical shift were needed to obtain superposition; however, when both were applied, a master curve could be obtained. The shift factors followed a WLF-type relation at low temperatures (up to ~20°C), whereas the high-temperature points (~20–90°C) could be better fitted by an Arrhenius relation. The WLF constants were 16 and 59.3, i.e., very close to the universal values if the tan δ_{max} of 217°K was chosen as the reference temperature; the apparent activation energy in the Arrhenius region was 28 kcal. Since data were taken over only three decades of time, no deviations from time–temperature superposition of the type observed for the styrenes [8, 9] could be detected.

In the dynamic study, it was found that a rubbery plateau was evident even for low nickel concentration. Also, a high temperature (approximately $T_g + 50$ to $T_g + 130$) peak was observed, which moved to higher temperatures with increasing ion concentrations, again paralleling the behavior of the styrenes [10].

Other elastomeric ion-containing systems have received attention. Dieterich *et al.* [40] investigated the mechanical properties of ionic polyurethanes. They prepared a polyurethane—poly(*N*-methylamino-2,2′-diethanol)—containing tertiary nitrogen in the backbone. Quaternization by methyl iodide resulted in pronounced elastomeric character between −40°C and 130°C, the height of the pseudoequilibrium rubbery plateau increasing with increasing degree of quaternization.

Somoano *et al.* [41] prepared a variety of cationic elastomers from butadiene-based polyurethanes. Polymers containing quaternary nitrogen in either the backbone or on pendant groups were obtained. Some of the polyurethanes were also covalently cross-linked through the pendant amino group. The effect of quaternization was seen in the modulus–temperature behavior as a rubbery plateau region from −30° to +40°C. With covalent cross-linking, the rubbery behavior extended to much higher temperatures.

Rembaum *et al.* [42] studied the properties of three elastomeric ionenes of different ion contents. The elastomers were prepared according to the reaction scheme outlined in Fig. 4.32. The reaction of Solithane 113 (I), a glycerol ester of ricinoleic acid, with tetramethylaminopropanol gives a urethane prepolymer containing tertiary amino groups (II), while *n*-bromopropanol with I gives a bromine-containing prepolymer (III); the latter was reacted with tetramethylamino propylene glycol to give an ionene product termed PPGS3. The former (II) was combined with two types of dibromides to yield two different ionene products: (1) with dibromopropane to give DPS4 and (2) with a dibromopolybutadiene prepolymer to give SPSA3. The glass-transition temperatures of these three materials correlate well with their ion contents, based on weight percent ionic bromide: SPSA3—2.6% Br^-, $T_g = -80$°C; PPGS3—4.3%, −20°C; DPS4—12.9%, +40°C.

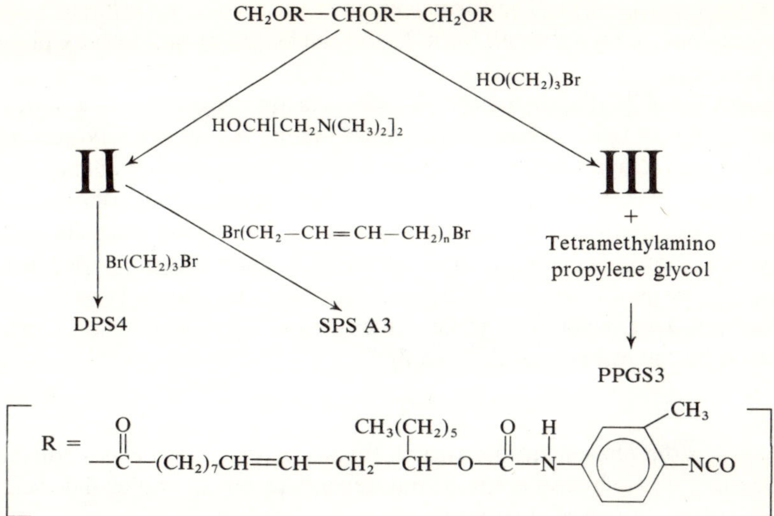

FIG. 4.32. Reaction scheme for synthesis of ionene elastomers [42].

The modulus–temperature behavior of the three materials is illustrated in Fig. 4.33. In this figure, one can see the effect of changing the length of the hydrocarbon chain in the dibromide, comparing SPSA3 and DPS4. Both the height of the rubbery plateau and the T_g are sharply increased. PPGS3, with intermediate ion content, falls in between. The comparison between a nonionically cured Solithane and the equivalent ionic cure DPS4

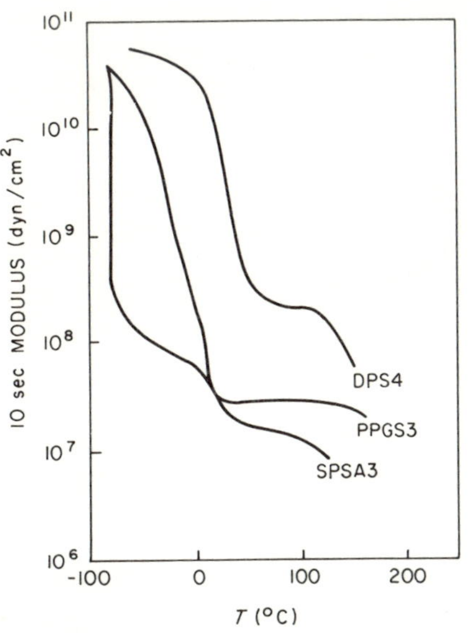

FIG. 4.33. Modulus–temperature curves for three ionene elastomers: DPS4, SPSA3, and PPGS3 [42].

can be seen in Fig. 4.34. The former exhibits ideal rubbery behavior, while in the ion-containing material, both T_g and the height of the rubbery plateau are increased.

Erhardt *et al.* [43] investigated the viscoelastic properties of a series of styrene–(*n*-butylmethacrylate)–(potassium methacrylate) terpolymers produced by partial hydrolysis of random S–*n*-BMA copolymers. Steady-state and dynamic viscosity measurements on the polymer melts showed that ionization produces a pronounced increase in zero-shear viscosity η_0 for any given molecular weight and S–*n*-BMA composition. Specifically, $d \log \eta_0/dc$ ranges from ~ 1.0–0.33, depending on styrene content and molecular weight, where c is mole percent ions. The viscosity data can be expressed by the following relationship

$$\log \eta_0 = \log \eta_{0,0} + K \cdot C$$

where $\eta_{0,0}$ is the zero-shear viscosity of the parent ester, and K is a constant, independent of molar ion content, but dependent on the molecular weight, the dependence being of the form

$$K = A + B/\text{molecular weight}$$

Plots of G' as a function of ω were made for materials formed from the copolymer containing 39 mole % styrene. Time–temperature superposition was found to be applicable. The shapes of the curves, as seen in Fig. 4.35, are very similar for ion concentrations up to 2.0 mole %. Above that concen-

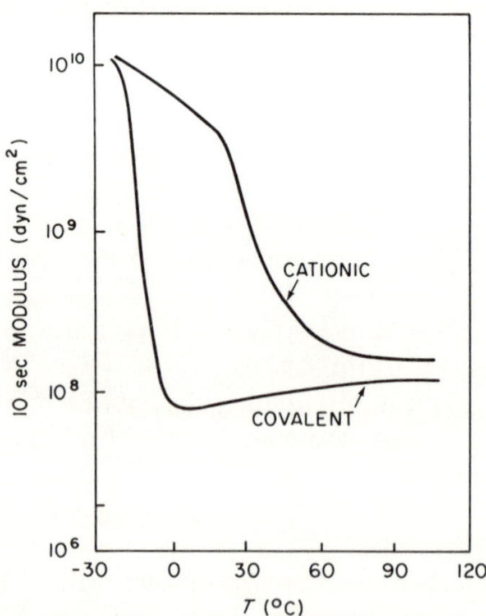

Fig. 4.34. Modulus–temperature curves for covalent and cationic solithane (DPS4) [42].

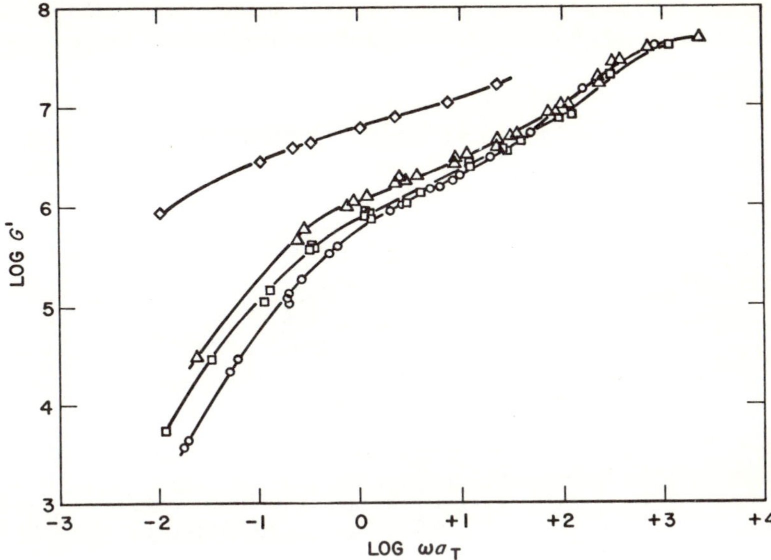

FIG. 4.35. Master curves of G' as a function of reduced frequency ωa_T for a series of ion-containing terpolymers formed from S–n–BMA copolymer containing 39 mole % S; M_n = 23,000. $T - T_g = 30$. ○ = 0.8 K^+, □ = 1.2 K^+, △ = 2.0 K^+, ◇ = 4.1 K^+ [43].

tration, however, appreciable deviations occur. Furthermore, a plot of the shift factors versus temperature shows great similarities in the low ion concentration region, with appreciable deviations for the 4.1 mole % K^+ sample. The authors suggest that this point may represent the onset of clustering.

Kresge [44] studied the properties of elastomers formed from ethylene–propylene (E–P) copolymers. Using a common polymeric backbone, various cross-link types were compared, including peroxide (C–C), sulfur (S_x), and metal carboxylate, the latter obtained through a (maleic anhydride)–E–P terpolymer. Both sulfur and ionic cross-links gave higher tensile strengths to rubbers than the equivalent peroxide cures. In each case, this was attributed to the relief of stress through the breaking and reforming of bonds. The ionically cross-linked materials also showed increased viscosity during extension due to the contribution of the ionic bonds, the increased energy dissipation leading to further suppression of rupture under dynamic conditions.

In the sulfur–cross-linked materials, compression set was found to be invariably high, while with ionically cross-linked systems, the compression set varied with the ion type and test conditions. In a $(COO)_2$–Ba system cured at 100°C, for example, a compression set test at 80°C showed recovery from 100 to 40% elongation. This appears to indicate the existence of a stable fraction of cross-links at this temperature, in line with the earlier observations on metal-oxide-cured butadiene rubbers.

C. CRYSTALLINE COPOLYMERS

1. Ethylene-Based Ionomers

At this time, ethylene-based ionomers represent perhaps the most extensively studied class of ion-containing polymers. The results of studies on these materials are presented in this section in six parts. The first part will reproduce data on modulus-temperature relations insofar as they have been obtained. The second will concentrate on stress relaxation results, while the third will be devoted to dynamic mechanical studies. Parts four and five will concern themselves with dielectric and NMR results, and the last part with optical and X-ray studies.

In many of their physical properties, ethylene ionomers are very similar to their parent acid copolymers. The most notable difference is the marked change in crystal morphology at low degrees of ionization, which manifests itself in the optical clarity of ionomer samples [45]. As will be described later, the development of spherulitic structure in ionomers is very dependent on annealing conditions. The degrees of crystallinity of well-annealed ionomers, however, are found to be very close to those of the corresponding unionized copolymers [46–48], decreasing in both cases with increasing carboxyl content, as indicated in Fig. 4.36. This decrease in crystallinity with increasing salt content, complicated by its dependence on the thermal history, has tended to obscure many of the changes in polyethylene viscoelasticity due to ions, as well as to make difficult the interpretation of structural changes which occur on ionization. However, as discussed in Chapter II, it appears certain that ion clustering occurs in these ionomers as well [15], and many of the viscoelastic properties to be discussed below reflect this fact.

(*a*) *Thermomechanical Behavior*

Ward and Tobolsky [49], in a study of the viscoelastic properties of a range of ionomers, investigated the 10-sec modulus as a function of temperature for the ionomer E–0.08MAA–0.47Na. The relationship obtained is illustrated in Fig. 4.37. It is evident that the breadth of the curve is reminiscent of a semicrystalline polymer; i.e., both the shape and the position are similar to those of a typical low-density polyethylene. X-ray studies on the sample, however, revealed a smaller percentage of crystallinity than in unmodified polyethylene and suggested that the effect of the ions, presumably in some aggregated state, might be similar to that of the crystallites or to strongly

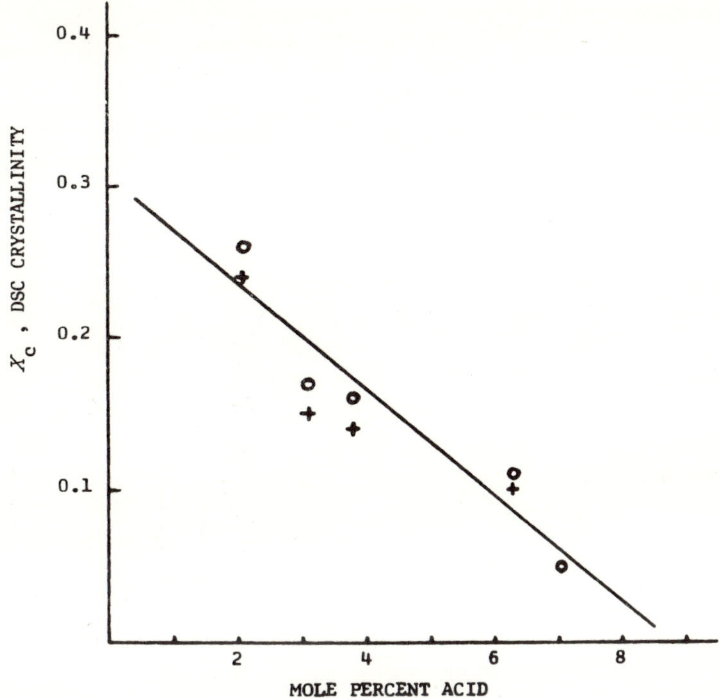

FIG. 4.36. Degree of crystallinity χ_c as measured by DSC versus carboxyl content of ethylene–(methacrylic acid) copolymers and salts; ○ = annealed copolymers, + = annealed cesium salts. [47].

bonded filler particles. Although the crystalline content may have been somewhat underestimated, as may well be the case in view of subsequent studies, it is still clear that the changes in the properties due to changes in both crystallinity and crystal morphology are compensated for by ionic aggregation. This reinforcing effect of ionic aggregation is similar to that encountered in the acrylates and in styrene ionomers, illustrating the universality of the observed phenomena.

In a study of the dynamic mechanical properties of the ionomer system E–0.041MAA, neutralized to varying degrees, MacKnight *et al.* [50] determined the modulus at ~1 Hz as a function of temperature, using a torsion pendulum. Their results, for both quenched and annealed samples, are shown in Fig. 4.38. It should be pointed out that the degrees of crystallinity varied, the percentages being ~15, 16, 12, and 7 for the samples of 0, 20, 60, and 78% ionization. It can be seen from Fig. 4.38 that, for the annealed samples, the low-temperature moduli decrease very slightly with increasing degree of

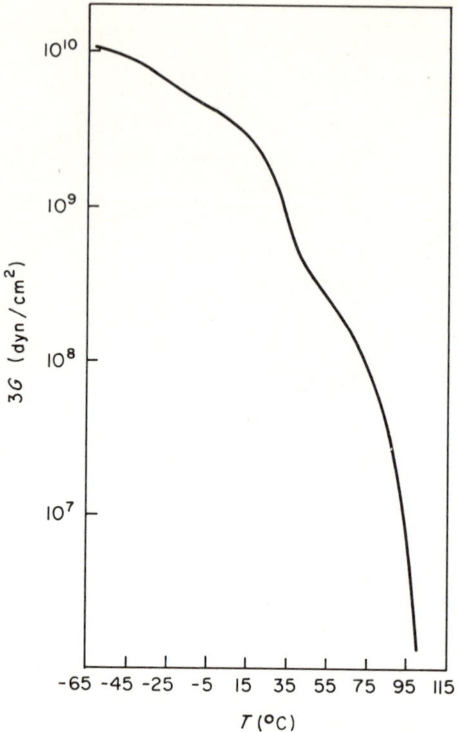

FIG. 4.37. Modulus–temperature behavior of quenched ionomer E–0.08MAA–0.47Na [49].

ionization; in the quenched samples, this decrease becomes more pronounced and indicates that the degree of crystallinity is primarily responsible for the level of the modulus. The other features of the curves resemble strongly those found by Ward and Tobolsky.

Otocka and Kwei [51] investigated the dynamic moduli of E–AA copolymers, fully neutralized by sodium and magnesium, the carboxyl contents varying from 0.66 to 3.40%. The results of this study, shown in Figs. 4.44 and 4.45 in Part (c) of this section, indicate two temperature regions in which the modulus drops quite rapidly as a function of temperature, i.e., at $\sim -20°C$ and at $\sim +50°C$, the latter temperature increasing with the ionic content of the material. The results also suggest that the nature of the counterion is not nearly as important in determining the value of the modulus as is the degree of ionization.

The very first study of the modulus of ethylene ionomers was performed by Rees and Vaughan [52], who obtained some preliminary data on one sample. Longworth and Vaughan [46] subsequently studied the modulus–temperature behavior of E–MAA ionomers, as part of a dynamic mechanical

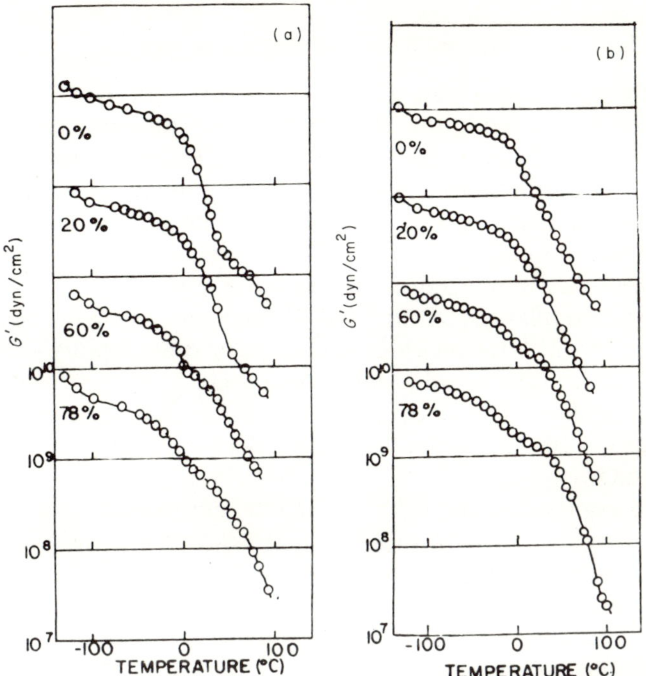

FIG. 4.38. G' at ~ 1 Hz versus temperature for (a) quenched and (b) annealed E–0.041MAA samples at various degrees of neutralization. Each curve is displaced by one decade above the preceding curve [50].

investigation. They found that at room temperature, a dry ionomer containing 5.4 mole % NaMA had a slightly higher modulus than low-density polyethylene, but that above 70 C and in the region -60° to 0°C, the reverse was true.

Bonotto and Bonner [15] extended the studies of ethylene ionomers to include the effect of water absorption on the thermomechanical behavior. They found that monovalent salts of an E–AA copolymer were considerably more sensitive to water than the divalent salts. For example, the secant modulus of an E–0.0635AA–0.7Na sample immersed in water for 24 hr was half that of the dry sample, while the 0.65Ca sample showed only a slight change in modulus upon water soaking.

Rafikov *et al.* [17] also studied the thermomechanical properties of ethylene ionomers. With random copolymers of ethylene and either MAA or AA (3–4 mole %), they found little change in the deformation–temperature curves of the neutralized copolymers in comparison with those of polyethylene or the unionized copolymers.

In summary, the following general observations can be made in regard to the modulus–temperature behavior of these polymers.

1. The low-temperature modulus decreases with increasing ionic content, probably paralleling the disappearance of spherulitic structure. This behavior contrasts markedly with that encountered in noncrystallizable copolymers of ionic materials, where the modulus increases with ion content.

2. As one might expect, quenching, which affects both crystallinity and crystal morphology, accentuates the differences in the low-temperature moduli, while annealing tends to deemphasize them.

3. For moduli between 10^{10} and 10^9 dyn/cm^2, there are two temperature regions in which the modulus tends to drop quite rapidly with temperature. At a low concentration of ions, these occur at $\sim -20°$ and $\sim +50°$C. Quenching tends to smooth out the curves somewhat, accentuating a sharp drop-off at $\sim -100°$C. The mechanisms underlying these phenomena will be discussed later.

4. In contrast to the behavior at low temperatures, the modulus in the temperature range $+40°$ to $+80°$ increases with increasing ion content, indicating that in that region the crystallites are no longer of primary importance and that ions contribute appreciably to the overall modulus.

5. In comparing the behavior of these materials with that of the styrene ionomers, one sees that the differences in properties between the ionic copolymers of ethylene and the parent polymer are not as drastic as those encountered in materials that are devoid of crystallinity, suggesting that the effect of crystallinity on mechanical properties is in some respects similar to the effect of ionic aggregates.

6. Substitution of a divalent cation for a monovalent one does not affect the properties of the polymer appreciably, at least in the modulus region above 10^9 dyn/cm^2. The total carboxylic content seems to be far more important.

(*b*) *Stress Relaxation Behavior*

In the study cited in Part (a), Ward and Tobolsky [49] also investigated the stress relaxation properties of the materials. In a preliminary study, they showed that both the thermal history and the water content affected the results appreciably. Specifically, they showed that, for the sodium salt mentioned above, in the vicinity of $E = 10^8$ dyn/cm^2, different thermal histories yielded curves which, while separated from each other by a factor of ~ 3, were practically parallel over the entire measured range (20 to 2×10^4 sec). Water soaking, on the other hand, yielded an initial (~ 100 sec) value of the modulus that was considerably lower than that for the dry samples; the rate of stress relaxation, however, was much slower than in the dry samples.

In the light of current understanding, it seems reasonable to suggest that there is a range of different aggregate structures in the material. Those that are highly susceptible to water (possibly the loose clusters) would probably relax over time periods much shorter than 100 sec, leading to a lower initial modulus. Those that are very resistant to water (tight multiplets) would relax very slowly, leading to only a minimal effect on stress relaxation over the time scale of the experiment.

Stress–strain experiments showed that up to 4% extension, the stress was proportional to $(\alpha - 1/\alpha^2)$, indicating that in that range the material is linearly viscoelastic; all the experimental results quoted subsequently were determined at 1% elongation.

Complete stress relaxation runs ($\sim -55°$ to $+85°$C) were performed on the parent acid copolymer, the sodium-containing sample discussed before, both quenched and annealed, and a quenched calcium salt E–0.065MAA–0.31Ca. The curves for the annealed sodium salt are shown in Fig. 4.39. It can be seen immediately, as the authors suggest, that the stress relaxation curves differ distinctly from those characteristic of a truly amorphous polymer, since the sharp glass-transition region is clearly absent, and the rate of stress relaxation is very low. It can also be seen that the results are very similar to those obtained for a completely noncrystalline ionic polymer,

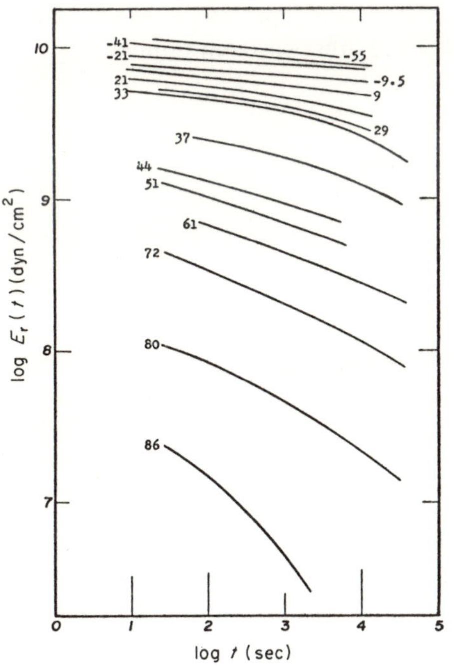

FIG. 4.39. Stress relaxation curves for the annealed ionomer E–0.08MAA–0.47Na [49].

such as plasticized poly(sodium acrylate), discussed in Chapter III. These facts tend to confirm that ion aggregation or clustering is important in these materials.

Attempts were made to superpose the individual stress relaxation curves to obtain a master curve. Reasonably good superposition could be obtained; however, it was observed that the shift factors were not of the WLF type. No master curves were plotted because it was evident that more than one relaxation mechanism was operative; thus, time–temperature superposition was assumed not to be applicable.

Several general trends were noted in the work; the most important of these are:

1. Incorporation of ions increases the modulus, but annealing is necessary to raise the modulus to the maximum attainable value.
2. Cations at the same concentration produce similar effects, independent of the valence of the ions. Thus, the ionomers E–0.065MAA–0.31Na and E–0.065MAA–0.31Ca gave identical stress relaxation curves over a wide range of temperatures. An example of this for 50°C is shown in Fig. 4.40, together with the results for the other quick-quenched samples at the same temperature. The similarity of curves *B* and *D* should also be noted. Curve *B* gives the results for a 3.76% Na sample, while curve *D* reproduces those for a 3.76% Ca sample. These curves do not superimpose exactly due to minor differences in stoichiometry, but are otherwise quite similar.

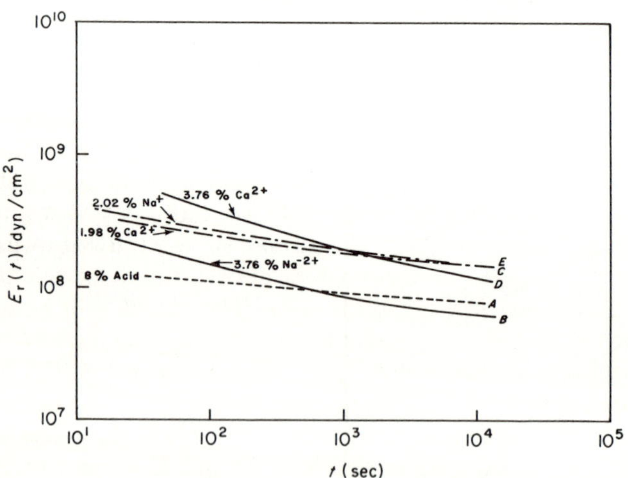

FIG. 4.40. Stress relaxation curves at 50°C for a quenched E–MAA copolymer (curve *A*) and its salts (*B* to *E*). The carboxylate ion content and counterion are indicated on each curve [49].

3. While partial neutralization of the acid tends to raise the initial (short-time) modulus, the rate of stress relaxation of the salt is faster in the rubbery and glassy regions, but not in the transition region. This factor was also noted for the ion–cross-linked elastomers, discussed in Section B of this chapter. In general, the higher the salt content, the higher the initial modulus and the faster the rate of stress relaxation.

(*c*) *Dynamic Mechanical Properties*

The first study of the dynamic mechanical properties was performed by Rees and Vaughan [52] on the ionomer system E–0.035AA. The writers showed that the main peak in the loss factor versus temperature plot is moved from ~25°C for the acid copolymer to ~60°C for the 70% ionic material. For polyethylene in the absence of ionic or acidic comonomers, the main peak also lies near 60°, indicating that, at least in this respect, the ionic polymer resembles pure polyethylene more than its parent acid; the mechanisms responsible for the high-temperature peaks in PE and the ionomer could hardly be the same, however.

It seems reasonable that in the unionized copolymer the peak is due to the normal α relaxation in PE, while in the ionomer it may be associated with the glass transition of the ionic region, as discussed in Chapter II. The crystallinity, as determined by X-ray diffraction in the annealed samples, was ~30%.

MacKnight *et al.* [50] studied the mechanical relaxations in an E–0.041AA copolymer and in sodium salts of that material ionized up to 78%. IR studies indicated that there were 5–10 short-chain branches per 1000 main-chain atoms. The crystallinity of the samples, as determined by DSC measurements, ranged from ~15% for the acid to 7% for the most highly ionized film. In the light of later studies [47], the decrease in crystallinity with increasing degree of ionization would seem to indicate incomplete annealing. The dynamic mechanical results for the annealed samples are shown in Figs. 4.41 and 4.42, which give, respectively, G'' and tan δ versus temperature for samples of degrees of neutralization 0, 20, 60, and 78%. The former curves tend to accentuate the low-temperature relaxations, while the latter emphasize the high-temperature transitions. The results will be discussed after the next two studies have been introduced.

Otocka and Kwei studied the dynamic mechanical properties of E–AA copolymers and their salts [51]. Their samples contained between 1.5 and 2.0 methyl groups per 100 backbone atoms. The mole percent acid was 0.66, 1.57, 2.78, and 3.40 for samples labeled B, C, D, and E, respectively, sample A being a low-density ethylene homopolymer. The results for the acid samples are shown in Fig. 4.43, while those for the sodium salts are given in

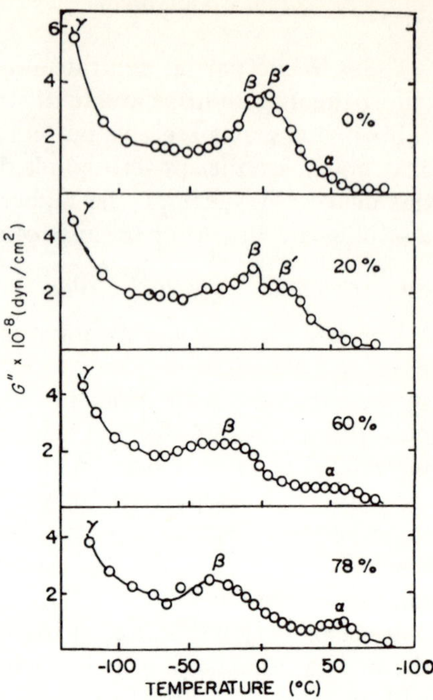

FIG. 4.41. G'' at ~1 Hz versus temperature for annealed E–0.041MAA at various degrees of neutralization [50].

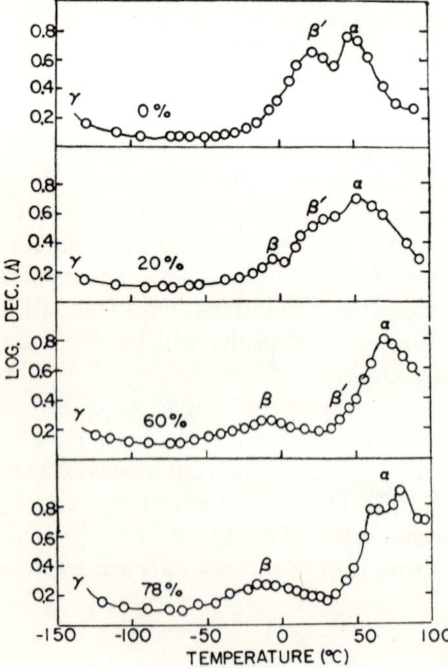

FIG. 4.42. Loss factor versus temperature for annealed E–0.041MAA at various degrees of neutralization [50].

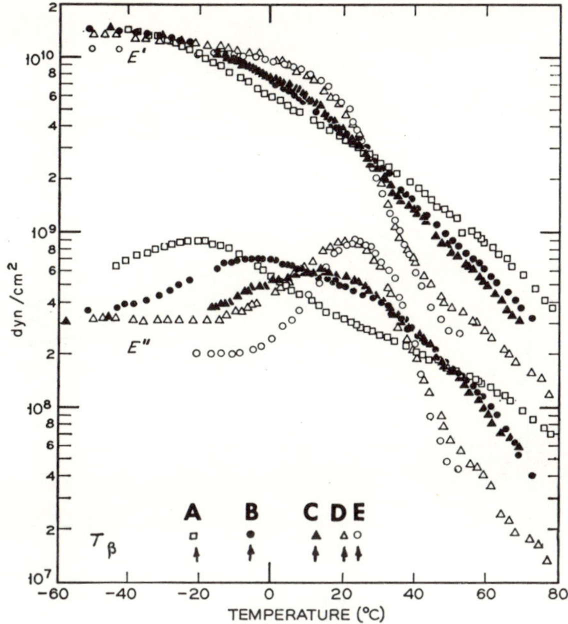

FIG. 4.43. Dynamic storage and loss moduli at 110 Hz for unneutralized E–AA copolymers. AA content for samples A to E, respectively: 0, 0.66, 1.57, 2.78, and 3.40 mole %. T_β for each material is indicated with the respective symbol [51a].

Fig. 4.44, in both cases as E' and E'' versus T at 110 Hz. The data for Mg salts are shown in Fig. 4.45. The crystallinity (as determined by DSC) for the acid samples ranged from 40% for sample B to 16% for sample E; while for the completely neutralized sodium salts, it ranged from 24 to 5%. Replacement of sodium by magnesium lowered the crystallinity appreciably.

Finally, Longworth and Vaughan [46] studied the dynamic mechanical properties of annealed E–MAA–Na ionomers containing 1.7 to 7.5 mole % NaMA. From G'' versus temperature plots, as shown in Fig. 4.46, they observed that the intensity of the peak at $\sim +20°C$ for the acid copolymers was drastically reduced upon ionization, and in its place, a peak appeared at lower temperatures, in the β relaxation region of low-density polyethylene ($\sim -20°C$). The intensity of this latter relaxation, as determined by the area under the peak, was found to increase linearly with increasing ion content.

Following the nomenclature of MacKnight *et al.* [50], four distinct relaxation regions, as indicated in Figs. 4.41 and 4.42, can be observed in these materials:

1. The α relaxation was observed by MacKnight *et al.* [50] in tan δ plots (Fig. 4.41); it lies at $\sim 50°C$ for the acid copolymer (4.1% MAA), and

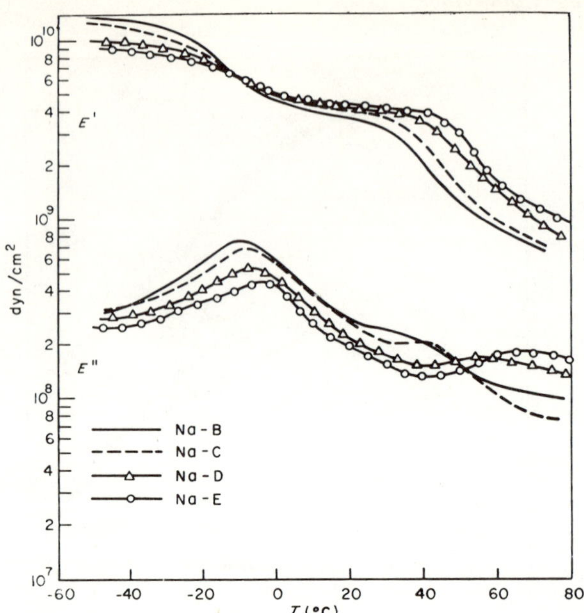

FIG. 4.44. Dynamic storage and loss moduli at 110 Hz for Na salts of the E–AA copolymers described in Fig. 4.43 [51b].

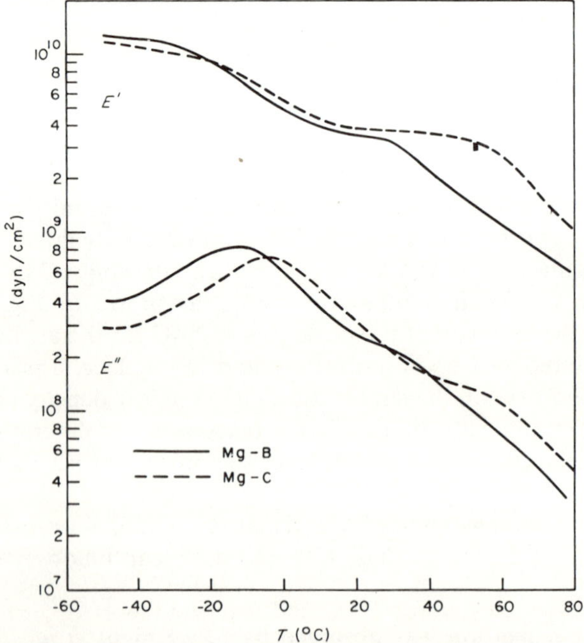

FIG. 4.45. Dynamic storage and loss moduli at 110 Hz for Mg salts of the E–AA copolymers described in Fig. 4.43 [51b].

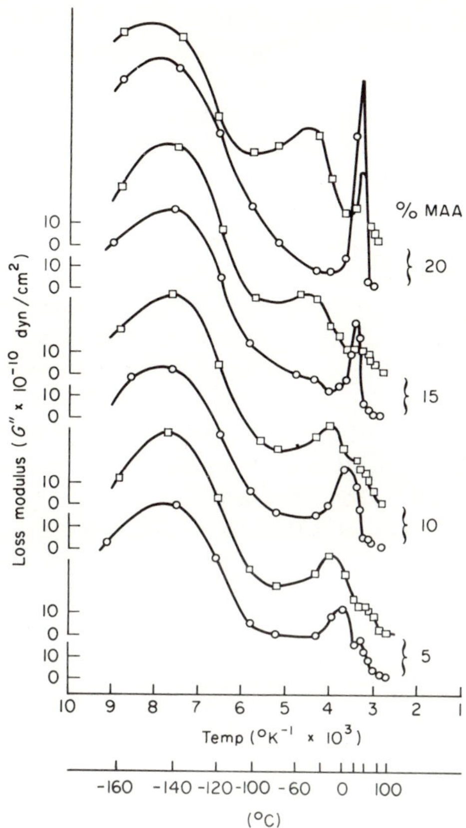

FIG. 4.46. G'' versus $1/T$ for E–MAA copolymers (○) and their Na salts (□). MAA content indicated as weight percent [52a].

moves to higher temperatures with increasing degrees of neutralization. Otocka and Kwei [51] and Longworth and Vaughan [46] also found that the α peak moves to higher temperatures with increasing ion content. This peak is attributed to motions of relatively large backbone segments, including the pendant carboxylate groups, and is thought to reflect the breakdown of the association between the salt groups. It can thus also be regarded as the glass transition of the ionic "phase." The most cogent reasons for this assignment are the facts that the peak increases in both magnitude and temperature with increasing degree of neutralization, and that divalent (Ca^{2+}) ions show a greater increase in both magnitude and temperature of the peak than do monovalent (Na^+ or Li^+) ions [53]. It may be recalled that a temperature of 50° was selected for T_c in the clustering theory described before; the reason for this selection was the assignment of the α peak as being associated with the breakdown of ionic aggregates.

2. The β' peak is also found in tan δ plots, but, for acid-containing materials, MacKnight *et al.* located it at $\sim +20$°C and found that it decreases in intensity with increasing ion content but stays at a constant temperature. In their study, Longworth and Vaughan found a similar effect of ionization for any given carboxyl content. There seems little doubt that this is also the peak observed by Otocka and Kwei for their acidic C, D, and E samples. Due to the approximate proportionality of the peak height and acid content, this peak is assigned to micro-Brownian motion of segments containing interchain hydrogen bonds. It can thus be regarded as the glass transition of the carboxyl-containing polyethylene. This assignment is supported by IR studies [54], which indicate that the carboxyl groups exist at low temperature almost entirely as hydrogen-bonded dimers, with free carboxyl groups beginning to appear at ~ 35°C. The dynamic nature of the hydrogen bond and the inherent limitation of the IR technique make it entirely reasonable that hydrogen-bonded segments may have a mobility which leads to a peak at ~ 20°C in the frequency region utilized here.

3. The β peak, in contrast with the α and β' peaks, is best observed in G'' or E'' plots. It occurs at ~ -10°C; its variation with ion content appears to depend on the total carboxyl content of the copolymer. Longworth and Vaughan observed a sharp increase in peak height and possibly even a decrease in the peak temperature with increasing ion content (1.7 to 7.5 mole % MAA), while Otocka and Kwei, in a lower carboxylate range (0.7–3.4%), observed a decrease in peak height and an increase in temperature. MacKnight *et al.* (4.1 mole % MAA) observed a slight increase in peak height but no change in peak temperature with increasing degree of ionization.

The β peak has been identified by MacKnight *et al.* with the glass transition of the noncrystalline, branched polyethylene fraction from which most of the ionic components have been excluded. This interpretation assumes that most of the ions, at least at higher carboxyl contents, are incorporated in clusters. This assignment is supported by the fact that the peak position does not change much with ion content, but that its intensity increases. As Longworth [45] points out, this indicates that the molecular structure of the material giving rise to the relaxation remains constant and increases only in extent. The increase in T_β at low carboxylate contents was attributed by Otocka and Kwei to the effect of copolymerization and, in their view, indicated a uniform dispersal of ions in the amorphous phase. The data, however, are also consistent with the incorporation of most of the ions in aggregates, the number of ions remaining in the amorphous polyethylene phase being small.

In summary, at very low ion contents [51], the rise in T_β suggests that clustering is incomplete, with a considerable fraction of ions outside the clusters, although the ratio of the number of ions outside the clusters to the number inside may not be changing much or may even be decreasing with increasing ion concentration in the range 0 to 3%. At slightly higher concentrations [50], a situation may be reached where the total number of dispersed ions in the bulk remains relatively constant, suggesting an increased clustering efficiency, since the total ion concentration is increasing. (In this study [50] partially neutralized materials were utilized, but unneutralized carboxylates are also presumably incorporated in the clusters [15], so that the total concentration of species which can be incorporated in clusters may be higher than the absolute ion concentration.) At still higher concentrations [46], due to the proximity of the ions to each other, the clustering efficiency may become so great that the total number of dispersed ions starts decreasing again, resulting in a reduction in T_β.

4. Finally the γ peak, which occurs below $-100°C$ and can be identified best in G'' or E'' plots, is given the same identification as the corresponding peak in ethylene homopolymers [55].

Variations in the detailed results by different groups may be due in part to the differences in sample composition, i.e., E–AA versus E–MAA, partial neutralization versus complete neutralization, etc.

Several brief studies have been performed on the variation of dynamic mechanical properties with the nature of the counterion [51, 53]. Only slight changes were observed, even if divalent ions were substituted for monovalent ones. The work was confined to the α and β regions.

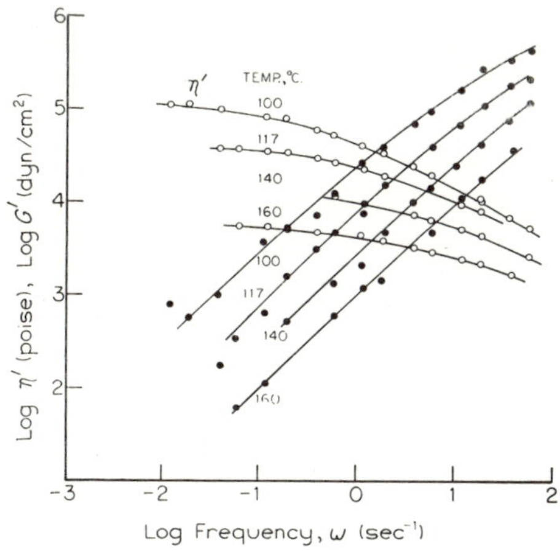

FIG. 4.47. Dynamic viscosity and modulus of E–0.041MAA as a function of angular frequency [56].

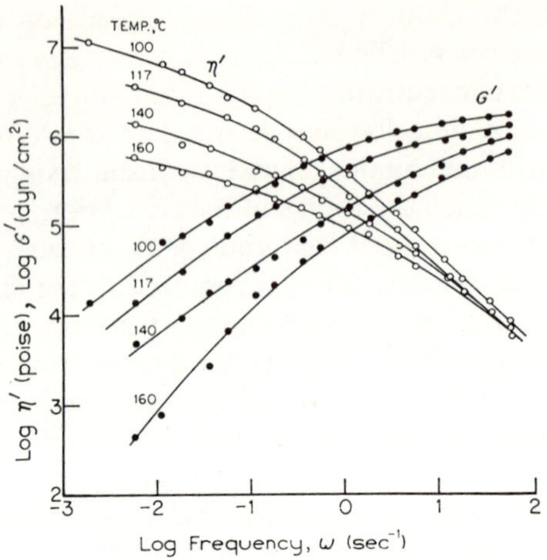

FIG. 4.48. Dynamic viscosity and modulus of E–0.041MAA–0.56Ca as a function of angular frequency [56].

A study of the dynamic melt rheology of an E–MAA copolymer was performed by Sakamoto *et al.* [56]. An angular frequency range of 10^{-2} to 10^{+2} was covered using a Weissenberg rheogonionmeter. The results for the acid form are shown in Fig. 4.47, while those for the 56% neutralized calcium salt can be seen in Fig. 4.48. A 59% neutralized sodium salt was also studied, the results being quite similar to those for the calcium salt.

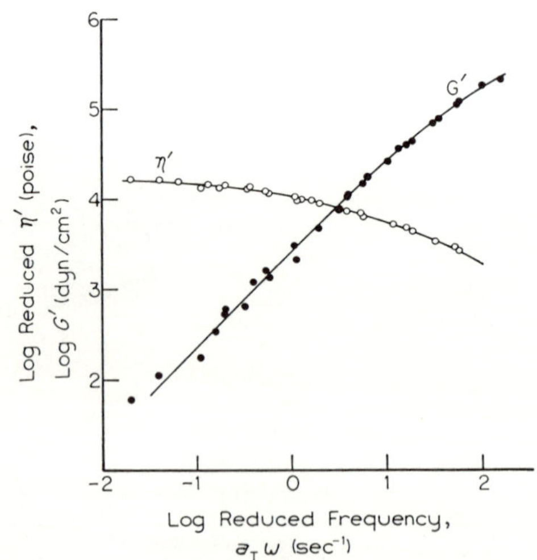

FIG. 4.49. Master curves of dynamic viscosity and modulus for E–0.041MAA. Reference temperature 140°C [56].

It was observed that time–temperature superposition is obeyed for the acid copolymer, the results for 140°C being shown in Fig. 4.49. The distributions of relaxation times thus can be expected to be independent of temperature except for a horizontal shift, and indeed this is the case. However, no superposition could be achieved for the salts, and, as might be expected, the distribution of relaxation times changes appreciably with temperature. The results for the calcium salt are shown in Fig. 4.50, where an indication of a rubbery plateau can be seen.

The authors conclude that these findings support a two-phase structure for the polymer (the crystals having disappeared at a lower temperature), and they note that the ionic domains become more mobile with increasing temperature; however, some ionic domains do persist to very high temperatures. Thus the ionic copolymers have effectively different structures at different temperatures, and time–temperature superposition is not to be expected. This has been observed in a number of the other ionic systems discussed in this book.

In terms of the multiplet–cluster theory described in Chapter II, the results can be interpreted in the following way: Most of the clusters probably break up at relatively low temperatures, between ~50 and 100°C depending on the composition of the material. Both the clusters and the multiplets affect the material properties below that temperature, giving the material the characteristics of a filled system. Above ~100°C, the existence of only a few clusters,

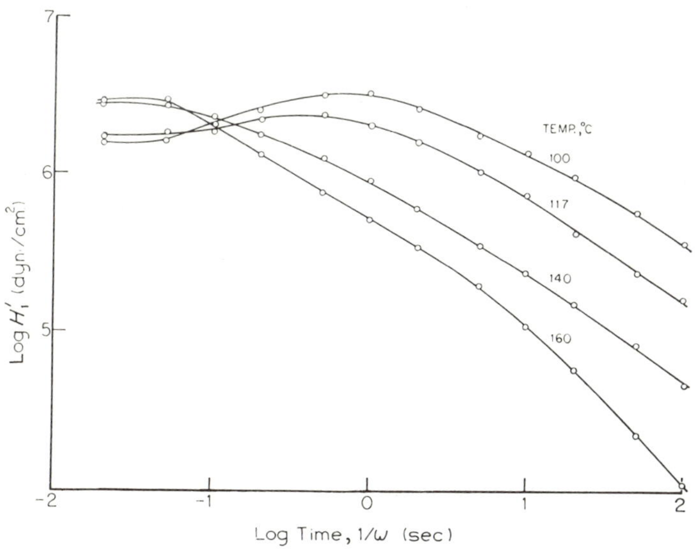

FIG. 4.50. Relaxation spectra of E–0.041AA–0.56Ca at various temperatures [56].

with the rest present as multiplets alone, would suffice to explain the phenomena, since presumably the multiplets (and clusters) do act as cross-links. Furthermore, the lifetime of the multiplets is long by comparison to the relaxation time of the system, but it is not infinite, and it decreases with increasing temperature. Thus, flow is retarded by the residence of any ion pair sequentially in a number of multiplets (or clusters), with the residence time decreasing with temperature. All this is superposed on the normal distribution of relaxation times for the polymer sample, which makes time–temperature superposition, at least over most accessible ranges, unfeasible.

(*d*) *Dielectric Properties*

The dielectric properties of ethylene-(methacrylic acid) copolymers and some of their salts were studied by Read *et al.* [57] and by Phillips and MacKnight [58]. The frequencies covered the range of 50 Hz–10 kHz, although a few results outside that range were also obtained. The first study concentrated on the acid copolymer (4.1% MAA) and one sodium salt which was 53% ionized, both dry and in the presence of varying amounts of water; the second explored the effect of varying the percent ionization and the cation.

Both the β' and the α dispersions (using the nomenclature of the preceding section) can be seen dielectrically in the acid copolymer. This fact suggests that both hydrocarbon segments and polar groups are involved in the processes. With regard to the β' process, it is characterized by a broad loss peak, a slight curvature in the plot of log ν_{max} versus $1/T$, and a high activation energy (50–60 kcal), all indicative of the type of motions associated with the glass transition. The dielectric results thus confirm the mechanistic assignments of the mechanical study.

In an attempt to understand the nature of the polar groups involved in the process, it should be recalled that most of the acid groups exist as hydrogen-bonded dimers in the temperature range studied here, and these would not be expected to participate in the dispersion. The equilibrium constant for the dimerization is known, so the total number N_t of single carboxyl groups that would participate can be calculated. This, together with the group moment of COOH (μ = 1.7D) allows a calculation of the magnitude of the dispersion of $N_t\mu^2$, which can be compared with experimental results. This comparison shows that groups other than monomeric acid segments must participate in the process; the authors suggest 0.2–0.4 mole % carbonyl groups would be sufficient to explain the discrepancy, as would also longer H-bonded structures such as acid trimers and tetramers. Both of these may well exist in the polymer. It would also seem possible to involve H-bonding between neighboring groups on the same chain yielding only one H-bond per pair, but resulting in a net dipole moment for the dimeric group. The γ

process, which is also encountered in homopolymeric ethylene, is of no interest in this context and will not be dealt with further.

Figure 4.51 shows the results for the dry E–0.041MAA–0.53Na sample at three frequencies over a wide temperature range. Three dispersion regions are visible, and in accordance with the nomenclature used for the description of the mechanical results, they are labeled α, β, and γ. The introduction of water affects the results profoundly, as is seen in Fig. 4.52. The α dispersion is moved to lower temperatures and a new peak appears at $\sim -40°C$. It is worthwhile to contrast the behavior of this material with that of the Nafions. In the latter case, water did not yield an independent peak, but rather shifted the position of the β peak to lower temperatures at a drastic rate. A plot summarizing all the results in the form of log ν_m versus $1/T$ is shown in Fig. 4.53; this plot also contains some mechanical results, which were discussed before, as well as some NMR results, which will be discussed later.

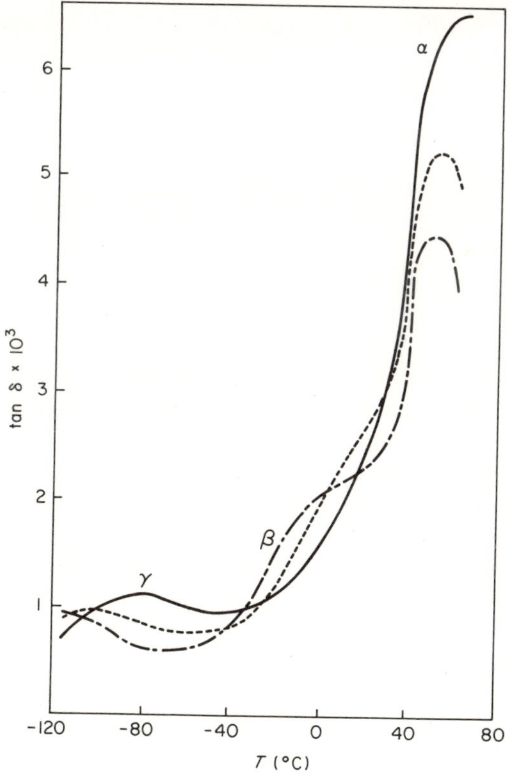

FIG. 4.51. Dielectric loss tangent of dry E–0.041MAA–0.53Na at 10 kHz (——), 1 kHz (····), and 100 Hz (– – –) [57].

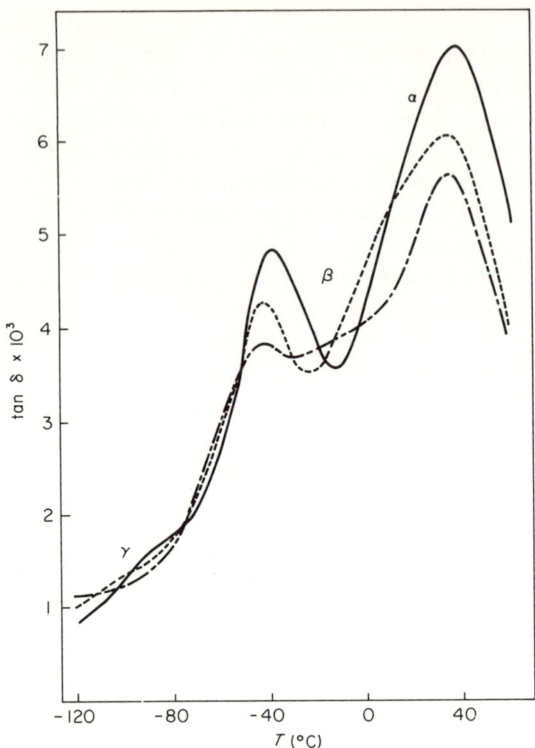

FIG. 4.52. Dielectric loss tangent of E–0.041MAA–0.53Na containing 1.1 wt % H_2O at 10 kHz (——), 1 kHz (····), and 100 Hz (– – –) [57].

It is evident that the dielectric α and β peaks correlate with the mechanical results and are therefore assigned to the same molecular mechanism. The β relaxation is considered to be associated with the glass transition in the amorphous polyethylene phase, an assignment supplrted by the curvature of the log ν versus $1/T$ plot. It is rendered dielectrically active by the presence of carbonyl groups on the chain. The α mechanism is assigned to the glass transition in the ionic regions. This assignment is supported by high magnitude of the dielectric α peak, which suggests that ionic groups may indeed be involved, by the curvature of the log ν versus $1/T$ plot and by the shift of the α peak to lower temperatures with increasing water content.

In regard to the water peak, the authors derive, from a calculation of $N\mu^2$, a value for μ of 1.5D, presumably by taking N to be the total number of water molecules. The dipole moment of water is 1.8D, indicating that the water molecule is, indeed, participating. Also from the much higher affinity

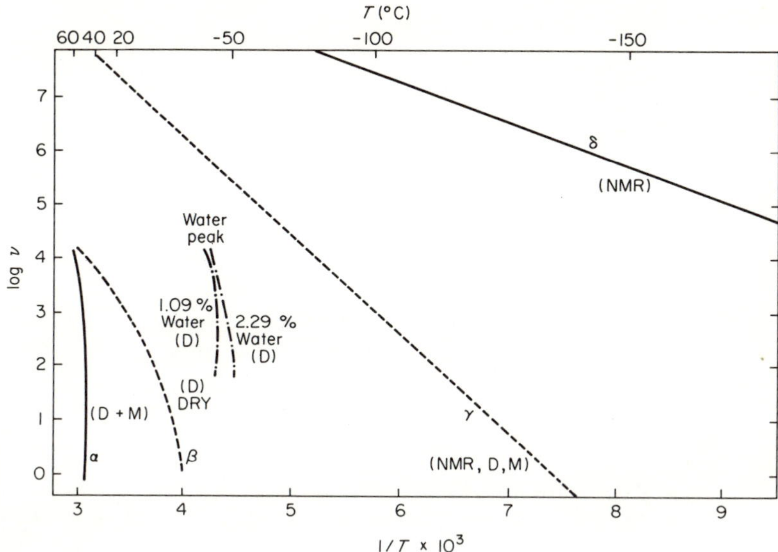

FIG. 4.53. Summary of loss tangent data for E–0.041MAA–0.53Na. Data obtained from mechanical (M), dielectric (D), and NMR studies for dry and wet samples [57].

for water of the sodium salt as compared with the acid copolymer, there seems little doubt that the water is preferentially located in the ionic phase.

The effects of the degree of ionization and of the nature of the counterion were investigated by Phillips and MacKnight [58]. Figure 4.54 shows the results for the calcium salt and includes the acid copolymer for comparative purposes. A low degrees of ionization there seems to be only a very slight counterion effect, since the Ca salt strongly resembles the corresponding Li or Na salt. As the degree of neutralization increases, the α peak height increases, and the peak also shifts to higher temperatures. In the two curves for 31 and 43% Ca, no α peak can be detected at all since it has moved beyond the range of measurement.

The activation energies for the various peaks were also obtained in that study, and they are given in Table I. It should be pointed out here that on a tan δ versus temperature plot, the β peak is not at all well resolved; however, on a plot of tan δ versus log ν, the resolution is excellent, and it is the latter procedure that allows a precise determination of the activation energies for the β and β' mechanisms.

While, in general, the dielectric results closely parallel the mechanical ones, the one serious discrepancy lies in the value of activation energy for the β process. Mechanically, it is of the order of 35 kcal, while dielectrically

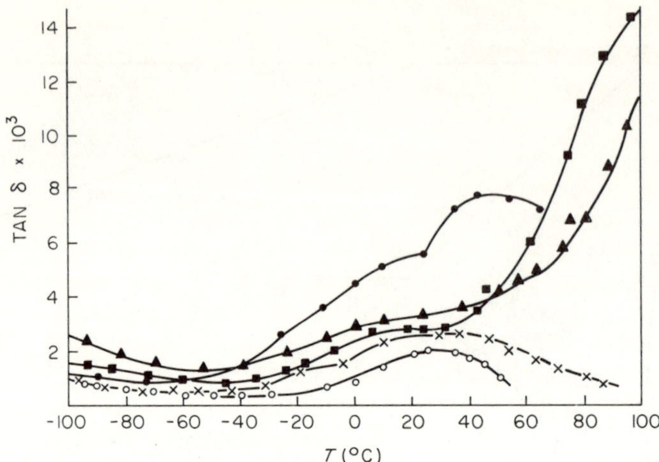

FIG. 4.54. Temperature dependence of dielectric loss factor at 1 kHz for various calcium salts of E–0.041MAA. Degrees of neutralization: (○) 0%; (×) 36%; (●) 46%; (■) 62%; (▲) 86% [58].

TABLE I
Activation Energies for E–MAA Dispersion Mechanisms[a]

Sample	Activation energies (kcal/mole)			
	α	β'	β	γ
E–0.041MAA				
100% Acid	—	56	—	8.9
0.41Na	120	~70	—	6.9
0.53La	99	—	16.4	8.3
0.19Li	85	—	~19	12.3
0.42Li	~80	—	12.4	12.8
0.72Li	—	—	8.6	11.4
0.36Ca	103	64	—	9.4
0.46Ca	88.5	—	—	12.4
0.62Ca	(~60)?	—	7.9	11.2
0.84Ca	—	—	12.6	12.2
E–0.083MAA				
0.95Na	—	—	13.7	16.0

[a] Phillips and MacKnight [58].

it has the value of 10–15 kcal. While the β process is due, as was pointed out, to the glass transition in homopolymeric ethylene from which ionic or carboxylic groups have been excluded, some undoubtedly must have remained to render the process dielectrically active.* The dielectric relaxation

* Carbonyl oxygens could also render the material dielectrically active.

would reflect the reorientation of only these groups, resulting from micro-Brownian motion of chain segments, while the mechanical relaxation would reflect all the processes occurring in the sample at the measuring frequency.

(e) NMR Studies

Read *et al.* [57] also studied the NMR relaxation of partially neutralized ionomers of E–0.041MAA. Using pulse methods, they obtained the spin-lattice relaxation time T_1 at 30 MHz and the correlation time $T_{1\rho}$ at 20–60 kHz. At low temperatures, these studies contribute to an understanding of the γ and δ processes in polyethylene, which are unrelated to the ionic character of the polymer. At higher temperatures for the acid polymer, a pronounced minimum can be seen for the $T_{1\rho}$ process at $\sim +40°C$, which correlates with the β' relaxation. For the salt, the α and β processes merge even in the 25-kHz region, yielding only a single $T_{1\rho}$ minimum for both processes. The NMR results for the salt are shown, together with the dielectric data, in Fig. 4.53.

In another study, Otocka and Davis [59] compared the linewidth in broad-line NMR spectra for ^7Li and ^1H for the ionomer E–0.049AA–Li. The results were shown in Fig. 2.13 of Chapter II, along with those for the acid copolymer. The authors take these results to indicate that the salt groups are physically well dispersed in the amorphous phase of the copolymer. As was pointed out in Chapter II, alternate interpretations cannot be excluded.

(f) Optical and X-Ray Studies

A number of additional studies on the ethylene-based ionomers have been performed using a variety of electromagnetic spectroscopic techniques. These studies have involved infrared spectroscopy and dichroism [51a,b, 54, 60], optical birefringence [61], X-ray diffraction [62], light scattering [63, 64], and far-infrared spectroscopy [65].

In initial studies, Rees and Vaughan [52] found that the carboxylic acid peak at 1700 cm^{-1} in the acid copolymer (E–0.017MAA) is replaced by carboxylate ion absorption at 1575 and 1400 cm^{-1} in the ionized species (E–0.071MAA–0.90Na). They also found that bands characteristic of the α-olefin crystalline regions, and which are present in the acid copolymer, are suppressed in the unannealed ionomer, but develop upon annealing.

As a result of subsequent analyses [51a,b, 54, 60], assignments, summarized in Table II, were made of the most important bands appearing in the acid copolymer and in the ionomer.

MacKnight *et al.* [54] studied the temperature dependence of the relative intensities of the 3540 and 1700 cm^{-1} bands in the acid copolymer E–0.041-MAA, from which the dissociation constant K for the carboxyl monomer–dimer association was obtained. They observed that at room temperature carboxyl dimerization is virtually complete. Free carboxyl groups became

TABLE II

IR Assignments for Ethylene Ionomers[a]

ν (cm^{-1})	Observed in	Polarization (w.r.t. chain axis)	Assignment
3540	Acid		Free OH stretching
2650	Acid		H-bonded OH stretching
1750 (shoulder)	Acid		Free C=O stretching
1700, 1705	Acid and salt	$\perp$	H-bonded C=O stretching
1560, 1575	Salt	$\perp$	Asymmetric vibration of COO^-
1470	Acid and salt	$\perp$	CH_2 bending
1400	Salt		Carboxylic deformation (?)
1263	Acid and salt	$\perp$	Carboxyl group deformation (?)
935, 940	Acid	$\parallel$	H-bonded OH bending
720, 730	Acid and salt	$\perp$	CH_2 rocking (doublet)

[a] Otocka and Kwei [51], MacKnight *et al.* [54], and Uemura *et al.* [60].

detectable only above 30°C, the concentration increasing with temperature. A plot of log K versus $1/T$ yielded a value of 11.6 kcal/mol for the heat of dissociation of the H-bonded dimer, which is well within the range observed for low-molecular-weight carboxylic acids [66].

Otocka and Kwei [51a,b] investigated the ethylene–acrylic acid copolymers and found a ΔH value of 11.5 kcal/mole for the association constant for the carboxylic acid. They also investigated the dichroic ratio as a function of elongation for the acids.

Uemura *et al.* [60] followed the variation in the dichroic ratio and the orientation function with sample elongation for various infrared wavelengths corresponding to hydrocarbon and carboxyl constituents in the copolymer E–0.041MAA and its 55% neutralized sodium salt. The orientation function f is related to the dichroic ratio D by

$$f = C[(D - 1)/(D + 2)] \tag{3}$$

where $D = A_{\parallel}/A_{\perp}$ is the ratio of the absorbances for radiation polarized parallel and perpendicular to the direction of elongation. The constant C is given by

$$C = (D_0 + 2)/(D_0 - 1) \tag{4}$$

with $D_0 = 2\cot^2\phi$, ϕ being the angle between the transition moment of the absorbing group and the segment axis.

The dichroic ratios were observed to decrease monotonously with increasing elongation at constant temperature, the only exception being the band at 935 cm^{-1}, associated with the H-bonded OH band of the unionized carboxyl group, for which $\phi = 0°$; for all the other bands studied, $\phi = 90°$.

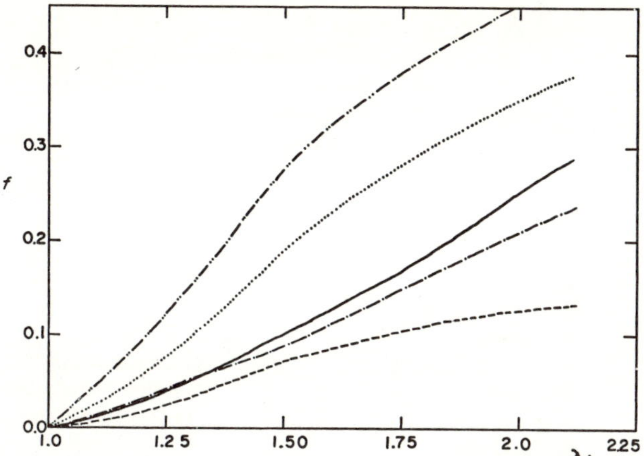

FIG. 4.55. Orientation function versus elongation at 40°C for various absorption bands of E–0.041MAA–0.55Na: 1700 cm^{-1} (——), 1560 cm^{-1} (---), 1470 cm^{-1} (····), 1263 cm^{-1} (—·—·), 720 cm^{-1} (—··—) [60].

As shown in Fig. 4.55, the value of f at each absorbance maximum shows a continuous increase with sample extension at constant temperature.

The variation in f^{50}, the orientation function at 50% elongation, for the acid, showed a sigmoidal decrease for all bands studied, with the exception of 935 cm^{-1}. This is illustrated in Fig. 4.56, the results for the 1263 cm^{-1} band having been omitted for clarity. By contrast, the values of f^{50} for the salt were all observed to pass through maxima in the region of 40°C. The results for two bands (the highest and lowest values of f^{50}) are shown in Fig. 4.57.

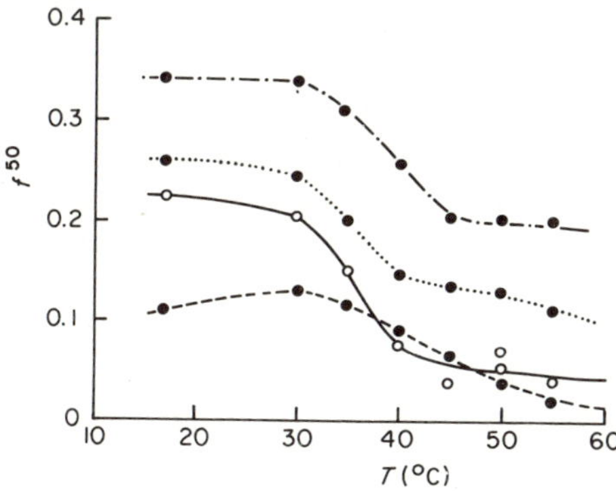

FIG. 4.56. Orientation function at 50% elongation versus temperature for E–0.041MAA at various absorption bands: 1700 cm^{-1} (——), 1470 cm^{-1} (····), 935 cm^{-1} (---), 720 cm^{-1} (—··—) [60].

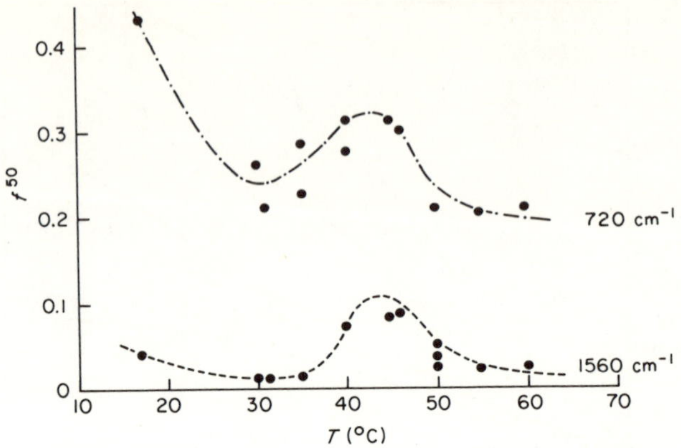

FIG. 4.57. Orientation function at 50% elongation versus temperature for E–9.041MAA–0.55Na at 1560 cm^{-1} (– – –) and 720 cm^{-1} (— · —) [60].

The maximum in orientability for the salt correlates with the α peak obtained in mechanical and dielectric studies of the same ethylene ionomer and can be considered further evidence for the onset of mobility in the ionic domains. It is very significant that both bands in Fig. 4.57 show peaks, i.e., increased orientability, between 40° and 50°C, since the 1560 cm^{-1} band is due to carboxylate ion, while the 720 cm^{-1} band is associated with the hydrocarbon phase. This suggests that the orientational changes in the two regions are coupled; i.e., the softening of the ion-containing domains is accompanied by an increase in the orientability of the hydrocarbon regions and implies a parallel connection between the two phases. Finally, as the authors point out [60], the very low values of orientability of the ionic regions below the softening temperature (Fig. 4.57) suggest that these are more nearly spherical than rodlike or sheetlike.

Kajiyama, Stein, and MacKnight [61] performed bifringence studies of the relaxation of E–MAA copolymers (4.1 and 8.2%) and their salts. The investigations included both the static and dynamic strain–optical coefficients as a function of composition and temperature. It was observed that for the salts the birefringence is linearly related to the elongation ratio, and that over the temperature range of ~5–90°C, the slope (i.e., the static strain–optical coefficient K_S) first decreases with increasing temperature, then increases to a maximum, and finally decreases again; the birefringence is only very slightly time dependent. The existence of the peak is characteristic only of the salts, the copolymer acid exhibiting a continuous, although nonuniform decrease, as shown in Fig. 4.58. A peak is also observed in the real part of the dynamic strain–optical coefficient K' at 1.02 Hz, shown in

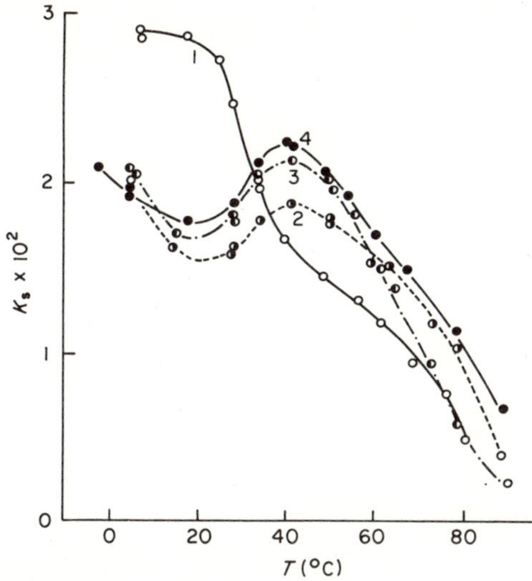

FIG. 4.58. Static strain–optical coefficient versus temperature for E–0.041MAA and three salts: (1) unionized, (2) 0.78 Na, (3) 0.72 Li, (4) 0.80 Ca [61].

Fig. 4.59, where it is compared with the static strain–optical coefficient and the mechanical loss tangent for the same materials.

The K' maximum is observed to lie at a somewhat higher temperature than that of K_S, probably because of the different time scale of the measurement. This difference also explains the difference in magnitude between K' and K_S; while K' contains only the reversible part of the birefringence change accompanying vibration, K_S contains both reversible and irreversible changes in birefringence.

The maxima for the salt are superficially similar to those found for pure polyethylene in the vicinity of the α loss peak. In the latter material, the maximum was ascribed to a competition between two processes involving the crystalline regions, one of which permits increasing orientation with increasing temperature while the other leads to a loss of orientation [55]. In the salts the mechanism for the maximum must also result from two factors, but they cannot be associated with the crystalline regions since the α loss process is directly attributable to the ionic component. Presumably, the increase in orientability is associated with a softening of the ionized regions, i.e., the glass transition of that component. The subsequent decrease of both K' and K_S at higher temperatures is due to a decrease in the effective degree of cross-linking (due to the ionic groups) with increasing temperature, i.e., a type of thermal dissociation of the cross-links. According to the theory

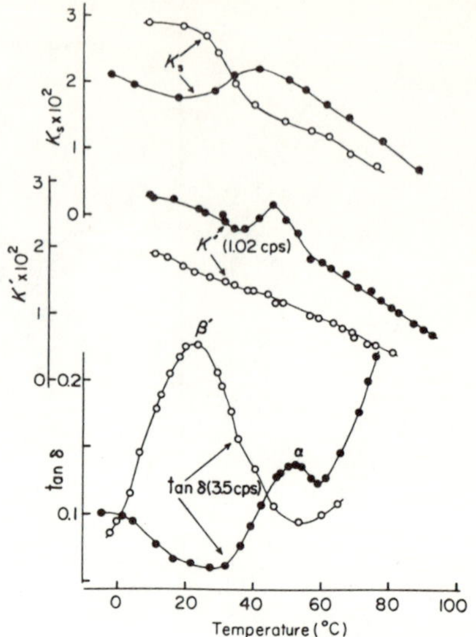

FIG. 4.59. Temperature dependence of static strain–optical coefficient K_s, dynamic strain–optical coefficient K', and mechanical loss tangent for E–0.041MAA and E–0.041-MAA–0.80Ca. ○ = unionized, ● = 0.80 Ca [61].

of Kuhn and Grün [67] the strain-optical coefficient of a cross-linked rubber is proportional to the degree of cross-linking, so that as the degree of cross-linking decreases (thermally), both K' and K_S decrease. This effect is predominant only above the glass-transition temperature.

Studies of the effect of the degree of ionization revealed that, as in the case of tan δ peaks, the K' peak position moves to higher temperature with increasing degree of ionization. Furthermore, the intensity of the K' peak increases also. Both of these effects can be seen in Fig. 4.60 and further confirm the assignment of the peak to the ionic component.

Two other noteworthy effects were observed by Kajiyama *et al.* It was found that a type of degree of ionization–temperature superposition exists for these materials. This is manifested by the fact that the plots for the static strain–optical coefficient versus the degree of ionization for the Na, Li, and Ca salts appear almost identical at the same temperatures. Also, as with polyethylene, the strain–optical coefficient follows temperature–frequency superposition. Both a horizontal and a vertical shift are required, indicating that both the nature of the process (which would require a vertical shift) and its time scale (which requires a horizontal shift) change with temperature. An increase in temperature decreases the height of the curve and shifts it toward higher frequencies, which means that for short-time-scale processes, a higher temperature is required to produce the same structural mobility that occurs for longer-time-scale processes at lower temperatures.

In an attempt to explain the vertical shift, two possible suggestions were offered. One suggests that the intrinsic birefringence values of the various

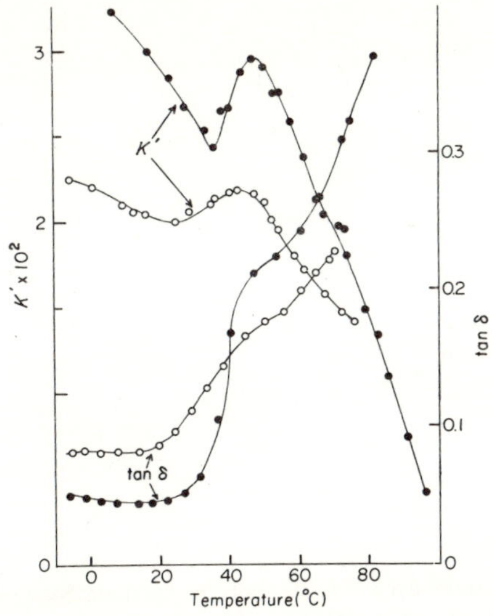

FIG. 4.60. Temperature dependence of K' and tan δ for E–0.041MAA–0.78Na (○) and E–0.082MAA–0.95Na (●) [61].

contributing regions depend upon temperature; the other is based on the possibility that the orientation mechanism is temperature dependent. The horizontal shift factors themselves are not of the WLF type, but rather are Arrhenius in nature with different low- and high-temperature activation energies, the change occurring at ~55°C for E–0.041AA–0.80Ca. The vertical shift factors show similar behavior. By contrast, the shift factors for the acid show a simple Arrhenius behavior over the entire temperature range. The existence of two sets of shift factors is consistent with the conclusion that competitive processes are responsible for the α peak in the copolymer salts.

In a subsequent publication, Kajiyama *et al.* [62] used X-ray diffraction to study not only the degree of crystallinity but also the crystal orientation of ethylene copolymers containing 4.1 mole % methacrylic acid in the acid state and ionized 78% with sodium. Orientation functions for the three crystal axes were determined as a function of elongation and temperature and were used to calculate the crystalline component of birefringence. In Figs. 4.61 and 4.62, the crystalline and amorphous contributions to the total strain–optical coefficient are shown for both the acid and the salt.

It is seen that the crystalline contribution decreases with temperature over the entire region; by contrast, the amorphous contribution in the salt goes through a maximum in the region of the α relaxation. Below this relaxation region, the amorphous material is considered to be interlaced with glasslike ionic domains. This gives the material a low compliance, so that the primary effect of an external strain is to orient crystals, leading to a high

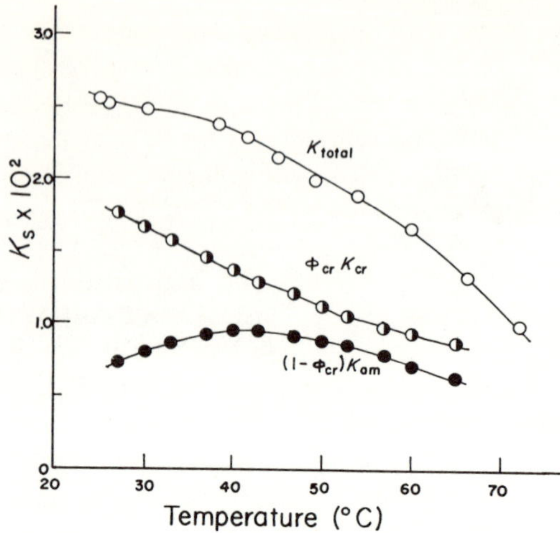

FIG. 4.61. Temperature dependence of total strain–optical coefficient, the crystalline contribution, and the amorphous contribution for annealed E–0.041MAA [62].

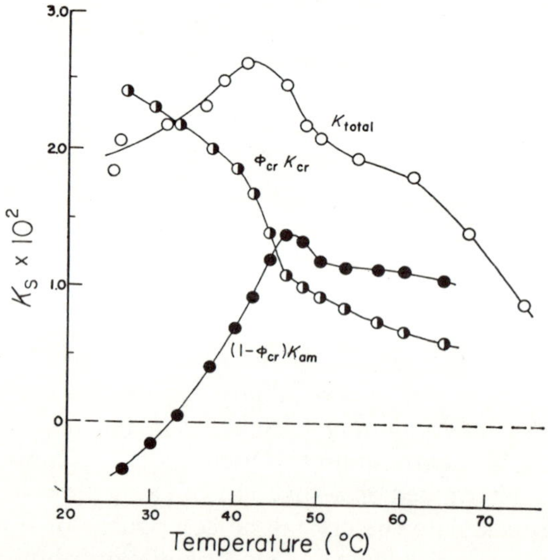

FIG. 4.62. Temperature dependence of total strain–optical coefficient, the crystalline contribution, and the amorphous contribution for annealed E–0.041MAA–0.78Na [62].

contribution of the crystals to the strain–optical coefficient. In the relaxation region itself, the ionic domains soften, permitting a higher orientation of the amorphous regions. This, in turn, leads to an increased amorphous contribution, but also results in a decrease in the force transmitted to the crystals, which explains the decrease in the crystalline contribution. The increase in birefringence with increasing temperature for the amorphous regions is greater than the decrease resulting from a decrease in crystal orientation, which leads to a net increase in the total value of K_S. At still higher temperatures, after the ionized regions have softened considerably, further heating leads to viscous flow with a concomitant decrease in K_S. The above combination of effects leads to the appearance of a maximum in K_S total.

Prud'homme and Stein [63] examined the light-scattering behavior of the copolymer E–0.041MAA and its 55% ionized sodium salt. The variation in scattered light intensity was followed as a function of the scattering angle θ, between the incident and scattered rays, and the azimuthal angle μ, made by the scattered ray in a plane perpendicular to the incident ray with respect to the stretching direction of the sample. Measurements were normally done with an H_v polarization mode (vertically polarized incident light and horizontally polarized analyzer), and the stretching direction was vertical. For pure polyethylene, which has a spherulitic structure, a four-leaf-clover pattern is obtained; i.e., the scattered intensity is at a maximum at $\mu = \pm 5°$ and at a scattering angle θ_m, which is related to the spherulite radius R by

$$4\pi(R/\lambda)\sin(\theta_m/2) = 4.1 \tag{5}$$

where λ is the light wavelength within the sample.

The unionized copolymer E–0.041MAA, which had been molded at 125° and then air-cooled, was found to exhibit the four-leaf-clover pattern typical of a spherulitic structure, with the value of θ_m corresponding to a spherulitic radius of 2.1 μm. The salt E–0.041MAA–0.55Na, with the same thermal history, showed only a continuous decrease in scattered intensity with increasing θ at $\mu = 45°$. The pattern in the latter case was more like the pattern observed for a rodlike lamellar aggregate structure, suggesting disorder in the orientation of crystals within the spherulites.

For samples prepared by quenching in dry ice/methanol, the spherulitic pattern was suppressed in both the acid and the ionomer. On the other hand, by annealing for 18 hr at 91°C, a four-leaf-clover pattern was established even for the ionomer, indicating that spherulites do develop under very mild conditions of sample preparation. The authors point out that the intensity of scattering from the rapidly cooled ionomers is comparable with that of the acid, which suggests that crystalline aggregates of comparable size are present, i.e., that lamellar morphology persists but that the lamellae are disordered with respect to each other. They conclude that it is thus not surprising that the α crystalline mechanical loss peak, associated at least in part with interlamellar motion, is not seen in the ionomers.

When the H_v light-scattering patterns of deformed samples were measured, it was found that the ionomer pattern changed in the same manner as that observed for polyethylene. No appreciable differences were seen on comparing the H_v scans upon deformation at 25 and 45°C.

In a subsequent study, Prud'homme and Stein [64] examined the dynamic light-scattering (DLS) behavior of the same ionomer used in the earlier work. They measured the variation in light-scattering intensity in phase and out of phase with an applied dynamic strain of 1% at 2.9 Hz superimposed on a static strain of 5%. For the acid copolymer the dynamic components $\Delta I'$ and $\Delta I''$ both increased continuously over the range 25 to 70°C, while, for the salt, the pattern was quite different, as seen in Fig. 4.63.

These results were explained in terms of the Van Aartsen–Stein theory for spherulitic deformation [68], which considers the contribution of deformation processes such as chain tilting with respect to the lamellae and lamellar twisting to the light-scattering pattern. The tilting process is associated with mobility within the crystalline regions and leads to negative values of the DLS components, while the twisting process, which would result from motion in the interlamellar regions, leads to positive values of $\Delta I'$ and $\Delta I''$.

It was concluded that the negative values of the DLS components of the acid sample at low temperatures indicate the predominance of the twisting process over the tilting process. As the temperature increases, the values of $\Delta I'$ and $\Delta I''$ become more positive, and the tilting process occurs more easily, as one might expect. In the salt, the values of $\Delta I'$ and $\Delta I''$ are positive at room temperature, indicating that the tilting deformation process is more important at low temperatures. This behavior is consistent with the existence

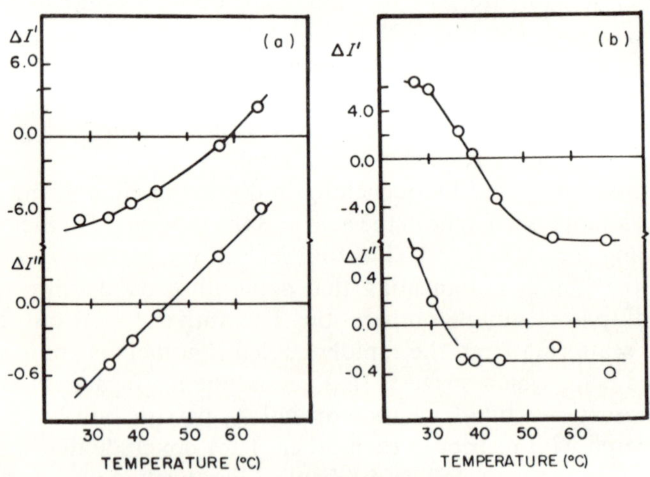

FIG. 4.63. Temperature dependence of the DLS components at $\theta = \theta_m$, $\mu = 45°$ for (a) E–0.041MAA and (b) E–0.041MAA–0.55Na [64].

of an ion-containing amorphous phase of very low compliance. When the sample is stretched, at room temperature, most of the strain is concentrated in the crystalline regions. As the α relaxation temperature is approached, the softening of the ionic domains permits a more uniform distribution of strain between the two phases of the polymer, and the DLS components become more negative.

Finally, cation motion within ionic domains has been observed in a study by Tsatsas *et al.* [65] involving far-infrared spectra of ethylene ionomers. in each of the ionomers studied, a broad, well-defined band in the region below 600 cm^{-1} was observed. This feature, which was absent in the spectrum of the acid form, was attributed to cation motion within an anionic field in the copolymer. The frequency variation of this band was found to correlate with the change in mass of the alkali metal counterion, taking the vibrational reduced mass to be approximately that of the cation.

Tsatsas *et al.* also observed several types of perturbations in the bands involving vibrational motion of the polyanion itself. These perturbations involved polymer backbone deformations as well as the vibrational modes localized on the carboxylate group. The authors found most appropriate a vibrational model in which the cation is in the field of both a carboxylate site and a hydrocarbon fragment. They suggest that such an aggregate may provide a molecular basis for interpreting cation-dependent rheological data.

2. Other Crystalline Ionomers

Only a small number of publications dealing with the mechanical properties of crystalline ionomer systems other than those based on polyethylene have appeared.

Wissbrun [69] investigated the properties of ionic oxymethylene copolymers. The materials were prepared by copolymerizing trioxane and epichlorohydrin and then ionizing the chlorinated fraction by reaction with disodium thioglycolate. Copolymers with ion contents of 1.4–8.5 mole % were obtained. As with ethylene ionomers, ionization was found to disrupt crystalline order, resulting in increased transparency. The tensile modulus and the maximum tensile strength at 23°C both decreased with increasing ion content. The strength properties were dependent on thermal history, as well as on atmospheric water. Ionization was found to increase the melt viscosity of the copolymers, as well as increase the temperature dependence of the viscosity. These latter findings are in qualitative agreement with those obtained in the melt rheology studies of both styrene and ethylene ionomers.

The viscoelastic properties of ionene polymers have also received attention. Tsutsui *et al.* [70] studied the properties of a system termed *x*,*y*-oxyethylene ionene (*x*,*y*-OEI) of the type (II), where *x* and *y* varied from 2 to 5.

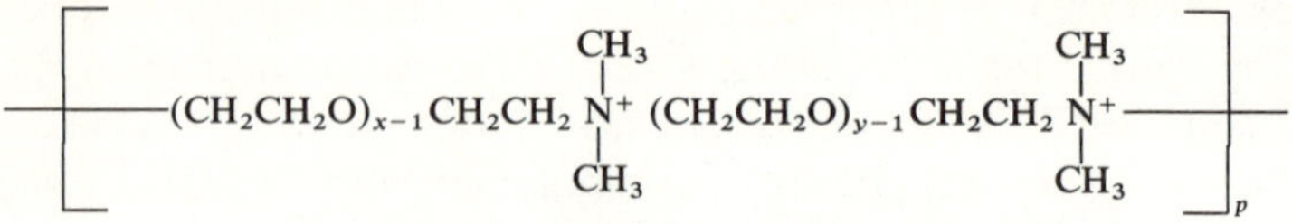

$$\left[\!\!-\!(CH_2CH_2O)_{x-1}CH_2CH_2\overset{\displaystyle CH_3}{\underset{\displaystyle CH_3}{\overset{|}{\underset{|}{N^+}}}}(CH_2CH_2O)_{y-1}CH_2CH_2\overset{\displaystyle CH_3}{\underset{\displaystyle CH_3}{\overset{|}{\underset{|}{N^+}}}}-\!\!\right]_p$$

(II)

Only some of the *x*,*y*-OEI polymers were crystalline, but the entire system is discussed here for convenience. The temperature dependence of the dynamic mechanical properties of these materials was studied by torsion braid analysis over the range -170 to $+150°C$. For each polymer, except 2,2-OEI, two dispersions were observed, as seen in the examples shown in Fig. 4.64. The temperature T_α of the primary dispersion was found to be linearly dependent on the charge density along the ionene backbone and was identified with the T_g of the amorphous phase of the polymer. The low temperature or β dispersion showed no regular temperature dependence. It was associated with a "local mode" dispersion. Both relaxation regions were greatly influenced by the presence of absorbed water.

D. POLYELECTROLYTE COMPLEXES

Polymeric complexes formed through the interaction of oppositely charged polyelectrolytes have received the attention of scientists for many years; however, industrial interest in such materials has been limited to more recent years [71].

In studies by Bungenberg de Jong and co-workers, it was found that certain naturally occurring polyelectrolytes could interact in aqueous media to form colloidal complexes, which were termed complex coacervates [72]. These complexes were found to be generally gellike or quasi-liquid phases of high water content and rather indefinite composition. Furthermore, the nature and extent of the reaction between natural polyelectrolytes were found to be highly dependent on such variables as pH, temperature, and ionic strength, because of the fact that natural polyelectrolytes are generally weak and often polyampholytic and strongly hydrogen bonded.

The interaction between strong, synthetic polyelectrolytes in aqueous solution, first reported by Fuoss and Sadek [73], yields a colloidal precipitate rather than the gelatinous coacervate obtained from weak polyelectrolytes. The reaction was found to occur rapidly and yield a product containing essentially stoichiometric equivalents of the component polyions. Michaels and Miekka [74] found that the polymeric precipitate formed in this way was almost exactly stoichiometric, regardless of the composition of the

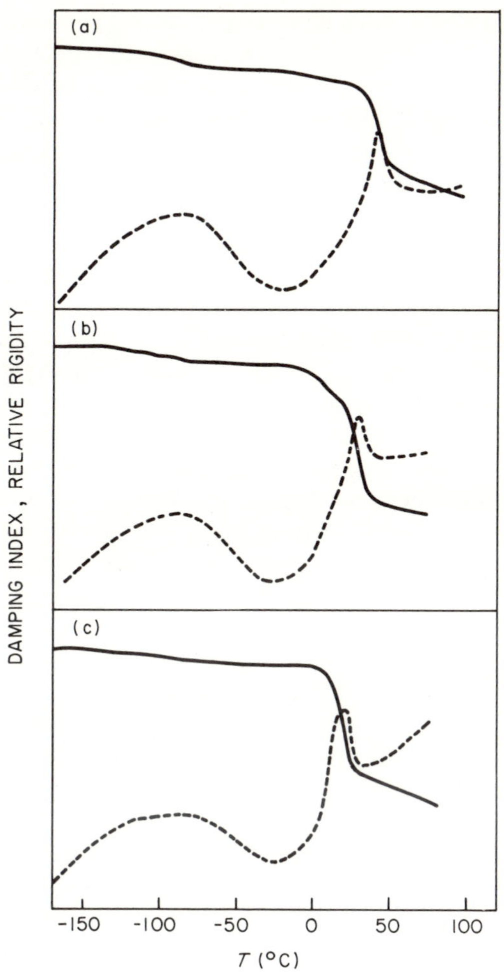

FIG. 4.64. Temperature dependence of relative rigidity (——) and damping index (– –) of *x,y*-oxyethylene ionenes: (a) 3,3-OEI; (b) 3,4-OEI; (c) 4,4-OEI [70].

starting material, and that virtually all of the counterions initially associated with the polymers were excluded from the precipitate. These polyelectrolyte complexes were found to be infusible and insoluble in all common solvents; they could, however, be dissolved in ternary solvents containing water, an electrolyte, and a polar organic solvent. In addition, it was noted that polyanions and polycations in any relative proportion could be codissolved without reaction in an appropriately chosen ternary solvent. Thus, by evaporating the water and organic solvent and leaching out the electrolyte, both stoichiometric and nonstoichiometric polyelectrolyte complexes can be prepared in mechanically useful bulk form from polyelectrolyte solutions in ternary solvents. The polysalts prepared in this way are found to be clear glassy solids, which swell and absorb salt in aqueous electrolyte solutions.

The dilute solution interactions of the systems poly(sodium styrene sulfonate)—PSSA–Na—and poly(vinyl benzyl trimethyl ammonium chloride)—PVBTACl—were investigated in detail by Michaels *et al.* [75], and the water and salt sorption and swelling of this system were studied by Gray [76]. In addition, a number of reports dealing with applications of these materials have been published. Many of these have been described in the reviews of Michaels [71], Michaels and Bixler [77], and Lysaght [78]. In the discussion to follow, only the studies relating to the viscoelastic properties of polyelectrolyte complexes will be described.

The dielectric behavior of PSSA–Na/PVBTACl complexes was examined by Michaels *et al.* [79] in a study which proved helpful in elucidating the structure of polyelectrolyte complexes. The stoichiometric NaSS–VBTACl complex, containing no microions, exhibits no particularly unusual dielectric behavior, the values of ε' and $\tan \delta$ decreasing monotonically over the frequency range 10^2 to 10^5 Hz. In this respect, the dielectric properties of the salt-free complex are similar to those observed in many nonionic, dipolar polymers above their glass transitions. With increasing concentration of water and low-molecular-weight salts, the dielectric constant of the stoichiometric complex increases by orders of magnitude and its frequency dependence in the 10^2- to 10^5-Hz range becomes marked, as illustrated in Fig. 4.65. Similar dielectric responses were also obtained for nonstoichiometric complexes, which contain an excess of one of the polymeric components and the accompanying microions [74]. Characteristic of these materials are the very broad dispersion curves illustrated in Fig. 4.66, even for the nominally salt-free material, in which the broadness is attributed to residual salt. The loss tangent values are quite high, and maxima in this frequency range are detectable only at high NaBr contents.

It was observed that the dc conductivities of polysalt complexes are very low, even at fairly high sorbed salt concentrations [71]; it was therefore concluded that dc conduction contributes little to the dielectric response, in contrast to what might be expected of materials so rich in ions. Thus, the anomalous dielectric properties of the salt-containing complexes were attributed primarily to polarization processes involving short-range microion movements. It should be pointed out that since the dielectric measurements were not made with a four-terminal bridge, the possibility of electrode polarization should be considered, which could account in part for the extremely high values of dielectric constant obtained.

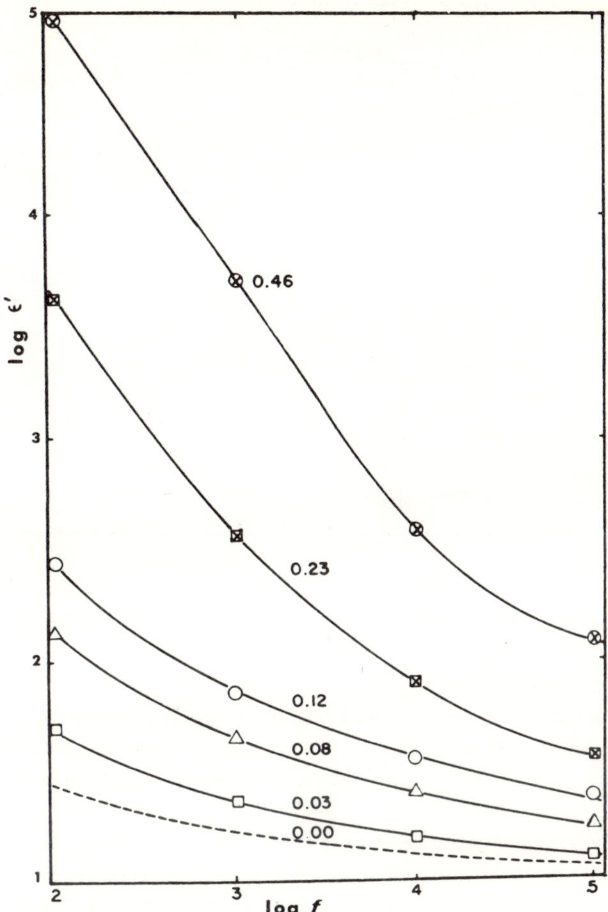

FIG. 4.65. Frequency dependence of storage dielectric constant of PSSA–Na/PVBTACl complex at various NaBr contents. $T = 27.5°C$; RH = 51%; equivalent NaBr per equivalent PS given on each curve [79].

Michaels *et al.* [79] explain the dielectric behavior of microion-containing complexes by postulating a structure in which noninterconnecting domains containing sorbed microions and associated polyions are set in a continuous matrix of neutral, microion-free polysalt. The formation of such a nonuniform structure is presumably favored for thermodynamic reasons similar to those that govern the formation of domain structures in ionomers, since the continuous matrix is of low dielectric constant. At the frequencies employed here, the polymeric mobilities are sufficiently low to consider the

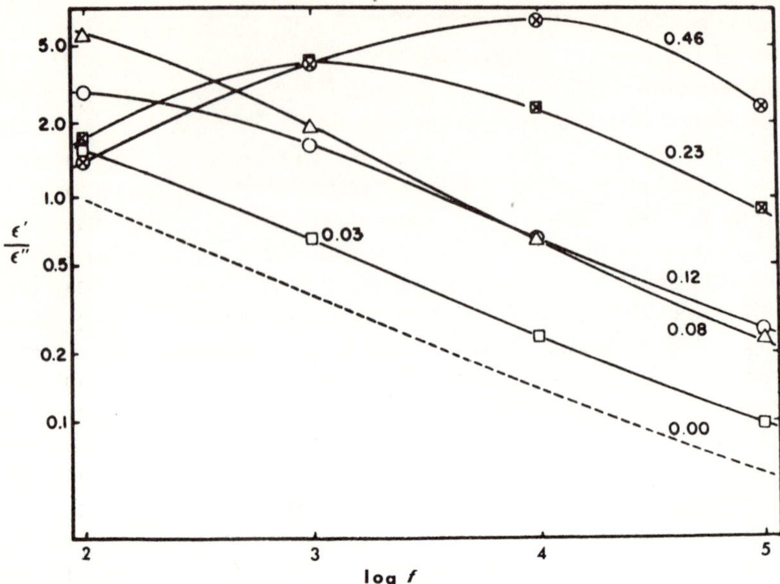

FIG. 4.66. Frequency dependence of dielectric loss tangent of PSSA–Na/PVBTACl complex at various NaBr contents; equivalent NaBr per equivalent PS given on each curve [79].

domain structure fixed with time. Thus the dielectric response observed upon application of an electric field is assumed to arise from the polarization of the ionic domains, which results from the internal migration of the mobile microions.

Variations in the water content of the salt-containing complexes appear to affect only the polarization time constant; that is, by increasing the water content, the dispersion shifts to higher frequency, and the dispersion curves obtained at different water contents can be superimposed by horizontal shifting along the log frequency axis [79]. This indicates that the main effect of water is that of a plasticizer, which loosens the structure. Variations in NaBr content at constant relative humidity, on the other hand, tend to change the shapes of the dispersion curves, with both ε_0 and ε_∞ increasing with increasing salt content. This is consistent with the supposition that the dielectric dispersion arises mainly from the polarization of the microions, since an increase in the concentration of polarizable elements would be expected to increase ε_0 and the dielectric increment.

According to the theories of O'Konski [80] and Schwartz [81], developed for a similar model, the dielectric increment per domain will increase with the surface charge density of the mobile ions and with the domain size, both of which are determined by the NaBr concentration. In addition, the relaxation time associated with the polarization is inversely proportional to the square of the domain radius. [See Eqs. (10)–(13) of Chapter III.] Thus, variations in ion mobilities and a distribution of domain sizes could lead to the broadened distribution of relaxation times observed in the polysalts.

The study by Gray [76], involving measurements of both mechanical properties and sorption, attempted to determine the nature of ionic interaction in polysalts. Two models were considered: a "diffuse interaction" model, analogous to molten salts, in which each ionic site shares its Coulombic interaction with its near neighbors, and an "ionic cross-link" model, which assumes that the interaction between ionic states is mainly pairwise. It was pointed out that the former model does not introduce a new mechanical transition, and, in fact, the shape of the relaxation curve should be the same as that of an amorphous polymer. On the other hand, with the latter model, an additional relaxation mechanism, analogous to chemical relaxation, due to the breaking and reforming of interchangeable cross-links should result. A possible modification of the latter model, not considered by Gray, but discussed above in connection with the dielectric study, would also require two relaxation mechanisms, corresponding to the yielding of the two phases, in the same manner as observed in styrene ionomers. If the discontinuous phase yielded at a rate much faster than the rate of yielding of the continuous phase, however, this contribution would not likely be observable by a static mechanical technique.

Stress relaxation curves (taken over ~3.5 decades of time) were prepared by Gray, applying a number of variables—temperature, salt content, moisture content—to the PSSA–Na/PVBTACl stoichiometric complex. Time–temperature superposition of the relaxation data was obtained, but only over a limited temperature range (2.0°C). In addition, material–time superposition, with respect to both water content and salt content, was obtained. The study remained inconclusive, however, in regard to the nature of the relaxation process in polysalts.

Otocka and Eirich [37] in their studies on the rheological behavior of butadiene ionomers, also investigated the behavior of copolyanion–copolycation mixtures. The viscoelastic properties of the parent B–MAA and B–MVP ionomers were described in Section B of this chapter. Films of the stoichiometric complexes were prepared by mixing solutions containing the appropriate concentrations of B–MAA–Li and B–MQMVP and evaporating the solvent. Gelation was too rapid to permit the preparation of films from the ionomers of highest ionic concentration. Also, efforts to prepare ionomer mixtures free of counterions were unsuccessful.

The T_g values of the butadiene ionomer mixtures were found to be intermediate between those of the parent ionomers, when compared at equivalent ion contents. The modulus-temperature curves of the two ionomer mixtures,

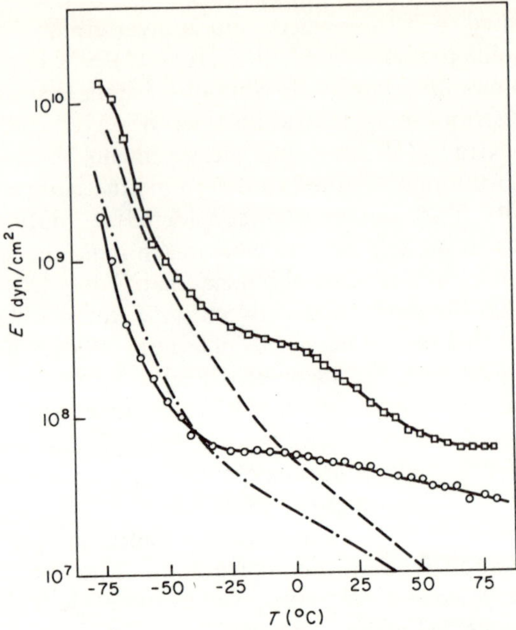

FIG. 4.67. Modulus temperature curves for B–MAA–Li/B–MQMVP mixtures and the corresponding B–MAA/B–MVP mixtures. Symbols: B–0.047MAA/B–0.018MVP: (—·—) acid–base, (○) salt–salt; B–0.077MAA/B–0.065MVP: (– – –) acid–base, (□) salt–salt [37, 82].

as shown in Fig. 4.67, are quite similar in appearance to those obtained for the B–MQMVP ionomers (Fig. 4.26). However, the height of the rubbery plateau or, equivalently, the apparent cross-link density was found to be higher for the mixtures than for either of the equivalent parent ionomers, confirming that a cross-linking effect due to pyridinium–carboxylate pairs persists above the glass-transition temperature.

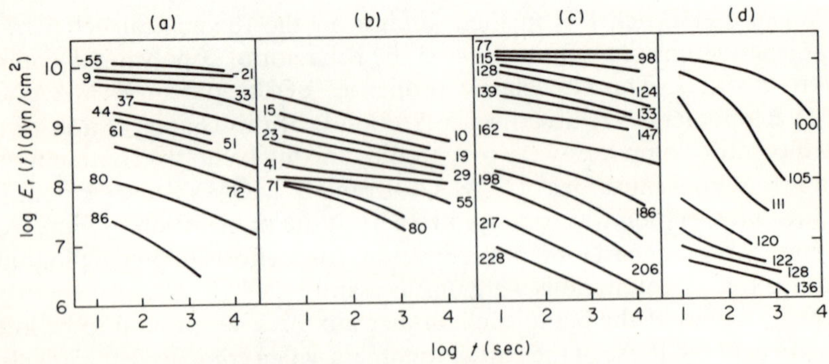

FIG. 4.68. Stress relaxation curves for: (a) E–0.08MAA–0.47Na, (b) EA–0.12AA–Na, (c) S–0.077MAA–Na, and (d) PS [83].

Otocka and Eirich [82] found no evidence for ionic cross-linking in the mixtures prepared from the parent acidic and basic forms of the butadiene copolymers. Although there was an increase in T_g and in the apparent cross-link density, the levels obtained were much less than those observed in any of the ionomers, as is evident from Fig. 4.67. Also, they observed no spectroscopic evidence for pyridium or carboxylate ion formation and concluded that the mechanical effect in this case was merely due to increased hydrogen bonding.

Michaels *et al.* [75] also observed sharp differences between the mixtures obtained from the reaction of the parent acidic and basic homopolymers and those prepared from their respective salt forms. In the former case, the reaction was found to be incomplete and nonstoichiometric; in the latter case, stoichiometric counterion-free products were obtained.

E. AN OVERVIEW

Before concluding this chapter, it might be useful to compare the behavior of several different families of materials, by summarizing the results of different types of studies, i.e., stress relaxation, the temperature dependence of the isochronal modulus, and the loss tangent.

Figure 4.68 shows the stress relaxation behavior of three ionomers and one of the parent polymers [83]. The ionomers are (a) an ethylene-(methacrylic acid) ionomer, E–0.08MAA–0.47Na, discussed in Section C1(b) (see Fig. 4.39); (b) an (ethyl acrylate)-(acrylic acid) ionomer, EA–0.12AA–Na [84]; (c) a styrene-(methacrylic acid) ionomer, S–0.077MAA–Na, discussed in Section A1 (see Fig. 4.2); and (d) a polystyrene of comparable molecular weight [85].

It is clear that all the ionomer samples show a great deal of similarity in their stress relaxation behavior in that the dip in modulus with time is relatively slow, and typical of a partially crystalline material. It should be noted, however, that only the PE ionomer has any crystalline content. The contrast between curves (c) and (d) is particularly noteworthy if it is recalled that the materials differ from each other only in the presence of 7.7 mole % NaMA, which raises the T_g by only $\sim 20°$. It is anticipated that all clustered systems will show similar behavior since, as was pointed out before, the clusters give the polymer many of the characteristics of a microphase-separated system.

Figure 4.69 shows the modulus–temperature curves for several ionomers along with the nonionic parent polymers, in addition to that for Nafion–K. Each of these materials was discussed separately in the appropriate section of Chapter III or IV. Only the general trends will be summarized here. In all the cases where comparisons can be made, T_g increases, and the slope of

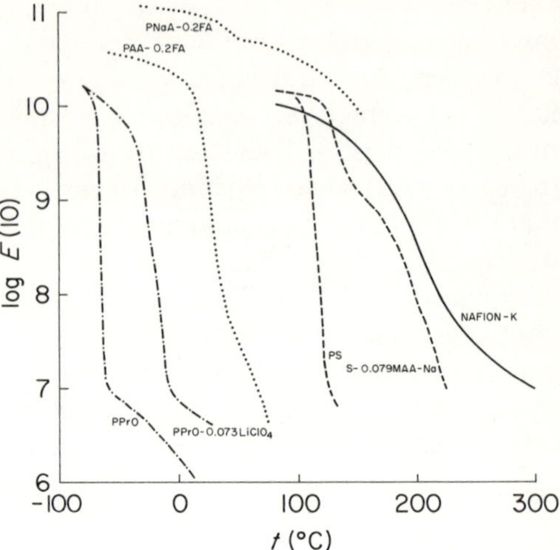

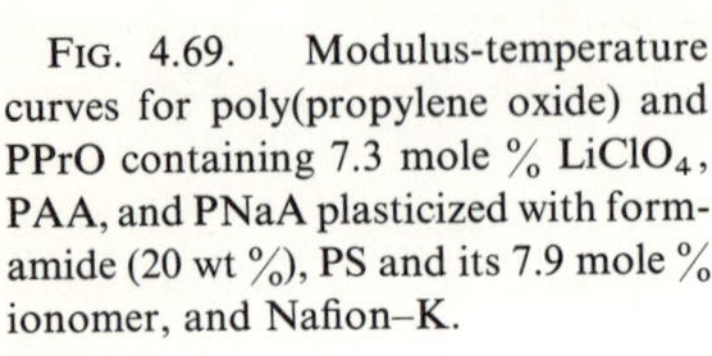
FIG. 4.69. Modulus-temperature curves for poly(propylene oxide) and PPrO containing 7.3 mole % $LiClO_4$, PAA, and PNaA plasticized with formamide (20 wt %), PS and its 7.9 mole % ionomer, and Nafion–K.

the modulus-temperature curve above T_g decreases over very wide ranges of modulus. This seems to be true for both plasticized and unplasticized systems.

Finally, Fig. 4.70 shows the loss tangent for three ionomer systems: PS (at an ion content of 0, 7.7, and 9.7 mole %), PEA (18.1%), and PE (0, 2.78, 3.40%). Again, it is clear that as the ion content increases, T_g increases, and that another high-temperature peak appears, which migrates to higher

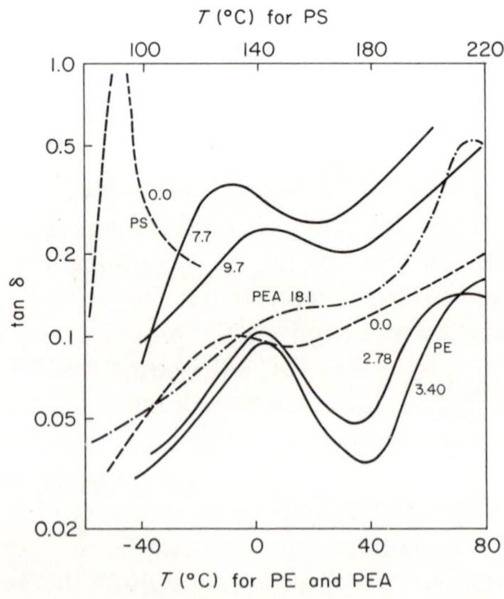

FIG. 4.70. Loss tangent versus temperature for PS (containing 0, 7.7, and 9.7 mole % ions), PEA (18.1%), and PE (0, 2.78, 3.40%) [83].

temperatures with increasing ion content. The similarities between PE and PS ionomers are striking (after the appropriate transposition due to changes in T_g). These materials resemble each other far more strongly than, for example, the styrene ionomers resemble polystyrene.

Not enough studies are available at this time to allow for meaningful comparisons involving other properties such as dielectric, stress–strain, and rheooptical behavior. Undoubtedly extrapolations could be attempted for some of these properties in view of the results that are available, but this would be too speculative for inclusion here. The only purpose of this section is to illustrate the similarity in the properties of some clustered systems independent of the parent polymers.

REFERENCES

1. W. E. Fitzgerald and L. E. Nielsen, *Proc. Roy. Soc.* (*London*) **A282,** 137 (1964).
2. N. Z. Erdi and H. Morawetz, *J. Colloid. Sci.* **19,** 708 (1964).
3. F. Ide, T. Kodama, and A. Hasegawa, *Kobunshi Kagaku* **26,** 863 (1969).
4. F. Ide, T. Kodama, A. Hasegawa, and O. Yamamoto, *Kobunshi Kagaku* **26,** 873 (1969).
5. S. Tamura, M. Fukada, H. Nakanishi, and K. Murakami, *Kobunshi Kagaku* **28,** 843 (1971).
6. J. D. Ferry, "Viscoelastic Properties of Polymers," 2nd ed., Chapter XI. Wiley, New York, 1970.
7. E. P. Otocka, Ph.D. Thesis, Polytechnic Inst. of Brooklyn (1966).
8. A. Eisenberg and M. Navratil, *J. Polym. Sci. Part B* **10,** 537 (1972).
9. A. Eisenberg and M. Navratil, *Macromolecules* **6,** 604 (1973).
10. A. Eisenberg and M. Navratil, *Macromolecules* **7,** 90 (1974).
11. M. Navratil and A. Eisenberg, *Macromolecules* **7,** 84 (1974).
12. J. J. Aklonis and V. B. Rele, *J. Polym. Sci. Polym. Symp.* **46,** 127 (1974).
13. A. Eisenberg, *J. Polym. Sci. Polym. Symp.* **45,** 99 (1974).
14. A. V. Tobolsky, "Properties and Structure of Polymers." Wiley, New York, 1960.
15. S. Bonotto and E. F. Bonner, *Macromolecules* **1,** 510 (1968).
16. L. L. Chapoy and A. V. Tobolsky, *Chem. Scripta* **2,** 44 (1972).
17. S. R. Rafikov *et al.*, *Vysokomol. Soyed.* **A15,** 1974 (1973); Engl. transl.: *Polym. Sci. USSR* **15,** 2225 (1973).
18. K. Gaertner, G. Pohl, H. Schade, and K. Schroeder, *Plaste Kautsch.* **21,** 736 (1974).
19. H. Schade and K. Gaertner, *Plaste Kautsch.* **21,** 825 (1974).
20. K. Iwakura and T. Fujimura, *J. Appl. Polym. Sci.* **19,** 1427 (1975).
21. E. Shohamy and A. Eisenberg, *J. Polym. Sci. Polym. Phys.* **14,** 1211 (1976).
22. F. B. Leitz, M. A. Accomazzo, and S. A. Michalek, Development of Electrochemical Hypochlorite Generator. Paper presented at the 141st Meeting of the Electrochem. Soc., Houston, Texas, 1972.
23. J. E. Currey, A. T. Emery, and C. S. McLarty, Future Trends in Chlorine Cells. Paper presented at the *Chlorine Bicentennial Symp., San Francisco* (1974).
24. L. J. Nuttall and W. A. Titterington, General Electric's Solid Polymer Electrolyte Water Electrolysis. Paper presented at the *Conf. Electrolytic Production Hydrogen, London* (1975).
25. W. G. F. Grot, *Chem. Ing. Tech.* **44,** 167 (1972); **47,** 617 (1975).
26. S. C. Yeo and A. Eisenberg, *J. Appl. Polym Sci.* (to be published).
27. A. Noshay and L. M. Robeson, *J. Appl. Polym. Sci.* **20,** 1885 (1976).
28. J. H. Wang, *J. Am. Chem. Soc.* **73,** 4181 (1951).

29. N. G. McCrum, *J. Polym. Sci.* **34,** 355 (1959).
30. H. P. Brown, *Rubber Chem. Technol.* **30,** 1347 (1957).
31. W. Cooper and T. B. Bird, *Ind. Eng. Chem.* **50,** 771 (1958).
32. W. Cooper, *J. Polym. Sci.* **28,** 195 (1958).
33. J. I. Cuneen, C. G. Moore, and B. R. Shephard, *J. Appl. Polym. Sci.* **3,** 11 (1960).
34. J. C. Halpin and F. Bueche, *J. Polym. Sci. Part A* **3,** 3935 (1965).
35. N. D. Zakharov, *Rubber Chem. Technol.* **36,** 568 (1963).
36. A. V. Tobolsky, P. F. Lyons, and N. Hata, *Macromolecules* **1,** 515 (1968).
37a. E. P. Otocka and F. R. Eirich, *J. Polym. Sci. Part A-2* **6,** 921 (1968).
37b. E. P. Otocka and F. R. Eirich, *J. Polym. Sci. Part A-2* **6,** 933 (1968).
38. C. K. Lim, R. E. Cohen, and N. W. Tschoegl, *Adv. Chem. Ser.* **99,** 397 (1970).
39a. M. Pineri, C. Meyer, A. M. Levelut, and M. Lambert, *J. Polym. Sci. Polym. Phys.* **12,** 115 (1974).
39b. C. T. Meyer and M. Pineri, *Polymer* **17,** 382 (1976).
40. D. Dieterich, W. Keberle, and H. Witt, *Angew. Chem. Int. Ed.* **9,** 40 (1970).
41. R. Somoano, S. P. S. Yen, and A. Rembaum, *J. Polym. Sci. Part B* **8,** 467 (1970).
42. A. Rembaum, S. P. S. Yen, R. F. Landel, and M. Shen, *J. Macromol. Sci. Chem.* **A4,** 715 (1970).
43. P. F. Erhardt, J. M. O'Reilly, W. C. Richards, and M. W. Williams, *J. Polym. Sci. Polym. Symp.* **45,** 139 (1974).
44. E. N. Kresge, Ionic Bonding in Elastomeric Networks. Paper presented at the 18th Canadian High Polymer Forum, Hamilton, Ontario, 1975.
45. R. Longworth, *in* "Ionic Polymers" (L. Holliday, ed.), Chapter II. Applied Science Publ., London, 1975.
46a. R. Longworth and D. J. Vaughan, *Amer. Chem. Soc. Polym. Preprints* **9,** 525 (1968).
46b. R. Longworth and D. J. Vaughan, *Nature* (*London*) **218,** 85 (1968).
47. W. J. MacKnight, W. P. Taggart, and L. McKenna, *J. Polym. Sci. Polym. Symp.* **46,** 83 (1974).
48. C. L. Marx and S. L. Cooper, *J. Macromol. Sci. Phys.* **B9,** 19 (1974).
49. T. C. Ward and A. V. Tobolsky, *J. Appl. Polym. Sci.* **11,** 2403 (1967).
50. W. J. MacKnight, L. W. McKenna, and B. E. Read, *J. Appl. Phys.* **38,** 4208 (1967).
51a. E. P. Otocka and T. K. Kwei, *Macromolecules* **1,** 244 (1968); errata: *Macromolecules* **2,** 110 (1969).
51b. E. P. Otocka and T. K. Kwei, *Macromolecules* **1,** 401 (1968).
52a. R. W. Rees and D. J. Vaughan, *Amer. Chem. Soc. Polym. Preprints* **6,** 287 (1965).
52b. R. W. Rees and D. J. Vaughan, *Amer. Chem. Soc. Polym. Preprints* **6,** 296 (1965).
53. W. J. MacKnight, T. Kajiyama, and L. McKenna, *Polym. Eng. Sci.* **8,** 267 (1968).
54. W. J. MacKnight, L. W. McKenna, B. E. Read, and R. S. Stein, *J. Phys. Chem.* **72,** 1122 (1968).
55. L. W. McKenna, T. Kajiyama, and W. J. MacKnight, *Macromolecules* **2,** 58 (1969).
56. K. Sakamoto, W. J. MacKnight, and R. S. Porter, *J. Polym. Sci. Part A-2* **8,** 277 (1970).
57. B. E. Read, E. A. Carter, T. M. Conner, and W. J. MacKnight, *Brit. Polym. J.* **1,** 123 (1969).
58. P. J. Phillips and W. J. MacKnight, *J. Polym. Sci. Part A-2* **8,** 727 (1970).
59. E. P. Otocka and D. D. Davis, *Macromolecules* **2,** 437 (1969).
60. Y. Uemura, R. S. Stein, and W. J. MacKnight, *Macromolecules* **4,** 490 (1971).
61. T. Kajiyama, R. S. Stein, and W. J. MacKnight, *J. Appl. Phys.* **41,** 4361 (1970).
62. T. Kajiyama, T. Oda, R. S. Stein, and W. J. MacKnight, *Macromolecules* **4,** 198 (1971).
63. R. E. Prud'homme and R. S. Stein, *Macromolecules* **4,** 668 (1971).
64. R. E. Prud'homme and R. S. Stein, *J. Polym. Sci. Polym. Phys.* **11,** 1347 (1973).
65. A. T. Tsatsas, J. W. Reed, and W. M. Risen, Jr., *J. Chem. Phys.* **55,** 3260 (1971).

66. G. C. Pimentel and A. L. McClellan, "The Hydrogen Bond." Freeman, San Francisco, California, 1960.
67. W. Kuhn and F. Gruen, *Kolloid-Z.* **101,** 248 (1942).
68. J. J. Van Aartsen and R. S. Stein, *J. Polym. Sci. Part A-2* **9,** 295 (1971).
69. K. F. Wissbrun, *Makromol. Chem.* **118,** 211 (1968).
70. T. Tsutsui, R. Tanaka, and T. Tanaka, *J. Polym. Sci. Polym. Phys.* **13,** 2091 (1975).
71. A. S. Michaels, *Ind. Eng. Chem.* **57,** 32 (1965).
72. H. G. Bungenberg de Jong, *in* "Colloid Science" (H. R. Kruyt, ed.), Vol. II. Elsevier, Amsterdam, 1949.
73. R. M. Fuoss and H. Sadek, *Science* **110,** 552 (1949).
74. A. S. Michaels and R. G. Miekka, *J. Phys. Chem.* **65,** 1765 (1961).
75. A. S. Michaels, L. Mir, and N. S. Schneider, *J. Phys. Chem.* **69,** 1447 (1965).
76. C. A. Gray, Ph.D. Thesis, Massachusetts Inst. of Technol. (1965).
77. A. S. Michaels and H. J. Bixler, *in* "Encyclopedia of Chemical Technology" (H. F. Mark, J. J. McKetta, Jr., and D. F. Othmer, eds.), Vol. XVI, 2nd ed., p. 117. Wiley, New York, 1968.
78. M. J. Lysaght, *in* "Ionic Polymers" (L. Holliday, ed.), Chapter VI. Applied Science Publ., London, 1975.
79. A. S. Michaels, G. L. Falkenstein, and N. S. Schneider, *J. Phys. Chem.* **69,** 1456 (1965).
80. C. T. O'Konski, *J. Phys. Chem.* **64,** 605 (1960).
81. G. Schwartz, *J. Phys. Chem.* **66,** 2636 (1962).
82. E. P. Otocka and F. R. Eirich, *J. Polym. Sci. Part A-2* **6,** 895 (1968).
83. A. Eisenberg, *Pure Appl. Chem.*, **46,** 171 (1976).
84. H. Matsuura, E. Polatajko, and A. Eisenberg (to be published).
85. M. Navratil, Ph.D. thesis, McGill Univ. (1973).

Chapter V

Configuration-Dependent Properties

A. POLYELECTROLYTE DIMENSIONS IN SOLUTION

The time-averaged distribution of the segments of a macromolecule about an internal reference point can be considered to determine its molecular configuration. The study of the structure of macromolecules in solution deals with both the mathematical description and the experimental determination of polymer configurations. Of interest is the way in which the polymer structure varies under the influence of such variables as the degree of polymerization, the extent of chain branching and stereoregularity, the type of solvent, temperature, polymer concentration, and the concentration of low-molecular-weight cosolutes.

The structure of nonionic, linear polymers in solution has received a great deal of attention and has been the subject of a number of books and reviews [1–7]. The knowledge gained in this general area of research, from both the theoretical and experimental viewpoints, in large part determines our knowledge of the structure of ion-containing polymers in solution, since ionic interaction is, in effect, just another variable which influences polymer configurations. The methods used in the study of structure in ion-containing polymers, both theoretical and experimental, are, in general, modifications of the basic approaches to the problem. Thus, it will be necessary in Section A1 of this chapter to include a brief review of the configurational properties of nonionic polymers, upon which the effect of ionic interactions can be added.

It is well known that linear macromolecules are normally more or less randomly coiled in solution; their configurations can be characterized by the mean-square displacement between the chain ends or the mean-square radius of gyration. Upon the introduction of ionized groups to the polymer chain backbone, the mean-square displacement between the chain ends may change, but the basic random-coil character of the configuration is retained. Based on this concept, the change in configuration is determined by changes in intramolecular interaction on a microscopic level. The ionized polymers also remain more or less spherical in shape and can be characterized by an equivalent hydrodynamic radius. In the latter case, the specific chainlike character of the polymer is not considered, in contrast to the former view.

Thus two main types of theory have been developed to describe the configurations of ion-containing polymers, based on models which emulate these two different physical views. One type of model employs a randomly coiled chain on which all of the fixed charges are arranged. The specific chainlike character of the polymer is taken into account in introducing the effect of the electrostatic charges. The other type of model represents the polymer as a sphere in which the fixed charges are assumed to be distributed continuously. The overall charge density and volume of the sphere are chosen to be equivalent to those of the real polymer.

The change in the polymer dimensions which results from the addition of charges can be calculated for both types of models by balancing the expansive force, which results from the placement of the charges, with the entropic retractive force. It is worth recalling at this point that most studies of polyelectrolyte solutions have been performed in aqueous systems, and under these circumstances of low ion concentration and high dielectric constant, the driving force for multiplet or cluster formation is minimal, in contrast with the situation in the solid state. The predicted chain dimensions vary considerably according to the various assumptions made about the distribution and effect of the fixed charges, counterions, and added simple electrolyte. The results obtained from both of the above approaches will be described in Section A2; in addition, comparisons with experimentally determined polymer dimensions will be made.

An alternate approach used to predict the dimensions of ionic polymers in good solvents is to adapt those theories which have been developed primarily to describe the expansion of nonionic polymers in good solvents. Most of these theories use a parameter which is dependent on the interaction energy per average pair of segments. The parameter that results from these energetic considerations has the dimensions of volume and can be physically associated with a mutual volume of exclusion per pair of segments; hence, it is commonly termed the excluded volume. This concept of excluded volume can also be applied to ion-containing polymers; it will be seen in Section A2(c) that equations, which successfully predict the expansion of nonionic polymers, are also applicable to ionic polymers. This implies that, although the origins of the forces that cause expansion in ionic polymers may be different, the dependence of the expansion on such molecular parameters as molecular weight and excluded volume is essentially the same as it is in nonionic polymers.

The expansion of polyelectrolytes in good solvents is largely the result of interactions between nonbonded chain segments, just as it is in the case of nonionic polymers. Theories that describe nonionic polymer expansion often consider the effect of long-range interactions to be superimposed on the effect of short-range interactions, which occur between neighboring chain segments. The rotational isomeric theory [2, 8] has been employed successfully to predict the unperturbed dimensions of nonionic polymers, in which only short-range interactions are important. The same theory has been applied with some success in the calculation of unperturbed dimensions of polyelectrolytes. It will be worthwhile to look first at the short-range effects of electrostatic charges on polymer dimensions, before considering the superimposable long-range effects of the same charges.

It should be stressed that the treatment given here is brief and is presented for the sake of completeness. More extensive works dealing with the configurational properties of polyelectrolytes exist, and the interested reader is referred to these [9–13].

1. Unperturbed Dimensions

Since the study of the structure of ion-containing polymers in solution builds on our knowledge of the configurational properties of nonionic polymers, it will be worthwhile to begin this section with a brief review of the theories that describe the configurations of nonionic polymers in solution. For a more complete treatment of this topic, the reader is directed to any of several existing books and reviews dealing with configurational statistics of polymers and polymer solutions [1–8].

The simplest case is that of the freely jointed chain. This is a chain consisting of N links, each of length l, and connected by universal joints. There are no steric or other energetic interactions between the links of this chain; thus all configurations are equally probable. This results in a rather tightly coiled chain in most configurations. The probability distribution function $W(h)$ of the displacement h between the ends of such a chain has been studied [1]. For normal extensions of high polymer chains, the distribution is Gaussian, as given by the expression

$$W_{0,0}(h)\,dh = (3/2\pi Nl^2)^{3/2} \cdot \exp(-3h^2/2Nl^2) \cdot 4\pi h^2\,dh \tag{1}$$

where the subscript 0,0 refers to the fact that there are neither short-range nor long-range effects that perturb this strictly random coil.

The average coil dimensions can be described in terms of $\overline{h^2}$, the mean-square displacement of the chain ends, or in terms of h^*, the most probable displacement. For the freely jointed chain, the following relationships hold:

$$\tfrac{3}{2}(h^*_{0,0})^2 = \overline{h^2_{0,0}} = Nl^2 \qquad (2)$$

The simplest modification of this model takes into account the fact that the bond angles between links are fixed, but free rotation is maintained. The expression for $\overline{h^2}$ becomes [2]

$$\overline{h^2_{\mathrm{fr}}} = Nl^2(1 + \cos\theta)/(1 - \cos\theta) \qquad (3)$$

where θ is the angle between successive bonds. For tetrahedral bond angles, $\overline{h^2_{\mathrm{fr}}}$ assumes the value $2Nl^2$.

The chain model becomes more complex when the effects of the short-range, steric interactions that result in hindered rotation about consecutive bonds of the chain backbone are considered. The resultant expansion of the polymer coil can be described in terms of the average angles of rotation about successive bonds along the polymer backbone. Several types of expressions have been developed, the simplest example being that for vinyl polymers with symmetric side groups. For this type, the expression [2]

$$\overline{h_0{}^2} = \overline{h^2_{\mathrm{fr}}}(1 + \overline{\cos\phi})/(1 - \overline{\cos\phi}) \qquad (4)$$

describes the expansion of the coil due to the steric hindrance of nearest neighbors. The term $\overline{\cos\phi}$ represents the average cosine of the angle of rotation of two successive bonds. For unsymmetrical, stereoregular polymers, more complex expressions, involving pairs of successive angles, can be written. Given a detailed knowledge of the potential functions governing bond rotation, the values of $\overline{\cos\phi}$ and hence $\overline{h_0{}^2}$ can be calculated, assuming a Boltzmann distribution for the angles of rotation. These potential functions are normally estimated from those corresponding to similar small molecules. In principle, second-nearest-neighbor and even higher interactions can be taken into account by this method [2].

Because of the difficulty in obtaining a detailed knowledge of the shapes of potential functions, the unperturbed dimensions of linear macromolecules are more readily estimated using the rotational isomeric state theory [2, 8]. The basic postulate of this theory is that each bond in the linear chain is considered to be in one of a small number of discrete rotational states. The relative probabilities of these rotational states are determined by the differences in the potential energy minima of the various conformations which result from bond rotation. This approach involves a great simplification since only the minima of the potential functions and not their shapes are considered.

The unperturbed dimensions of a number of stereoregular, nonionic polymers have been calculated, using potential energy curves derived from related small molecules [8]. These calculated values, of course, apply to the case where short-range interactions are the only factors that contribute to coil expansion. This condition can be realized experimentally by measuring polymer dimensions in θ solvents, in which the effect of long-range interactions is eliminated or, more correctly, compensated for. The polymer dimensions calculated from rotational isomeric theory have been found to agree quite well with experimental values [8].

This approach can, at least in principle, be extended to linear polyelectrolytes. In this case, the potential energies of the various conformations will depend on the Coulombic interactions. Given the geometry of each conformational state, the potential energy can be evaluated as the sum of contributions from van der Waals and Coulombic interactions, and from the potential energy, a relative probability can be assigned to each configuration. However, attempts to calculate electrostatic contributions to the conformational potential energies have not been particularly successful, mainly because of the difficulty in choosing an appropriate value of the local dielectric constant. This was observed by Semlyen and Flory [14], who applied the rotational isomeric theory to the linear polyphosphates. The geometry and distribution of charge in the model they used were based on crystallographic and chemical data from a number of simple phosphates. They considered the rotational states resulting from the various 3-fold rotations about a consecutive pair of bonds in the phosphate chain [i.e., the various combinations of trans and gauche ($\pm$) conformations]. In this respect, the same techniques were employed for the polyphosphates. However, the potential energies of the conformational states could not be evaluated with confidence because of the great uncertainty in estimating the value of the local dielectric constant. Since this led to a corresponding uncertainty in the values of the relative probabilities of each state, Semlyen and Flory evaluated the characteristic ratio for a range of possible conformational energies. They found that the values of conformational energy which resulted in a value of σ^2 (σ = characteristic ratio) in line with the experimentally observed values (see below) also led to a temperature coefficient of σ^2 in agreement with the observed value. This result indicates that the application of rotational isomeric theory to polyelectrolytes is basically correct, but actual *ab initio* calculations of polyelectrolyte dimensions require more sophisticated methods of estimating values of local electrostatic energy than are presently available.

The experimental determination of polyelectrolyte unperturbed dimensions is relatively straightforward, involving the same general techniques used for other chain polymers. The unperturbed dimensions, which are usually expressed in terms of the characteristic ratio $\sigma^2 = \overline{h_0{}^2}/\overline{h_{fr}^2}$, are usually estimated from viscosity measurements in θ solvents with the aid of the Flory–Fox equation [1], which may be written

$$\overline{h_0{}^2} = (M[\eta]_\theta/\Phi)^{2/3} \tag{5}$$

In Eq. (5), M is the polymer molecular weight, $[\eta]_\theta$ is the intrinsic viscosity as determined in a θ solvent, and $\Phi = 2.6 \times 10^{21}$ is the "universal" constant. The value of $\overline{h_{fr}^2}$ for a polymer of given molecular weight is readily calculated from the local geometry—the fixed bond lengths and valence angles.

θ solvents for polyelectrolytes are generally concentrated salt solutions in which the long-range interactions between the fixed charges on the polymer are effectively screened by the added electrolyte. Normally, the θ condition for a given polyelectrolyte-solvent system is defined in thermodynamic terms, such as the temperature and electrolyte concentration at which the value of the second virial coefficient becomes zero. Viscosity measurements in θ solvents have been reported for only a few polyelectrolytes. The values of σ^2 for these and for some related polymers are listed in Table I.

It can be seen that the values of σ^2 for a given polyion are not independent of the type of counterion. The counterion effect on polyelectrolyte unperturbed dimensions has been studied by Allegra *et al* [21], whose findings indicate that the counterion contribution to the local electrostatic energy should be included in the calculation of the unperturbed dimensions. This

TABLE I

Characteristic Ratios of Various Polymers

Polymer	θ condition	σ	Ref.
PE	Various	1.63 ± 0.08	[4]
PS	Various	2.22 ± 0.05	[4]
PAA	Dioxane (30°C)	1.85 ± 0.05	[15]
PAA–Na	1.5 *M* NaBr (15°C)	2.38	[16]
PVSA–K	1.001 *M* KCl (44.5°C)	2.81	[17]
PVSA–Na	1.003 *M* NaCl (32.4°C)	2.97	[17]
	1.008 *M* NaBr (40.1°C)	2.96	
PSSA–K	3.1 *M* KCl (25°C)	1.98	[18]
$LiPO_3$	1.80 *M* LiBr (25°C)	1.88	[19]
$NaPO_3$	0.415 *M* NaBr (25°C)	1.82	[20]

seems reasonable in view of the fact that, for highly charged polymers, a significant fraction of the counterions are effectively bound to the polyion, and that the extent of counterion binding is dependent on the nature of the counterion.

It is of interest to note that the unperturbed dimensions of polyelectrolytes, as measured experimentally, do not differ greatly from those of nonionic polymers. This indicates that most of the polymer coil expansion, which results from the addition of charges to the chain, is due to the long-range interactions of electrostatic forces rather than to the short-range interactions which affect unperturbed dimensions.

2. Dimensions in Good Solvents

(*a*) *Discrete Charge Models*

In order to study the dimensions of polyelectrolyte coils where significant long-range electrostatic interaction occurs, as is normally the case with ionic polymers in good solvents, a different approach must be taken. In one such approach [22], the hypothetical, freely jointed chain in which all the short-range, nonionic interactions are grouped together is considered; the number of links Z and the length of each link A in this equivalent chain are chosen to make h_{max}, the maximum extension, and $\overline{h_0^2}$ the same as in the real case; i.e., $AZ = h_{max}$ and $A^2Z = \overline{h_0^2}$. Thus, the real polymer chain can be described with the same statistics as this hypothetical, equivalent chain, whose probability density function $W_0(h)$ is Gaussian, and can be expressed for $h \ll h_{max}$ as

$$W_0(h)\,dh = (3/2\pi\overline{h_0^2})^{3/2} \exp(-3h^2/2\overline{h_0^2})4\pi h^2\,dh \tag{6}$$

Equation (6) is obtained from Eq. (1) by replacing Nl^2 with $\overline{h_0^2}$, its equivalent value,

With the introduction of charges to this chain, the effect of electrostatic repulsion is superimposed on the equivalent chain, resulting in further expansion. Now the actual probability density function can be expressed as the *a priori* probability function of the unionized, equivalent chain multiplied by the statistical weight due to the average repulsion energy of a given chain-end displacement; i.e.,

$$W(h)\,dh = W_0(h) \exp(-E_{el}/kT)\,dh \tag{7}$$

The term E_{el} represents the average repulsion energy corresponding to each value of h. For any given conformation, E_{el} may be written

$$E_{el} = \sum_{i>j}^{N} q^2/\varepsilon r_{ij} \tag{8}$$

It is assumed that each charge is of magnitude q; r_{ij} is the separation of the ith and jth charges. ε, the dielectric constant within the polymer coil, may differ from the macroscopic value ε_0.

In order to evaluate the polymer coil expansion, as measured by $\overline{h^2}$ or h^*, the electrostatic potential must be expressed as a function of chain-end displacement. It is generally easier to evaluate h^* instead of the more customary $\overline{h^2}$. Since only a relative value of the chain expansion is sought, it does not matter which is calculated. h^* may be obtained by setting

$$(dW/dh)_{h=h^*} = 0 \tag{9}$$

which is equivalent to

$$kT\,\frac{d\ln W_0}{dh} = \frac{d(E_{\mathrm{el}})}{dh} \tag{10}$$

From Eqs. (1), (2), and (10), one obtains

$$(h^*/h_0{}^*)^2 - 1 = -(h^*/2kT)(dE_{\mathrm{el}}/dh)_{h=h^*} \tag{11}$$

The evaluation of the term for the electrostatic potential energy depends on the model assumed. Kuhn, Künzle, and Katchalsky [22] considered a hypothetical polymer chain composed of Z links, with $s\lambda$ charges of magnitude e, the electronic charge, located at each junction point. λ is the degree of ionization of the corresponding real polymer and s is the number of monomer units per statistical link. Kuhn *et al.* first calculated E_{el} for the case where all the counterions are removed. The result of the summation required by Eq. (8), taken over the gaussian distribution of chain links, is given by [22]

$$E_{\mathrm{el}} = (2\nu^2 e^2/\varepsilon h)\ln(h/h_0{}^*) \tag{12}$$

where $\nu = \lambda Z$, and ε is taken as the macroscopic dielectric constant. Substitution of Eq. (12) in Eq. (11) gives

$$(h^*/h_0{}^*)^2 - 1 = (\nu^2 e^2/\varepsilon kTh^*)[\ln(h^*/h_0{}^*) - 1] \tag{13}$$

or in terms of the expansion ratio α

$$\alpha^2 - 1 = (\nu^2 e^2/\varepsilon kTh_0{}^*\alpha)(\ln\alpha - 1) \tag{14}$$

The expression thus obtained greatly overestimates the magnitude of expansion, even for very dilute solutions. For a typical ionic polymer, such as poly(sodium methacrylate), Eq. (14) predicts expansions of greater than 90% of the maximum extension. Observed chain expansions, even at extreme dilutions, are far less than this. This would indicate, and it has been shown, that even in very dilute polyelectrolyte solutions, a substantial portion of the counterions are held closely within the polymer coil and, as a result, are effective in shielding some of the charge.

If it is now assumed that the counterions are present and provide an ionic atmosphere to shield the fixed ions, then, as shown by Katchalsky and Lifson [23], the expression for the pairwise potential energy can be modified according to the Debye–Hückel approximation for dilute solutions. This gives rise to a shielding factor $\exp(-\kappa r)$, which is the factor by which the electrostatic potential at a distance r from a fixed ion is reduced by the screening effect of the mobile ions. The constant κ is given by the expression [9]

$$\kappa^2 = 4\pi \sum q_i{}^2 c_i/\varepsilon kT \tag{15}$$

q_i is the magnitude of the charge of each free ion of species i and c_i is its concentration. Employing the shielding factor, Eq. (8) becomes

$$E_{\text{el}} = \sum_{i>j}^{N} q^2 \exp(-\kappa r_{ij})/\varepsilon r_{ij} \tag{16}$$

By summing over all pairs of charges, using the Gaussian distribution of chain links defined in Eq. (6), the total electrostatic interaction energy is obtained as

$$E_{\text{el}} = (v^2 e^2/\varepsilon h) \ln(1 + 4h/\kappa h_0^{*2}) \tag{17}$$

Substitution of Eq. (17) in Eq. (11) gives

$$(h^*/h_0{}^*)^2 - 1 = (v^2 e^2/2\varepsilon kTh^*)[\ln(1 + 4h^*/\kappa h_0^{*2}) - 4h^*/(\kappa h_0^{*2} + 4h^*)] \tag{18}$$

In terms of α, assuming that $(4\alpha/\kappa h_0{}^*) \ll 1$, and expanding the terms within square brackets in Eq. (18), one obtains

$$\alpha - 1/\alpha = 4v^2 e^2/\varepsilon kTh_0^{*3}\kappa^2 \tag{19}$$

Equation (19) can be used to describe polymer coil expansion in solutions containing added low-molecular-weight salts. For solutions containing excess uni-univalent electrolyte in molar concentration c_s, Eq. (15) becomes

$$\kappa^2 = 8\pi e^2 N_{\text{Av}} c_s/\varepsilon kT \times 10^3 \tag{20}$$

This results in the following expression for polyelectrolyte expansion:

$$\alpha - 1/\alpha = 10^3 v^2/2\pi N_{\text{Av}}\, c_s h_0^{*3} \tag{21}$$

For large values of α, Eq. (21) predicts that the expansion ratio is proportional to $Z^{1/2}$ (since $v \propto Z$ and $h_0{}^* \propto Z^{1/2}$). As discussed by Morawetz [24], this half-power dependence of α on chain length represents an overestimate of the equilibrium chain expansion, since the overall dependence of $\overline{h^2}$ on chain length would then be first power, which is greater than that normally observed.

Rice and Harris [25] considered a different polyelectrolyte model, based on an equivalent chain, with the total backbone charge per unit located in the center of each link. As the chain expands, the average angle between adjacent links of the equivalent chain increases from the unperturbed value of 90°. By considering only the interactions between adjoining chain segments, they obtained a chain expansion factor α independent of chain length for a given charge density, a result appropriate only for expansion under θ conditions for the polyelectrolyte. For these circumstances, the values of α predicted by the Rice–Harris treatment are also somewhat greater than observed.

The main reason that the equivalent chain model fails is that it overestimates the electrostatic energy available for expansion. A number of explanations for this fact have been offered, and various modifications to the equivalent chain model have been introduced to provide a better estimate of the expansion.

One explanation that can be offered for the overestimate of the energy available for expansion derives from the fact that ionization is normally accompanied by an increase in chain stiffness, which can be attributed to the short-range effect of the electrostatic forces. The nature of this effect and its magnitude were discussed in Section A1 of this chapter. The important point at this stage is that part of the electrostatic energy is used to promote increased chain stiffness. In terms of the equivalent chain model, this means that the length of the statistical element increases with the increase in the unperturbed dimensions that normally accompanies ionization. Katchalsky and Lifson [23] point this out in accounting for observed values of polyelectrolyte expansion. They assumed that the length of the statistical element varies linearly with α. In this way they accounted for the expansion of PMAA upon ionization by allowing the statistical element to increase from 9 to 24 monomer units as ionization proceeds from 0 to 100%.

Noda, Tsuge, and Nagasawa [26] obtained an expression for polymer expansion in which the gradual lengthening of the statistical chain element is expressed in a quantitative manner. They assumed that the most probable chain-end separation in the function $W(h)$ is given not by $h_0{}^*$ but by $h^* = \alpha h_0{}^*$. This changes the expression for the electrostatic potential, Eq. (17), to

$$E_{\rm el} = (\nu^2 e^2/\varepsilon h) \ln(1 + 4h/\alpha^2 h_0^{*2} \kappa) \qquad (22)$$

Insertion of this equation in Eq. (11) results in the expression

$$\alpha^5 - \alpha^3 = 4v^2e^2/\varepsilon kTh_0^{*3}\kappa^2 \tag{23}$$

For subsequent purposes, Eq. (23) can be recast in terms of the factor z_{el}, commonly referred to as the excluded volume parameter due to electrostatic interaction. The use of this parameter will be dealt with later, in subsection (c). For now, it may be stated that it assumes the value

$$z_{el} = \pi^{-1/2} \cdot 4v^2e^2/\varepsilon kTh_0^{*3}\kappa^2 \tag{24}$$

under normal circumstances. Thus, Eq. (23) can be written in the simple form

$$\alpha^5 - \alpha^3 = 1.77z_{el} \tag{25}$$

In subsection (d), Eq. (25) will be compared with other expressions used to describe polyelectrolyte expansion.

Another reason put forward to explain the overestimate of the electrostatic energy available for expansion is the fact that a significant fraction of the counterions must be considered to be held tightly bound or immobilized in the vicinity of the fixed ions. This phenomenon, commonly referred to by a number of roughly equivalent expressions—association, immobilization, binding, or condensation, has been investigated from both the theoretical and phenomenological points of view [9–13, 27].

In theoretical treatments of counterion condensation, the most successful model employed to describe the polyelectrolyte solution is the cylindrical cell model, in which the polyions are represented as long parallel cylinders, which contain a surface charge of uniform density. The numerical values of the parameters that describe the model are chosen to correspond with the values obtained for a real polyelectrolyte solution. A theoretical basis for an effective charge parameter was discussed by Manning [13]. For the cylindrical cell model, the linear charge density can be written

$$\zeta = e\lambda Z/h \tag{26}$$

where h is the contour length of the polymer chain represented by the charged cylinder, and λZ is the number of fixed charges on the chain. The model assumes the dielectric constant to be that of the pure bulk solvent and that the interactions between polyions are negligible if a mobile electrolyte such as NaCl is present in excess. A charge density parameter ξ appears in the statistical–mechanical phase integral for an infinite line-charge model. This parameter is given by

$$\xi = \zeta/\varepsilon kT \tag{27}$$

For all values of $\xi > 1$, the phase integral diverges; hence it is proposed that the linear charge density must be reduced to a value which will make $\xi < 1$. This gives a theoretical prediction of an effective charge if ξ is set equal to 1 for the particular polymer in question. Since $\xi = 2.85\lambda$ for polyphosphates or vinyl polymers in aqueous solution (25°C), where λ is the fraction of monomer units with charges, then the charge would have to be reduced to $1/\xi$ of its value to make $\xi = 1$. For a fully charged vinyl polyelectrolyte, the model thus predicts an effective charge parameter of $1/2.85 = 0.35$. This value is quite similar to experimentally determined values of counterion activity coefficients in systems with added simple electrolyte [10, 13].

In terms of the prediction of polyelectrolyte dimensions based on the equivalent chain model, the use of an effective charge, based on the counterion activity coefficient or the osmotic coefficient, would reduce the values of α predicted by Eqs. (19) and (21).

(*b*) *Models with Continuous Charge Distribution*

Another method of approach to the problem of polymer coil expansion on addition of electrostatic charge is to consider as the polymer model a spherically symmetrical charged cloud rather than a coiled chain. The polymer expansion can then be considered to result from the electrostatic repulsion of such a cloud which will modify the assumed end-to-end distribution function, as formulated in Eq. (7).

The basic ideas of this model are as follows: The polymer molecule is envisaged as a charged sphere whose hydrodynamic radius R_s is equivalent to that of the real polymer coil. The overall concentration of backbone charges is the same as in the coil and is given by

$$c_f = 3v/4\pi R_s^3 \tag{28}$$

where v is the total charge per polymer molecule. The concentration of mobile electrolyte is given by c_s; if $c_s > c_f$, then the charge density at any point can be taken to be

$$\rho = e \cdot [c_f + c_s \exp(-e\psi/kT) - c_s \exp(+e\psi/kT)] \tag{29}$$

The factor ψ is the electrostatic potential at a distance r from any given backbone ion. Because the backbone ions are fixed, their concentration cannot be multiplied by an exponential weighting factor as is done with the mobile ions. Thus, Poisson's equation

$$\Delta\psi = -4\pi\rho/\varepsilon \tag{30}$$

which relates the electrostatic potential to the local charge density and becomes, on combining Eqs. (29) and (30), the Poisson–Boltzmann equation

$$\Delta\psi = (4\pi e/\varepsilon)[-c_f + 2c_s \sinh(e\psi/kT)] \tag{31}$$

Hermans and Overbeek [28] provided a solution for this equation by applying the Debye–Hückel approximation that $\sinh(e\psi/kT) \simeq e\psi/kT$, for $e\psi/kT \ll 1$. They obtained an approximate expression for E_{el}, which can be substituted into Eq. (11), giving for $\kappa R_s \gg 1$

$$(h^*/h_0^*)^2 - 1 = 3v^2/8V_s c_s = (3v/8)(c_f/c_s) \tag{32}$$

where V_s is the volume of the equivalent sphere. In terms of the expansion factor, Eq. (32) becomes

$$\alpha^5 - \alpha^3 = 3v^2/8(V_s)_0 c_s \tag{33}$$

The disadvantage of the above approach is that it is limited to the case where the potentials within the charged sphere are low enough that the Debye–Hückel approximation can be employed. This condition limits the use of theory to small concentrations of added electrolyte and could even be contradictory to the requirement that κR_s be large, which implies that there be a substantial concentration of free charges and/or a moderately large polymer coil size.

It was demonstrated by Lifson [29] that a better approach to the problem is to use the Donnan assumption of electroneutrality within the polymer coil. With an excess concentration of added neutral electrolyte, this assumption appears to be especially valid. Hence, for $\rho = 0$

$$\sinh(e\psi/kT) = c_f/2c_s \tag{34}$$

This leads to an expression for electrostatic energy

$$E_{el} = vkT \int_0^1 \sinh^{-1}(c_f\lambda/2c_s)\, d\lambda \tag{35}$$

As before, substitution in Eq. (11) gives an expression for the expansion ratio:

$$(h^*/h_0^*)^2 - 1 = 3c_s V_s\{[1 + (v/2c_s V_s)^2]^{1/2} - 1\} \tag{36}$$

Where the concentration of added electrolyte is sufficient to make $v/2c_s V_s \ll 1$, Eq. (35) becomes identical to Eq. (32). That is, the Donnan approximation of electroneutrality leads to the same ultimate prediction of chain expansion as the Debye–Hückel approximation of low potentials, regardless of their conceptual difference.

Flory [30] used a somewhat different approach in predicting polyelectrolyte expansion. He calculated the free energy of formation of the charged sphere as the sum of two terms, one of which is due to the difference in free energy of solvent molecules and mobile ions inside and outside the polymer sphere, and the other due to the ordinary thermodynamic interaction between nonionic polymers and solvent. That is, the free energy is composed of an osmotic term and an excluded volume term in the nonionic sense.

As before, the expansive force dA/dh is equated to the elastic retractive force. In this case, the electrostatic energy E_{el} is replaced by A, the sum of the ionic and nonionic free energies. Flory also employed a Gaussian charge distribution instead of the assumed spherical distribution of the previous cases. The expression thus obtained for polymer expansion is

$$(\overline{h^2}/\overline{h_0{}^2}) = (\overline{h_1{}^2}/\overline{h_0{}^2}) + v^2/2.32(\overline{h^2})^{3/2}c_s \tag{37}$$

This applies in the case that the concentration of the added uni-univalent electrolyte is much greater than the polymer charge density. $(\overline{h_1{}^2})$ is the mean-square end-to-end distance of the polymer in the given solvent, if it were uncharged. This nonionic term $(\overline{h_1{}^2}/\overline{h_0{}^2})$ is often quite considerable, because good solvents for ionic polymers are generally good solvents for the corresponding nonionic polymers, as for example aqueous simple electrolyte which is a good solvent for both poly(acrylic acid) and its sodium salt. In the case where the solvent medium is a theta solvent for the uncharged polymer, the nonionic term becomes equal to unity, and Eq. (37) takes the same form as Eq. (33).

Equation (37) can be recast in terms of the expansion factor α. The nonionic part of the expansion has been given previously by Flory [1]. With the ionic and nonionic parts combined, the expression has the form

$$\alpha^5 - \alpha^3 = 2C_M(1 - \theta/T)M^{1/2} + 2(C_I/c_s)M^{1/2} \tag{38}$$

C_M and C_I are calculable quantities, which are constants for any given polymer–solvent system. The expression for C_I is given by

$$C_I = (10^3/4.67)(M/\overline{h_0{}^2})^{3/2}(1/N_{Av}m_s) \tag{39}$$

Flory's expression for the electrostatic expansion factor is identical, except for a numerical factor, with the expression for expansion given in Eq. (25). By recasting the last term in Eq. (38) in terms of the parameter z_{el} given in Eq. (24), Flory's equation becomes

$$\alpha^5 - \alpha^3 = 2.60z_{el} \tag{40}$$

In the following section, theories that relate polyelectrolyte expansion to an ionic excluded volume concept will be discussed, and the various theories will be related to this factor.

(c) *Excluded Volume Theories*

In most of the theories describing the expansion of nonionic polymers, the expansion factor α is related to a parameter z, as for example in Flory's expression [1]

$$\alpha^5 - \alpha^3 = 2.60z \tag{41}$$

which is formally identical with Eq. (40) for polyelectrolyte expansion. In simplest terms, the parameter z is given by

$$z = N\beta/V_0 \tag{42}$$

where N is the number of segments per polymer coil, V_0 the unperturbed volume ($=4\pi(\overline{s_0{}^2})^{3/2}/3$), and β the excluded volume per segment, characterizing the interaction between a pair of segments. Taking $N = M/m_s$, as defined previously, and assuming that $\overline{h_0{}^2} = 6\overline{s_0{}^2}$, the excluded volume parameter assumes its more familiar form

$$z = (3/2\pi)^{3/2} m_s^{-2} (M/\overline{h_0{}^2})^{3/2} \beta M^{1/2} \tag{43}$$

A number of functional relationships between α and z have been derived, including those by Flory [1], Ptitsyn [31], Fixman [32], Kurata *et al* [33], and Alexandrowicz [34], among others.

The excluded volume approach can be used to describe the expansion of ionic polymers if it is considered that chain expansion in any good solvent is conceptually the same whether it arises from ionic or nonionic interactions or both. Thus by replacing β in Eq. (43) with the factor $10^3\lambda^2/2N_{Av}c_s$, which has the dimensions of volume, the electrostatic excluded volume parameter z_{el} is obtained

$$z_{el} = (3/2\pi)^{3/2} m_s^{-2} (M/\overline{h_0{}^2})^{3/2} (10^3\lambda^2/2N_{Av}c_s) M^{1/2} \tag{44}$$

Combining this equation with Eqs. (2) and (20) results in the form of z_{el} given in Eq. (24). It will be seen subsequently that most of the theories on polyelectrolyte expansion can be expressed in terms of this parameter. It should be noted that the degree of ionization λ can serve as an adjustable parameter if the counterion activity coefficient is included in it.

The theory of Fixman [32], with respect to the expansion of nonionic polymers, is essentially a modification of Flory's. The difference lies in the manner in which the segment density varies with the distance from the center of mass. In Flory's theory, the segment density is taken to be that given by a Gaussian distribution. If it had been assumed to be constant, the expression for expansion would only have differed by a numerical factor, as evidenced by the theory of Hermans and Overbeek. Fixman obtains a segment density which is, for any radius of gyration R, given by

$$\rho(r) \simeq (N/2\pi^2)[r^2(2R^2 - r^2)^{1/2}]^{-1}H(2R^2 - r^2) \tag{45}$$

where H is the step function. The use of this segment density function, rather than a continuous one, results in the relationship between α and z given by

$$\alpha^3 - 1 = 2z \tag{46}$$

In his treatment of polyelectrolyte expansion, Fixman does not begin with an excluded volume approach [35]. The essence of his theory is the assumption that repulsive forces between segments in the interior of the molecule, although large, make no contribution to the expansion. The polymer model is the so-called fuzzy sphere model, which superimposes a continuous distribution of polymer segments on a background cloud of charge. The major cause of expansion is attributed to repulsion between segments outside and inside the background cloud.

The equation for the expansion ratio, which is obtained from this approach, is given by

$$\alpha^3 - \alpha \simeq \alpha^3 - 1 = \tfrac{1}{2}(\varepsilon_0/\varepsilon)(M/m_s)^{1/2}(\lambda/b_0\kappa) \tag{47}$$

The factor $\varepsilon_0/\varepsilon$ introduces an effective dielectric constant into the theory, which is equivalent to introducing ion-binding or an effective charge parameter, since the value of ε is adjustable. The factor b_0 is defined by $b_0 = R_s/N^{1/2}$, where R_s is the hydrodynamic radius and N is the number of segments.

Although Fixman does not explicitly use an excluded volume approach, it is certainly interesting to compare his theory with others that do. The comparison can be readily forced by assuming that $R_s{}^2 = \overline{h_0{}^2}/6$ or $b_0 = L/6^{1/2}$, where $L = (\overline{h_0{}^2}/N)^{1/2}$ is the segment length. Thus, if β in Eq. (43) is replaced by $(\pi^{3/2}/3)(\lambda\varepsilon_0/\varepsilon)(L^2/\kappa)$, which has the dimensions of volume, one obtains

$$z'_{\text{el}} = \tfrac{1}{4}(\varepsilon_0/\varepsilon)(M/m_s)^{1/2}(\lambda/b_0\kappa) \tag{48}$$

and Eq. (47) for polyelectrolyte expansion becomes formally identical to Eq. (46), Fixman's general equation for polymer expansion.

The factor β in the parameter z'_{el}, as shown above, is proportional to $L^2\kappa^{-1}$. This implicit excluded volume can be rationalized from Fixman's polyelectrolyte model. According to his key assumption, the spherical interior of the polyelectrolyte coil is essentially unperturbed. Hence, the radius of this inner sphere should be proportional to L, the link length of the unperturbed chain. Fixman's view is that repulsion between segments in the inner sphere and those surrounding it is the cause of expansion. The excluded volume must then be a concentric shell surrounding this inner sphere, the thickness of which should be proportional to κ^{-1}, the thickness of an ionic atmosphere. Hence, β must be proportional to the volume of such a concentric atmosphere, and thus $\beta \propto L^2\kappa^{-1}$, as given in Fixman's theory. This model, when subjected to this physical interpretation of excluded volume, does not seem reasonable for any but the lowest expansions, because of the assymmetry of this type of expansion. However, this does not rule out the possibility that such a physical interpretation of excluded volume is incorrect. The agreement with experimental results, which is the ultimate test of any model, will be discussed later.

Kurata [36] obtained an expression for polyelectrolyte expansion by employing an equivalent ellipsoid model for the polyelectrolyte. This model, employed originally in predicting the excluded volume effect of nonionic polymers [33], does not assume isotropic expansion of the polymer conformation in minimizing the free energy of intersegmental interaction. Rather, it assumes the expansion ratio to be different along one axis of the polymer coil than along the orthogonal axes. For nonionic polymers, the theory gives a result similar to that of Fixman, i.e., Eq. (46). For ionic polymers, the Donnan assumption of electroneutrality [Eq. (32)] was employed in solving the Poisson–Boltzmann equation; the charge distribution along each axis was considered to be Gaussian. The theory predicts for polyelectrolytes,

$$\alpha^3 - 1 \simeq 2z_{el} \tag{49}$$

where z_{el} is given by Eq. (44), which, it should be noted, differs from z'_{el} in Eq. (48).

Kurata's theory predicts that the polyelectrolyte conformation is highly rodlike under conditions of high charge. This is qualitatively consistent with the models used to explain the high-frequency dielectric relaxation of polyelectrolytes in dilute solution, as discussed in Chapter III, Section C3.

Another excluded volume approach, perhaps more reasonable, from the point of view of a physical interpretation, is that taken by Alexandrowicz [37]. He arrives at a factor β, which is proportional to $L\kappa^{-2}$, for all but extremely low ionic strengths. His model is a polymer coil in which the polymer filament is covered by a sheath from which other portions of the filament are excluded. The radius of such a sheath should be proportional to κ^{-1}, as long as L is somewhat greater than κ^{-1}. This will result in an excluded volume, $\beta \propto L\kappa^{-2}$, since $\pi L\kappa^{-2}$ is the volume of the ionic sheath surrounding each chain link.

Actually, the excluded volume must be a net value which is related to an effective radius of exclusion, which Alexandrowicz terms $\chi = (b^2 - b_\theta^2)^{1/2}$. The parameter $b \simeq \kappa^{-1}$, and b_θ is the value of b at the theta point. Since the theta point for ionic polymers is reached at some finite salt concentration, in general, then the value of $b(\simeq\kappa^{-1})$ must be greater than zero at this point. The excluded volume must, of course, be equal to zero at the theta temperature; this means that there must be a negative contribution to the radius of exclusion because of nonionic interactions. Hence, this justifies the form of the parameter χ.

Thus Alexandrowicz obtains for the excluded volume, $\beta = k_e \pi \chi^2 L$, where k_e is a constant of the order of 1, and which is in fact an adjustable parameter. The parameter β can then be used in Eq. (43) for z_{el}, which is then applied to Alexandrowicz' general expression for expansion [34], given by

$$\alpha^2 = 1.7z^{1/2} + 0.35 \qquad \text{for} \quad \alpha^2 > 2 \tag{50}$$

At extremely low values of c_s, the concept of excluded volume must be modified to account for the case where the ionic sheath radius b becomes larger than the segment length L. In this case, the size of the freely jointed link is determined by b, which means that the segment length must increase as c_s decreases. For this case, then, the segment length is denoted as L^*, and the excluded volume becomes $\beta^* = k_e \pi \chi^2 L^* = \beta(\chi/L)$, since L^* is taken to be equal to χ in this situation. The unperturbed mean-square end-to-end distance $\overline{h_0^2}$ must also change as L^* increases; i.e., $(\overline{h_0^2})^* = N^* L^{*2} = N(L/\chi) \cdot \chi^2 = \overline{h_0^2}(\chi/L)$. Thus the excluded volume parameter becomes $z^* = z(L/\chi)^{1/2}$. A single unified expression will result if one introduces an additional parameter γ which has the values $\gamma = 1$ for $b < L$ and $\gamma = (L/\chi)^{1/2}$ for $b > L$. Then z can be replaced by $z\gamma$ in Eq. (50) to give the general expression for polyelectrolyte expansion:

$$\alpha^2 = 1.7(z\gamma)^{1/2} + 0.35 \qquad \text{for} \quad \alpha^2 > 2 \tag{51}$$

For small expansions, the simple equation shown above does not hold, and an exact solution of Alexandrowicz' equation must be supplied [37].

Finally, Kurata [38] and, independently, Alexandrowicz [39] provided another expression for polyelectrolyte expansion in terms of excluded volume. The method, which employs an equivalent Gaussian chain, is based on a uniform expansion model of perturbed chains developed by Fujita *et al.* [40], which leads to a general equation for the excluded volume effect, which is expressed in terms of probabilities for segment contacts in the real chain. The equation obtained is

$$(\alpha^3 - 1) + (3/8)(\alpha^5 - \alpha^3) = (5/2)z \tag{52}$$

or

$$\alpha^5/5 + \alpha^3/3 = (4/3)z + 8/15$$

and numerically is in rather close agreement [38] with the general equation developed earlier by Ptitsyn [31]

$$\alpha^2 = (1/4.68)[3.68 + (1 + 9.36z)^{2/3}] \tag{53}$$

It can be seen that Eq. (52), the Kurata–Alexandrowicz equation, is really a hybrid of the Flory and Fixman equations [(41) and (46), respectively], and as a result, predicts intermediate values of polymer expansion.

The following is a summary of polyelectrolyte expansion relationships in terms of excluded volume, arranged in order of decreasing dependence of α on z_{el} or z'_{el}. In each case, exact relationships are given, but it should be recalled that most theories contain an adjustable parameter, so that the form of the equation is of greater significance than the absolute value.

(i) Katchalsky–Lifson

$$\alpha - 1/\alpha = 1.77z_{el} \tag{19}$$

(ii) Fixman

$$\alpha^3 - 1 = 2z'_{el} \tag{47}$$

(iii) Kurata

$$\alpha^3 - 1 \simeq 2z_{el} \tag{49}$$

(iv) Ptitsyn

$$\alpha^2 = (1/4.68)[3.68 + (1 + 9.36z_{el})^{2/3}] \tag{53}$$

(v) Alexandrowicz

$$\alpha^2 = 1.7(z_{el}\gamma)^{1/2} + 0.35 \qquad \text{for} \quad \alpha^2 > 2 \tag{51}$$

(vi) Kurata–Alexandrowicz

$$\alpha^5/5 + \alpha^3/3 = (4/3)z_{el} + 8/15 \tag{53}$$

(vii) Hermans–Overbeek–Lifson

$$\alpha^5 - \alpha^3 = 7.4z_{el} \tag{33}$$

(viii) Flory

$$\alpha^5 - \alpha^3 = 2.60z_{el} \tag{40}$$

(ix) Noda *et al.*

$$\alpha^5 - \alpha^3 = 1.77 z_{el} \tag{25}$$

In all of the above equations, the excluded volume parameter is proportional to $M^{1/2}/c_s$, except for the Fixman equation (47), where $z'_{el} \propto M^{1/2}/c_s^{1/2}$, and the Alexandrowicz equation (51) at low ionic strengths, where $z_{el}^* \propto M^{1/2}/c_s^{3/4}$.

(*d*) *Comparison with Experiments*

Some work has been done to test the theories that employ excluded volume considerations in predicting the expansion of polyelectrolytes. Ideally, polymer dimensions in dilute solution should be obtained by an absolute method, which gives the coil radius directly from an experimentally measurable quantity. Dyssymmetry measurements in light scattering give the radius of gyration of high polymers directly [1]. However, light scattering is an exacting and time-consuming experimental procedure, so the experimentally simpler measurement of intrinsic viscosity is generally used to determine polymer expansion ratios. If one is only interested in relative expansion, then absolute values of polymer dimensions are not important. For high polymers, the intrinsic viscosity $[\eta]$, is given by the relationship [1]

$$[\eta] = \Phi(\overline{h^2})^{3/2}/M \tag{54}$$

where Φ is a "universal parameter," whose value is only approximately known. Since $(\overline{h^2}) = \alpha^2(\overline{h_0{}^2})$, then one can write

$$[\eta] = \Phi(\overline{h_0{}^2/M})^{3/2} M^{1/2} \alpha^3 \tag{55}$$

Since $(h_0{}^2/M)$ is a constant for a given polymer, then Eq. (55) becomes

$$[\eta] = K M^{1/2} \alpha^3 \tag{56}$$

where $K = \Phi(\overline{h_0{}^2/M})^{3/2}$. At theta conditions, $\alpha = 1$, of course, and $[\eta]_\theta = KM^{1/2}$. Hence one obtains

$$\alpha^3 = [\eta]/[\eta]_\theta \tag{57}$$

The value of α obtained from intrinsic viscosity measurements in good solvents and in a theta solvent should be denoted as α_η. This viscosity expansion ratio is only equal to the true expansion ratio if the value of Φ remains a true constant with expansion. Φ is obtained by considering that the polymer coil acts as a hydrodynamically equivalent sphere of radius R_s, which is proportional to $(\overline{h^2})$. However, this proportionality could very well change as the expansion ratio becomes significant. For very large expansions, as one finds with dilute polyelectrolyte solutions with no added salt, one would expect the hydrodynamic properties of the polymer to be quite different from the unperturbed coil. In fact, the highly expanded polymer molecule could be expected to act more like a free draining coil than as a hydrodynamically equivalent sphere.

The results of viscometric measurements on the PAA–Na system with added simple electrolyte have been used to compare several theories which predict polyelectrolyte expansions in terms of electrostatic excluded volume. Alexandrowicz [37] used the viscometric data of Takahashi and Nagasawa [41] on fully neutralized PAA. It should be recalled that z_{el} is proportional to $M^{1/2}$ [cf. Eq. (44)]. Since the theory predicts α^2 to be linear with $z^{1/2}$, he therefore plotted α^2 versus $M^{1/4}$ for PNaA/NaBr, as shown in Fig. 5.1. The linearity of the plots at each of the ionic strengths and their fit to a common intercept (0.35) indicate an agreement with the predicted dependence of α on M, even at low ionic strengths.

If α^2 is then plotted against the parameter $M^{1/4}\chi\gamma^{1/2}$, the relationship shown in Fig. 5.2 is obtained. [In Eq. (51), $z \propto \beta M^{1/2} \propto \chi^2 M^{1/2}$.] The values of χ versus c_s were found from the slopes of the straight lines in Fig. 5.1; this procedure, in effect, evaluates the adjustable parameter k_e, which relates β and χ. For $\alpha^2 > 2$, the plot of α^2 versus $M^{1/4}\chi\gamma^{1/2}$ is linear, while the line predicted by Fixman's equation (47) diverges from the observed dependence at higher values of α. It should be pointed out that by choosing a different value of the adjustable parameter k_e, a much better

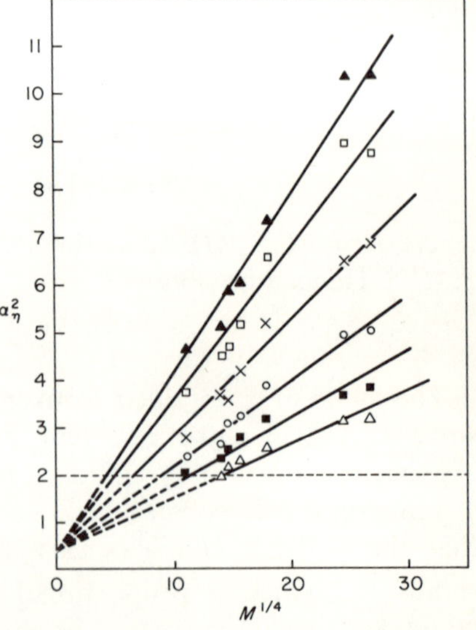

FIG. 5.1. α_η^2 versus $M^{1/4}$, for PNaA ($\alpha^2 > 2$). $c_s = 0.0025$ (▲); 0.005 (□); 0.01 (×); 0.025 (○); 0.05 (■); and 0.1 (△) mol/liter added NaBr [37a].

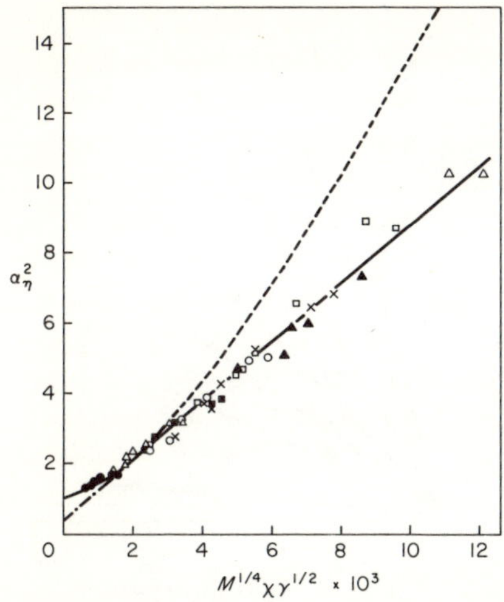

FIG. 5.2. α_η^2 versus $M^{1/4}\chi\gamma^{1/2}$. The solid line represents Eq. (51), which extrapolates to 0.35 for $\alpha^2 < 2$. The dashed line represents Eq. (47). The values of c_s are as in Fig. 5.1, with $c_s = 0.5$ (●) included as well [37a].

fit with the Fixman equation could be obtained, although the plot would still diverge at very high expansion ratios. The Flory equation (40) diverges in the opposite sense, as pointed out in the original paper by Takahashi and Nagasawa [41].

Noda *et al.* [26] showed that the Fixman equation (47) qualitatively describes the expansion of partially and fully neutralized polyacrylic acid in aqueous NaBr, in the case where degrees of expansion are fairly small. Equation (47) predicts that α^2 varies linearly with z and therefore with $M^{1/2}$. Accordingly, they plotted $[\eta]/M^{1/2}$, which is proportional to α_η^3, versus $M^{1/2}$ and obtained a series of straight lines for various degrees of ionization and ionic strength of added NaBr, as illustrated in Figs. 5.3 and 5.4. The former figure, pertaining to fully neutralized PAA, was taken from the study of Takahashi and Nagasawa [41].

The electrostatic part of the expansion factor $\alpha_{\eta e}$ is defined by [26]

$$\alpha_{\eta e}^3 = \alpha_\eta^3 - (\lim_{c_s \to \infty} \alpha_\eta^3 - 1) \tag{58}$$

This parameter, $\alpha_{\eta e}^3$, was plotted against $(M/c_s)^{1/2}$ for various degrees of neutralization and ionic strength of NaBr. $\alpha_{\eta e}^3$ was also plotted against $M^{1/2}/c_s$. Examples of these plots are shown in Figs. 5.5 and 5.6. Fixman's excluded volume β, it should be recalled, is proportional to $\kappa^{-1} \propto c_s^{-1/2}$. Hence, Eq. (47) predicts that α^3 varies as $z \propto (M/c_s)^{1/2}$. The linearity and the common slope in Fig. 5.5 indicate clearly the qualitative agreement with the theory of Fixman, while the noncoincidence of the curves in Fig. 5.6 must be considered a serious drawback to the theories that predict $\beta \propto M^{1/2}/c_s$.

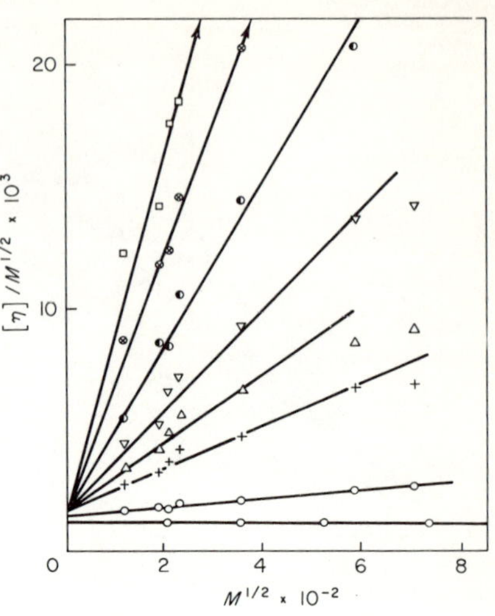

FIG. 5.3. $[\eta]/M^{1/2}$ versus $M^{1/2}$ for PNaA. $c_s = 1.506$ (●); 0.502 (○); 0.100 (+); 0.0502 (△); 0.0251 (▽); 0.0100 (◐); 0.00502 (⊗); 0.00251 (□) mol/liter NaBr [41].

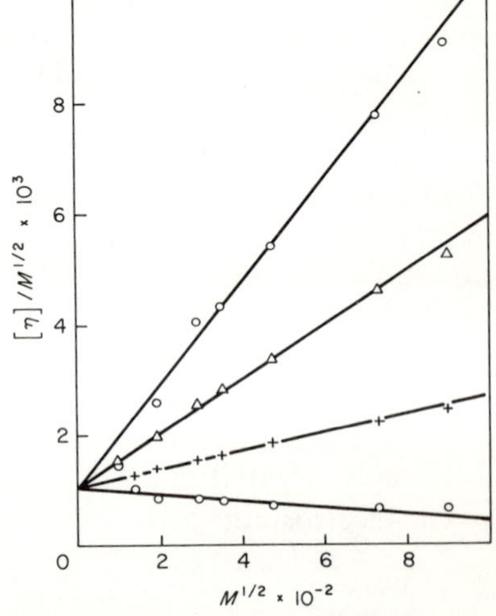

FIG. 5.4. $[\eta]/M^{1/2}$ versus $M^{1/2}$ for PAA–0.103Na. $C_s = 0.01$ (○); 0.025 (△); 0.1 (+); 0.5 (◑) [26].

This seeming contradiction that two different theories of polyelectrolyte expansion (i.e., those of Fixman and Alexandrowicz) equally well support the same set of data may not actually present any problem. It should be noted that the expansion ratios used in support of Fixman's theory are mainly the lower values which occur at the higher strengths of added electrolyte. The expansions used in support of Alexandrowicz theory are all greater than $\alpha^2 = 2$, since his theory converges with that of Fixman at low values of expansion.

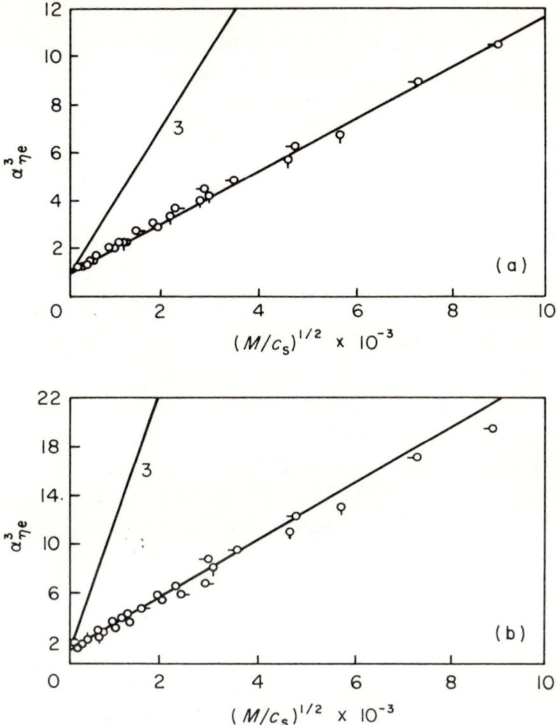

FIG. 5.5. $\alpha^3_{\eta e}$ versus $(M/c_s)^{1/2}$ (a) for PAA–0.103Na and (b) for PAA–0.6Na. $c_s = 0.5$ (○ with tick up); 0.1 (○–); 0.025 (○ with tick down); 0.01 (–○) mol/liter NaBr. Line 3 represents Eq. (47) [26].

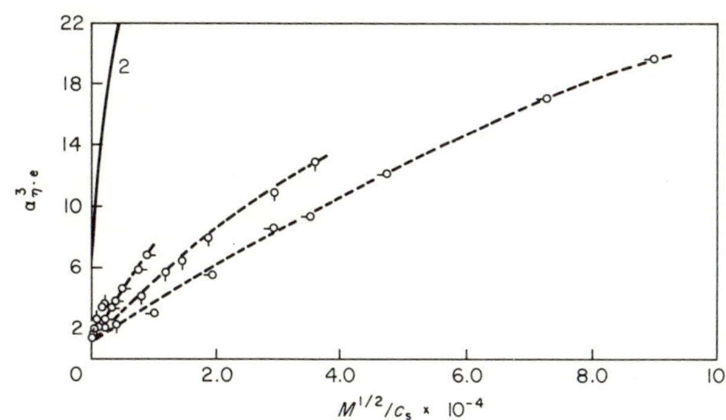

FIG. 5.6. $\alpha^2_{\eta e}$ versus $M^{1/2}/c_s$ for PAA–0.6Na. c_s values same as in Fig. 5.5. Line 2 represents Eq. (40) [26].

As was pointed out earlier, the uncertainties of using expansion ratios calculated from viscosity measurements may make any or all of the theories presented invalid. It has been estimated that in the free-draining limit [4]

$$\alpha_\eta{}^3 = \alpha^{5/2} \tag{59}$$

This would change Fixman's equation, for example, to the form

$$\alpha_\eta^{3.6} - 1 = 2z_{\mathrm{el}} \tag{60}$$

at high expansion ratios, and bring it more in line with the observed dependence of α on M. A more direct measurement of at least some of these expansion ratios would help to confirm or deny the validity of using expansion ratios obtained from intrinsic viscosities.

Finally, another form of "experimental" testing of excluded volume theories should be mentioned. Computer studies of non–self-intersecting walks on a lattice have also been used to test the applicability of the various theories. Alexandrowicz and Accad [42] used a Monte Carlo technique to generate simulated polymer chains of high molecular weight. The computed expansion ratios of the simulated chains were then compared for their fit to four theoretical dependences on chain length, i.e., Fixman ($\alpha^3 - 1 \propto N^{1/2}$), Flory ($\alpha^5 - \alpha^3 \propto N^{1/2}$), Alexandrowicz ($\alpha^2 - 0.035 \propto N^{1/2}$) and Kurata–Alexandrowicz ($\alpha^5/5 + \alpha^3/3 - 8/15 \propto N^{1/2}$), by fitting the predicted lines with the Monte Carlo results at $\alpha = 2$. As shown in Fig. 5.7, the best fit was obtained with the latter equation, that of Kurata–Alexandrowicz.

B. RUBBER ELASTICITY

The effect of ions on the elasticity of polymer networks is in general twofold: direct cross-linking effects resulting from interchain ion association (i.e., multiplet or cluster formation) and indirect effects through changes in chain configuration. Ionic cross-linking is considered to be a time-dependent phenomenon, as discussed in Chapters III and IV, although for practical purposes, the time constant for cross-link breakdown may be considerably longer than conceivable experimental times. In order to study the effects of ions not related to cross-linking, systems must be studied in which ionic cross-linking is absent or can be suppressed.

An ionizable network of convenience can be readily obtained from mixtures of poly(acrylic acid) and poly(vinyl alcohol), which can be cross-linked thermally in a wide variety of relative concentrations [43]. Ion aggregation effects can be eliminated by allowing the gel to swell in aqueous media. Two experimental situations for determining the modulus of elasticity will be considered here: one in which there is free exchange with the

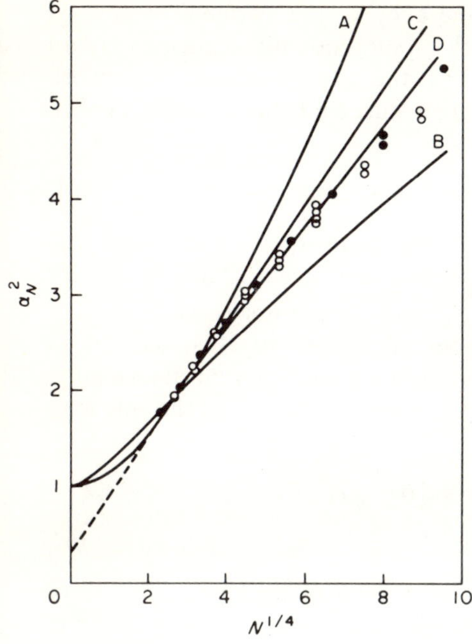

FIG. 5.7. The expansion coefficient of the chain end-to-end distance α_N as a function of the chain length N. Lines A, B, C, and D are derived from Eqs. (47), (40), (51), and (52), respectively, [42].

embedding medium during stretching, hence constant chemical potential, and one in which no material exchange takes place, i.e., constant volume, or more correctly constant mass.

1. Elasticity at Constant Volume [44]

The first part of the subsequent discussion will be devoted to the derivation of a theory for the elasticity of a two-component rubber, one component of which is ionizable. Subsequently, the experimental aspects will be discussed, and the theoretical predictions compared with experimental results. Finally, several reasons will be suggested which might explain the discrepancy between theory and experiment.

(*a*) *Theory*

Consider a unit cube of a rubber which consists of a mutually interpenetrating network of two types of chains. In the present case, the two types of chains are PAA and PVA. The nature of the cross-link in the PAA–PVA rubber system is taken to be a simple esterification. It is assumed, for the purposes of this calculation, that each cross-link is tetrafunctional and that from each cross-link emanate two chains of PAA and two of PVA.

Suppose that a unit cube of such a polymer undergoes isotropic swelling in the ratio $1/V_r$, where V_r is the total volume fraction of polymer in the mixture. The length of each edge of the cube in the swollen, unstrained state is given by $1/V_r^{1/3}$. If the network is subsequently deformed by an external force to the dimensions $\gamma_1, \gamma_2, \gamma_3$, the total difference in network entropy between the unstrained, unswollen state and the strained, swollen state for each type of chain is given by

$$\Delta S_p' = -\tfrac{1}{2} N_p k(\overline{r_p^2}_0/\overline{r_p^2}_f)(\gamma_1^2 + \gamma_2^2 + \gamma_3^2 - 3) \tag{61}$$

The expressions $\overline{r_p^2}_0$ and $\overline{r_p^2}_f$ represent the mean-square end-to-end distances of the chains in the unstrained, unswollen network and in the free state, respectively. It is assumed that $\overline{r_p^2}_0$ and $\overline{r_p^2}_f$ are not equal.

For a network consisting of two types of chains, the total entropy of deformation of the network $\Delta S_0'$, including both swelling and deformation resulting from the externally applied force, is simply the sum of the entropies of the components

$$\Delta S_0' = \Delta S_1' + \Delta S_2' = -\tfrac{1}{2} k[(\overline{r_1^2}_0/\overline{r_1^2}_f) N_1 + (\overline{r_2^2}_0/\overline{r_2^2}_f) N_2](\gamma_1^2 + \gamma_2^2 + \gamma_3^2 - 3) \tag{62}$$

where the subscripts 1 and 2 on S, r, and N refer to the two components. The change in entropy due to the isotropic swelling in the ratio $1/V_r^{1/3}$ in the absence of stress is

$$\Delta S_0 = -\tfrac{1}{2} k[(\overline{r_1^2}_0/\overline{r_1^2}_f) N_1 + (\overline{r_2^2}_0/\overline{r_2^2}_f) N_2](3V_r^{-2/3} - 3) \tag{63}$$

Therefore, the entropy of deformation of the swollen network ΔS is simply the difference of $\Delta S_0'$ and ΔS_0

$$\Delta S = \Delta S_0' - \Delta S_0 = -\tfrac{1}{2} k[(\overline{r_1^2}_0/\overline{r_1^2}_f) N_1 + (\overline{r_2^2}_0/\overline{r_2^2}_f) N_2](\gamma_1^2 + \gamma_2^2 + \gamma_3^2 - 3V_r^{-2/3}) \tag{64}$$

It is more convenient to express ΔS in terms of the extensions $\lambda_1, \lambda_2, \lambda_3$, referred to the swollen unstrained network, where $\lambda_1 = \gamma_1 V_r^{1/3}$, etc. Thus

$$\Delta S = -\tfrac{1}{2} k[(\overline{r_1^2}_0/\overline{r_1^2}_f) N_1 + (\overline{r_2^2}_0/\overline{r_2^2}_f) N_2](\lambda_1^2 + \lambda_2^2 + \lambda_3^3 - 3) V_r^{-2/3} \tag{65}$$

If changes in the internal energy of the system on deformation of the swollen network are neglected, the force of extension f of a swollen rubber of length L and volume V is given by

$$f = -T(\partial \Delta S/\partial L)_{T,V} \tag{66}$$

For simple extension of a rubber, $\lambda_1 = L/L_1$, where L_1 is the value of L

in the unstrained, swollen state. Therefore,

$$f = -(T/L_1)(\partial \Delta S/\partial \lambda_1)_{T,V} \tag{67}$$

If the rubber is considered to be incompressible, then $V = V_1$, the unstrained, swollen volume, and

$$\lambda_2 = \lambda_3 = (\lambda_1)^{-1/2} \tag{68}$$

Hence, the relationship

$$\lambda_1{}^2 + \lambda_2{}^2 + \lambda_3{}^2 - 3 = \lambda_1{}^2 + 2/\lambda_1 - 3 \tag{69}$$

is obtained. By substituting Eq. (69) in Eq. (65) and differentiating the result in Eq. (67), the expression for the force is obtained

$$f = (kT/L_1)[(\overline{r_1{}^2{}_0}/\overline{r_1{}^2{}_f})N_1 + (\overline{r_2{}^2{}_0}/\overline{r_2{}^2{}_f})N_2](\lambda_1 - 1/\lambda_1{}^2)V_r^{-2/3} \tag{70}$$

The Young's modulus E is then given by the ratio of stress to strain

$$\begin{aligned} E &= (f/A_0)/(L/L_1 - 1) \\ &= (kT/V_1)[(\overline{r_1{}^2{}_0}/\overline{r_1{}^2{}_f})N_1 + (\overline{r_2{}^2{}_0}/\overline{r_2{}^2{}_f})N_2] \\ &\quad \times [(\lambda_1 - 1/\lambda_1{}^2)/(\lambda_1 - 1)]V_r^{-2/3} \end{aligned} \tag{71}$$

In the limit of small strains, this becomes

$$E = (3kT/V_1)[(\overline{r_1{}^2{}_0}/\overline{r_1{}^2{}_f})N_1 + (\overline{r_2{}^2{}_0}/\overline{r_2{}^2{}_f})N_2]V_r^{-2/3} \tag{72}$$

If V_0 is the unswollen, unstrained volume, then $V_1 = V_0/V_r$, and the Young's modulus becomes

$$E = (3kT/V_0)[(\overline{r_1{}^2{}_0}/\overline{r_1{}^2{}_f})N_1 + (\overline{r_2{}^2{}_0}/\overline{r_2{}^2{}_f})N_2]V_r^{1/3} \tag{73}$$

It is clear, then, in the simplest approximation where $\overline{r_0{}^2}/\overline{r_f{}^2}$ is considered to be unity and $(N_1 + N_2)$ becomes the total number of chains, the equation for the Young's modulus assumes the familiar form

$$E = E_0 V_r^{1/3} \tag{74}$$

where E_0 is the modulus of the unswollen network.

Now consider the specific case of a rubber formed from a mixture of PAA (monomeric mole fraction m) and PVA (monomeric mole fraction $[1 - m]$). The mixture is cast as a film from aqueous solution, cross-linked by heating, and subsequently swollen in water. Since it was assumed that each cross-link is the junction of two chains of PAA and two chains of

PVA, it must hold that

$$N_{\mathrm{PAA}} = N_{\mathrm{PVA}} = N/2 \tag{75}$$

Therefore, the degrees of polymerization between cross-links P_{p} are in the ratio of the monomeric mole fractions; that is,

$$P_{\mathrm{PVA}}/P_{\mathrm{PAA}} = (1 - m)/m \tag{76}$$

The relative values of $\overline{r_{\mathrm{PVA}}{}^2{}_0}$ and $\overline{r_{\mathrm{PAA}}{}^2{}_0}$ can be deduced by balancing the force on each cross-link, keeping in mind that each cross-link must be the junction of two PAA chains and two PVA chains. The retractive force exerted by each chain is proportional to the end-to-end distance r as given by [45]

$$f = 3kT\, r/\overline{r^2}_{\mathrm{f}} \tag{77}$$

Since the average tension exerted on the PAA chains must be equal and opposite to the average tension exerted on the PVA chains in any one direction, and considering that the sample is isotropic, the relationship

$$(\overline{r_{\mathrm{PAA}}{}^2{}_0})^{1/2}/\overline{r_{\mathrm{PAA}}{}^2{}_{\mathrm{f}}} = (\overline{r_{\mathrm{PVA}}{}^2{}_0})^{1/2}/\overline{r_{\mathrm{PVA}}{}^2{}_{\mathrm{f}}} \tag{78}$$

can be written. A similar relationship holds for the ionic form

$$(\overline{r_{\mathrm{PNaA}}{}^2{}_0})^{1/2}/\overline{r_{\mathrm{PNaA}}{}^2{}_{\mathrm{f}}} = (\overline{r_{\mathrm{PVA}}{}^2{}_0})^{1/2}/\overline{r_{\mathrm{PVA'}}{}^2{}_{\mathrm{f}}} \tag{79}$$

where the symbol PVA′ refers to the basic medium, and PNaA is poly(sodium acrylate). It should be recalled, however, that it is not the mean-square end-to-end distances of the two types of chains that are equal, but only the average tension under which the chains find themselves.

It is more convenient to rewrite Eq. (73) using Eq. (75), with PAA representing component 1 and PVA component 2

$$E = (3NkT/2V_0)[(\overline{r_{\mathrm{PAA}}{}^2{}_0}/\overline{r_{\mathrm{PAA}}{}^2{}_{\mathrm{f}}}) + (\overline{r_{\mathrm{PVA}}{}^2{}_0}/\overline{r_{\mathrm{PVA}}{}^2{}_{\mathrm{f}}})]V_{\mathrm{r}}^{1/3} \tag{80}$$

Substitution of Eq. (78) in Eq. (80), for the polymer in the acidic form, gives

$$E_{\mathrm{acidic}} = (3NkT/2V_0)[(\overline{r_{\mathrm{PAA}}{}^2{}_0}/\overline{r_{\mathrm{PAA}}{}^2{}_{\mathrm{f}}}) + (\overline{r_{\mathrm{PVA}}{}^2{}_{\mathrm{f}}}/\overline{r_{\mathrm{PAA}}{}^2{}_{\mathrm{f}}})^2(\overline{r_{\mathrm{PAA}}{}^2{}_0}/\overline{r_{\mathrm{PVA}}{}^2{}_{\mathrm{f}}})]V_{\mathrm{r}}^{1/3} \tag{81}$$

or

$$E_{\mathrm{acidic}} = (3NkT/2V_0)(\overline{r_{\mathrm{PAA}}{}^2{}_0}/\overline{r_{\mathrm{PAA}}{}^2{}_{\mathrm{f}}})(1 + \overline{r_{\mathrm{PVA}}{}^2{}_{\mathrm{f}}}/\overline{r_{\mathrm{PAA}}{}^2{}_{\mathrm{f}}})V_{\mathrm{r}}^{1/3} \tag{82}$$

Similarly, for the polymer in the ionic form

$$E_{\mathrm{ionic}} = (3NkT/2V_0)(\overline{r_{\mathrm{PNaA}}{}^2{}_0}/\overline{r_{\mathrm{PNaA}}{}^2{}_{\mathrm{f}}})(1 + \overline{r_{\mathrm{PVA'}}{}^2{}_{\mathrm{f}}}/\overline{r_{\mathrm{PNaA}}{}^2{}_{\mathrm{f}}})V_{\mathrm{r}}^{1/3} \tag{83}$$

The ratio $\overline{r_{\text{PNaA}}^2}_0/\overline{r_{\text{PAA}}^2}_0$ can be obtained by equating the total root-mean-square distances between the cross-links for both components in acidic and basic forms, at constant volume fraction of polymer

$$(\overline{r_{\text{PAA}}^2}_0)^{1/2} + (\overline{r_{\text{PVA}}^2}_0)^{1/2} = (\overline{r_{\text{PNaA}}^2}_0)^{1/2} + (\overline{r_{\text{PVA}'}^2}_0)^{1/2} \tag{84}$$

Combining Eq. (84) with Eqs. (78) and (79) yields

$$\overline{r_{\text{PAA}}^2}_0(1 + \overline{r_{\text{PVA}}^2}_{\text{f}}/\overline{r_{\text{PAA}}^2}_{\text{f}})^2 = \overline{r_{\text{PNaA}}^2}_0(1 + \overline{r_{\text{PVA}'}^2}_{\text{f}}/\overline{r_{\text{PNaA}}^2}_{\text{f}})^2 \tag{85}$$

From Eqs. (82) and (83), the ratio of the Young's moduli at constant V_{r} is obtained

$$\frac{E_{\text{acidic}}}{E_{\text{ionic}}} = \frac{(\overline{r_{\text{PAA}}^2}_0/\overline{r_{\text{PAA}}^2}_{\text{f}})(\overline{r_{\text{PVA}}^2}_{\text{f}}/\overline{r_{\text{PAA}}^2}_{\text{f}} + 1)}{(\overline{r_{\text{PNaA}}^2}_0/\overline{r_{\text{PNaA}}^2}_{\text{f}})(\overline{r_{\text{PVA}'}^2}_{\text{f}}/\overline{r_{\text{PNaA}}^2}_{\text{f}} + 1)} \tag{86}$$

With Eq. (85) the ratio simplifies to

$$\frac{E_{\text{acidic}}}{E_{\text{ionic}}} = \frac{\overline{r_{\text{PNaA}}^2}_{\text{f}} + \overline{r_{\text{PVA}'}^2}_{\text{f}}}{\overline{r_{\text{PAA}}^2}_{\text{f}} + \overline{r_{\text{PVA}}^2}_{\text{f}}} \tag{87}$$

Since the present system deals with equal numbers of chains of each component, this ratio is simply the ratio of the average mean-square end-to-end distances of the free chains

$$E_{\text{acidic}}/E_{\text{ionic}} = (\overline{r^2}_{\text{f}})_{\text{basic}}/(\overline{r^2}_{\text{f}})_{\text{acidic}} \tag{88}$$

It is evident that Eq. (88) is formally the same as that obtained for an ionizable rubber consisting of a single component. The determination of the mean-square end-to-end distances of the free chains and hence the prediction of the ratio of the moduli will be discussed subsequently.

(b) *Comparison of Experiment with Theory*

The values of $\overline{r_{\text{PAA}}^2}_{\text{f}}$ and $\overline{r_{\text{PVA}}^2}_{\text{f}}$ in the polymer network are unknown; however, two estimates of their values were provided. In one [44], they were taken to be the unperturbed dimensions of the corresponding free polymer chains. The characteristic ratios σ for PAA and PNaA were given in Table I; for PVA, the estimated value is $\sigma = 2.04 \pm 0.10$. Since $\overline{r^2_{\text{fr}}}/P$ is a constant (9.49×10^{-16} cm^2) for vinyl polymers, then

$$\overline{r^2} \propto \sigma^2 P$$

and the following relationship is obtained from Eqs. (87) and (76)

$$\frac{E_{\text{acidic}}}{E_{\text{ionic}}} = \frac{\sigma_{\text{PVA}}^2 \times (1 - m) + \sigma_{\text{PNaA}}^2 \times m}{\sigma_{\text{PVA}}^2 \times (1 - m) + \sigma_{\text{PAA}}^2 \times m} \tag{89}$$

For a 50:50 mixture of PVA and PAA (by weight), $m = 0.38$, and the above ratio has the value 1.22.

A second estimate of the ratio $E_{\text{acidic}}/E_{\text{ionic}}$ was made using polymer dimensions determined in model solvents [46]. For the acidic form, the solvent was a mixture (by volume) of propionic acid (15%), ethyl alcohol (15%), and water (70%); for the ionic form, a mixture of sodium propionate (15%), ethyl alcohol (15%), and water (70%) was chosen. The volume percentages were selected to give the same volume percent of water as the point of comparison of the swollen polymers and also the same ratio of alcohol to acid as in the polymer mixture. Intrinsic viscosity/molecular-weight relationships established in these two media yielded a predicted value of $E_{\text{acidic}}/E_{\text{ionic}} = 1.06$ for the same 50:50 mixture of PVA and PAA.

Experimentally, the Young's modulus of PAA–PVA (50:50) rubber, was determined as a function of the degree of swelling in the unionized and ionized forms. The results are shown in Fig. 5.8. Comparison of moduli at $V_r = 0.3$ yields the value

$$E_{\text{acidic}}/E_{\text{ionic}} = 1.43$$

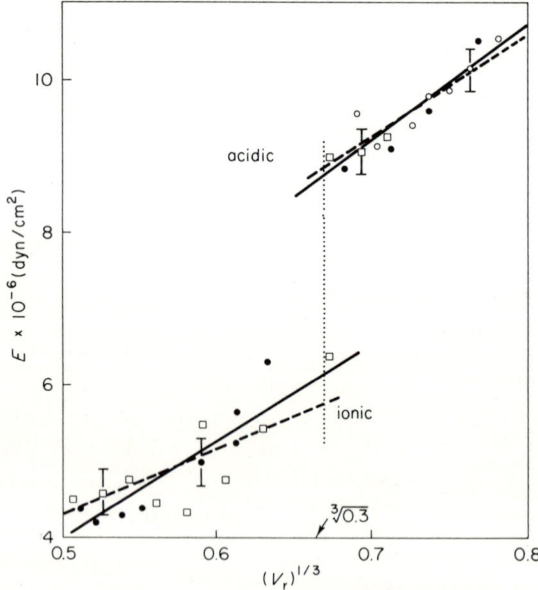

FIG. 5.8. Young's modulus versus $V_r^{1/3}$ for rubber in acidic and ionic forms. Data points indicated: sample 1 (○), sample 2 (□), sample 3 (●); best fit to $E = A + BV_r^{1/3}$ (———), best fit to $E = CV_r^{1/3}$ (— — —), comparison at $V_r = 0.3$ (····) [44].

Although the predicted values of $E_{\text{acidic}}/E_{\text{ionic}}$ are both smaller than the experimental value, they can be considered qualitatively correct, since they correctly predict the result that the modulus of the ionized rubber is less than that of the nonionic rubber. Several possible explanations could be offered to explain the discrepancy. These include the effect of imperfect network formation (the existence of PAA–PAA cross-links could raise the ratio, for example, since $E_{\text{acidic}}/E_{\text{ionic}}$ for a pure PAA network is predicted to be 1.65) and the contribution of internal energy to rubber elasticity. It is interesting to note that the prediction obtained from the unperturbed polymer dimensions is much closer to the experimental value than that obtained from dimensions in low-molecular-weight analogs. This finding is in line with the prediction of Flory [47a] that the degree of chain expansion for a polymer dissolved in its oligomer is inversely proportional to the degree of polymerization of the oligomer; i.e., where the solvent environment for a polymer chain is the polymer itself, the unperturbed dimensions provide the best estimate of free polymer dimensions. This latter fact has since been confirmed experimentally [47b].

It should be made clear that the observed effect of a ratio of $E_{\text{acidic}}/E_{\text{ionic}} > 1$ should be generally observable for all ionizable rubbers where effects of aggregation can be removed. The reason for this is the same reason that ionization of a polymer generally produces an increase in the unperturbed dimensions of a polymer chain. If ionization caused a decrease in polymer dimensions, then one would expect to see an increase in modulus on ionization.

2. Elasticity at Constant Chemical Potential

Consider the case where elongation or contraction of a polyelectrolyte gel takes place under conditions in which chemical exchange between the embedding medium and the gel is allowed to occur. This open system is the well-known pH muscle [43], of interest for its superficial similarity with the contraction and elongation of natural muscle systems and also as a mechanochemical converter, reversibly transforming chemical and mechanical energy. The properties of these and other mechanochemical systems were studied extensively in several laboratories and reviewed in 1960 [48, 49]. Only the homogeneous pH muscle, the PAA–PVA rubber, will be considered here.

One is interested in the elongation or contraction of an ionizable gel upon a change in the pH of the embedding medium, or conversely, in the change in pH that occurs when such a gel is elongated or contracted. The expansion that occurs upon neutralization of a gel is mainly due to the Donnan osmotic force and is balanced by the decrease in free energy due to the deformation of the polymer network from its equilibrium state. The changes in length that occur upon successive pH changes in the homogeneous PAA–PVA system were found to be completely reversible. Kuhn *et al.* [48] point out that the mechanical energy of deformation is obtained from the energy of neutralization of the added base, and that the converse, the conversion of mechanical to chemical energy, is also possible.

The change in pH of the embedding fluid produced by a mechanical change in the dimensions of a ionizable gel was studied for three experimentally feasible situations: for an embedding fluid containing (a) added salt, (b) no added salt, and (c) buffer. In the first case, consider an embedding fluid containing a small amount ($\sim 10^{-3}\ M$) of NaCl in equilibrium with an ionizable gel of length L and fixed cross-sectional area A. If $[\mathrm{AA}]_{g,L}$ is the total concentration of acrylic acid (ionized and unionized) in the gel, then

$$[\text{—COO}^-]_{g,L} = \alpha[\mathrm{AA}]_{g,L} \tag{90}$$

In the gels used by Kuhn *et al.*, $[\mathrm{AA}]_{g,L}$ was normally in the range 0.5–1.0 mol/liter. The Donnan condition requires that electroneutrality be maintained within the gel. Thus, for significant degrees of neutralization ($\alpha > 0.1$), the total ion concentration in the gel will be very much greater than in the embedding fluid, resulting in the chloride ion concentration $[\mathrm{Cl}^-]_{g,L}$ in the gel being negligible with respect to that in the external medium $[\mathrm{Cl}^-]_{e,L}$. If the dissociation constant K_{PAA} is of the order 10^{-5}–10^{-6}, then $[\mathrm{H}^+]_{g,L}$ and $[\mathrm{OH}^-]_{g,L}$ will be small with respect to $[\mathrm{Na}^+]_{g,L}$, and therefore

$$[\mathrm{Na}^+]_{g,L} \simeq [\text{—COO}^-]_{g,L} = \alpha[\mathrm{AA}]_{g,L} \tag{91}$$

$[\mathrm{H}^+]_g$ and $[\mathrm{OH}^-]_g$ are derived from K_{PAA} as follows:

$$[H^+]_{g,L} = K_{\mathrm{PAA}} \frac{[\text{—}COOH]_{g,L}}{[\text{—}COO^-]_{g,L}} = K_{\mathrm{PAA}} \frac{1-\alpha_1}{\alpha_1} \tag{92}$$

and

$$[\mathrm{OH}^-]_{g,L} = \frac{K_w}{[\mathrm{H}^+]_{g,L}} = \frac{K_w}{K_{\mathrm{PAA}}} \cdot \frac{\alpha_1}{1-\alpha_1} \tag{93}$$

where α_1 is the degree of neutralization of the unstretched swollen gel. Since the Donnan condition [1] requires that

$$[\mathrm{Na}^+]_{g,L}[\mathrm{OH}^-]_{g,L} = [\mathrm{Na}^+]_{e,L}[\mathrm{OH}^-]_{e,L} \tag{94}$$

one obtains

$$[\mathrm{H}^+]_{e,L} = \frac{K_w}{[\mathrm{OH}^-]_{e,L}} = \frac{K_w[\mathrm{Na}^+]_{e,L}}{[\mathrm{Na}^+]_{g,L}[\mathrm{OH}^-]_{g,L}} = \frac{[\mathrm{H}^+]_{g,L}[\mathrm{Na}^+]_{e,L}}{[\mathrm{Na}^+]_{g,L}} \tag{95}$$

With Eq. (92), this becomes

$$[H^+]_{e,L} = K_{PAA} \frac{1-\alpha_1}{\alpha_1} \cdot \frac{[Na^+]_{e,L}}{[Na^+]_{g,L}} \tag{96}$$

If the gel is stretched from L to λL, then, assuming that K_{PAA} is independent of swelling, one can also write

$$[H^+]_{e,\lambda L} = K_{PAA} \frac{1-\alpha_2}{\alpha_2} \frac{[Na^+]_{e,\lambda L}}{[Na^+]_{g,\lambda L}} \tag{97}$$

where α_2 is the degree of ionization in the stretched swollen state. Since A is held constant during stretching, the volume also increases by λ, the increase occurring practically completely by the uptake of water from the embedding solution, with $[Cl^-]_{g,\lambda L}$ and $[OH^-]_{g,\lambda L}$ still small. Thus, the dilution by water decreases $[AA]_g$, $[—COO^-]_g$, and $[Na^+]_g$ all by the same factor λ:

$$\begin{aligned} [—COO^-]_{g,\lambda L} \simeq [Na^+]_{g,\lambda L} &= \lambda^{-1}[Na^+]_{g,L} \\ [AA]_{g,\lambda L} &= \lambda^{-1}[AA]_{g,L} \end{aligned} \tag{98}$$

Therefore,

$$\alpha_2 = \frac{[Na^+]_{g,\lambda L}}{[AA]_{g,\lambda L}} = \alpha_1 \tag{99}$$

i.e., the degree of neutralization in the gel remains unchanged on stretching. This means that $[H^+]_g$ and $[OH^-]_g$ also remain unchanged.

In the embedding solution, from Eqs. (97) and (99), it is obtained that

$$[H^+]_{e,\lambda L} = K_{PAA} \frac{1-\alpha_1}{\alpha_1} \frac{[Na^+]_{e,\lambda L}}{[Na^+]_{g,\lambda L}} \tag{100}$$

Thus, assuming that the NaCl concentration in the embedding medium is maintained during the stretching of the gel, Eqs. (96), (98), and (100) give

$$[H^+]_{e,\lambda L} = [H^+]_{e,L} \frac{[Na^+]_{g,L}}{[Na^+]_{g,\lambda L}} = \lambda [H^+]_{e,L} \tag{101}$$

i.e., the pH of the embedding medium decreases upon stretching of the gel.

By similar reasoning, the ion concentrations inside and outside the gel were established for the second and third cases, i.e., where the embedding medium is pure H_2O and buffer, respectively. In the second case, a decrease in pH in the embedding medium also occurs upon stretching, but in this instance, the fractional change in $[H^+]_e$ is given by $\sqrt{\lambda}$. In the third case, where buffer is present in the external medium, the system behaves in the same way as in the second case as far as fractional changes in ion concentrations are concerned, although for different reasons. In each of the three situations, within the gel $[Na^+]_g$ decreases by $1/\lambda$ (due to water uptake) while α and $[H^+]_g$ remain constant. Kuhn *et al.* [48] were able to establish the quantitative validity of these mechanochemical relationships in the cross-striated PAA–PVA system, in which the cross-sectional area of the gel is constrained.

The pH change upon stretching was also observed in the homogeneous PAA–PVA system, and the same arguments were used to predict its magnitude, except that the factor by which $[H^+]$ changes is V/V_0, rather than λ. It should be noted that a significant volume change must occur upon stretching the homogeneous system in order to produce a change in pH in the external medium. Kuhn [50] pointed out that the Young's modulus measured during stretching of a homogeneous gel in equilibrium with an embedding medium is not the same as the usual Young's modulus of a gel in which no exchange with the surrounding medium occurs. If the polyelectrolyte gel is removed from the medium prior to testing, the Young's modulus E is determined in the usual way, i.e.,

$$E = FL_0/A_0{}^2\,\Delta L \tag{102}$$

where F is the force required to produce a small deformation ΔL, and L_0 and A_0 are the original length and cross-sectional area of the test specimen. In this case, the volume change upon stretching is negligible, and Poisson's ratio $\mu = L_0\,\Delta A/A_0\,\Delta L \simeq 0.5$. In the former case, the volume does not remain constant, and some of the mechanical energy of stretching is converted to chemical energy. By analogy with the preceding, the apparent Young's modulus in contact with the medium E' is

$$E' = F'L_0/A_0{}^2\,\Delta L \tag{103}$$

The Poisson's ratio μ' determines the relationship between E' and E, which is given by

$$\frac{E'}{1+\mu'} = \frac{E}{1+\mu} \tag{104}$$

For nonelectrolyte gels it is found that $\mu' = 0.3 - 0.4$, and for polyelectrolyte gels $\mu' \simeq 0$. In the latter case (assuming $\mu = 0.5$), $E' \simeq \frac{2}{3}E$. The relationship predicted by Eq. (104) was confirmed experimentally in the PAA–PVA system [51].

The thermoelasticity of the PAA–PVA gel system at low degrees of ionization was subsequently studied by Smith [52], who found that the internal energy component of the force of network deformation was significant and dependent on polymer–diluent specific interactions. The energy component f_e at temperature T is related to the total force f by the expression

$$\frac{f_e}{f} \equiv 1 - \frac{T}{f}\left(\frac{\partial f}{\partial T}\right)_{VL} = -T\left[\frac{\partial \ln(f/T)}{\partial T}\right]_{VL} \tag{105}$$

where subscripts V and L denote constant volume and length. Since the experimental variables are normally constant pressure and length, and since deformation occurs while the sample is in contact with the external medium, both volume and chemical composition must be considered variables; thus the following relationship for the equilibrium force of extension can be obtained:

$$\frac{f_e}{f} = -T\left[\frac{\partial \ln(f/T)}{\partial T}\right]_{PL} - \left(\frac{\partial f}{\partial V}\right)_{TLN}\left(\frac{\partial V}{\partial T}\right)_{PL} - \sum_j\left(\frac{\partial f}{\partial N_j}\right)_{TLVN'}\left(\frac{\partial N_j}{\partial T}\right)_{PL} \tag{106}$$

where the subscript N denotes fixed total composition, and N' constant composition of all species except the jth.

The coefficients $(\partial f/\partial N_j)_{TLVN'}$ in Eq. (106) are indicative of the extent of polymer–diluent interaction on chain conformation, since the force of extension is dependent on $\overline{r_f^2}$ as given in the following equation

$$f = (nkT/L_0)(\overline{r_0^2}/\overline{r_f^2})v_0^{-2/3}[\lambda - (v_0/\lambda^2 v)] \tag{107}$$

where $\lambda = L/L_0$. v_0 is the volume fraction of polymer in the swollen unstretched gel and v is the volume fraction of polymer in the stretched state. Substitution of Eq. (107) in Eq. (106) leads to

$$\frac{f_e}{f} = -T\left[\frac{\partial \ln(f/T)}{\partial T}\right]_{PL} - \frac{T\beta}{[\lambda^3(v/v_0) - 1]} - T\sum_j \sigma_j\left(\frac{\partial N_j}{\partial T}\right)_{PL} \tag{108}$$

$\beta = (\partial \ln V/\partial T)_{PL}$ is the thermal expansion coefficient of the swollen stretched gel at constant deformation and $\sigma_j = (\partial \ln \overline{r_f^2}/\partial N_j)_{LTVN'}$. For a two-component diluent (H_2O plus salt), Eq. (108) becomes, for $N_{H_2O} \gg N_s$,

$$\frac{f_e}{f} = -T\left[\frac{\partial \ln(f/T)}{\partial T}\right]_{PL} - \frac{T\beta}{[\lambda^3(v/v_0) - 1]} - TV\left[\frac{\sigma_{H_2O}}{\overline{V}_{H_2O}}(\beta - \overline{\beta}) + \sigma_s c_s(\beta - \beta^*)\right] \tag{109}$$

where c_s is the molar salt concentration in the gel, $\bar{V}_{H_2O}$ the partial molar volume of H_2O, and $\bar{\beta}$ and β^* are, respectively, the thermal expansion coefficients of the gel at constant composition and of the external solution. At low ionization, the Poisson number $\mu \simeq 0.25$ for deformation at constant chemical potential [1]. Thus, $V/V_0 \simeq \lambda^{1/2}$ and $\beta/\beta_0 \simeq 5/6$. In addition, $c_s \simeq c_s^*$, the external salt concentration. (At high degrees of ionization, $\mu \simeq 0$, and $V/V_0 \simeq \lambda$, $\beta/\beta_0 \simeq 2/3$; also $c_s \simeq c_s^*/\lambda$, as discussed previously in connection with the work of Kuhn *et al* [48].) Finally, with the approximation that $\bar{\beta} \simeq \beta^* < |\beta|$, Eq. (109) simplifies to

$$1 - \frac{T}{f}\left(\frac{\partial f}{\partial T}\right)_{PL} = \frac{f_e}{f} + \frac{5\beta_0 T\lambda^{1/2}V_0}{6}\left(\frac{\sigma_{H_2O}}{\bar{V}_{H_2O}} + c_s^*\sigma_s\right) + \frac{5\beta_0 T}{6(\lambda^{5/2} - 1)} \quad (110)$$

Equilibrium stress–temperature measurements at constant strain were made on PAA–PVA gels (85:15 by weight) immersed in various swelling media. Plots of $[1 - (T/f)(\partial f/\partial T)_{PL}]$ versus $1/(\lambda^{5/2} - 1)$ at fixed T were found to be linear, indicating the approximate strain independence of the polymer–diluent interaction term in Eq. (110). The values of the intercept (f_e/f + specific interaction) were -0.75 in pure H_2O, -0.85 in 0.01 *M* $CaCl_2$, and -5.7 in 0.1 *M* $CaCl_2$. From these values, Smith obtained $f_e/f = -1.32$, $\sigma'_{H_2O} = \lambda^{1/2}V_0\sigma_{H_2O} = 0.0105$ liter/mol, and $\sigma_s' = \lambda^{1/2}V_0\sigma_s =$ 21.4 liter/mol. This study shows that the equilibrium rubber elasticity of an ionizable gel is greatly affected by changes in the unperturbed dimensions of the polymer segments. This finding is in agreement with the conclusions of the study of Eisenberg and King [44] for rubber elasticity at constant volume. It should be noted that the study by Smith did not consider any possible effect of the addition of $CaCl_2$ on the unperturbed dimensions of the PVA component of the gel, since the change in elasticity depends on the change in unperturbed dimensions of both components supporting the stress.

Hasa *et al.* [53, 54] studied the deformational behavior of ionized poly(methacrylic acid) gels. They considered two types of deviation from ideal rubber elasticity theory: (1) a finite deformation contribution, taken into account by employing an inverse Langevin function to describe the distribution of chain-end displacements at finite extension ratios in place of the ideal Gaussian distribution normally used to obtain the entropic component of rubber elasticity and (2) an electrostatic contribution based on the Katchalsky–Lifson model of polyelectrolyte expansion (see Section A2), in which the length of the statistical chain element (or equivalently, the value of $\overline{h_0^2}$) increases with increasing degree of ionization. The former contribution was found to add to the rubbery modulus and be independent of α, while the latter was found to diminish it, the effect increasing with α.

Experiments with PMAA gels, subjected to strain in equilibrium with various swelling media, showed satisfactory agreement with the prediction of increased modulus with increased deformation. However, the predicted decrease in modulus with increasing degree of ionization was found to be valid only for the range $0 < \alpha < 0.5$, and was consistent with a gradual increase in the value of $\overline{h_0^2}$. Above $\alpha = 0.5$, the values of modulus were observed to increase.

This work can also be compared to the previously described study of Eisenberg and King [44], in which a decrease in the rubbery modulus was predicted and observed for an ionizable gel, based on the increase in $\overline{h_0^2}$ that normally accompanies ionization in aqueous media.

C. DILUTE-SOLUTION VISCOSITY

In this section, the variation of polyelectrolyte solution viscosity with molecular weight, concentration, and solvent environment (pure water, or water containing added small electrolytes, or nonaqueous solvents) will be considered. This brief review is not intended to be exhaustive since the body of literature pertaining to this subject is considerable, and more extensive treatments of the topic exist [26, 27, 55–57]. Also, this section will not deal directly with the related topics of sedimentation, electrophoresis, and diffusion, which also relate polymer mobility to configuration, since these aspects have been well covered in other books and reviews [9–13, 27, 56, 57]. The main rationale for including a discussion of dilute-solution viscosity here is that it, like rubber elasticity, represents one of the limiting aspects of viscoelastic behavior, which is the main thrust of the book.

1. Dependence on Polymer Concentration

The reduced viscosity or viscosity number of a polymer solution is defined by

$$\eta_{sp}/C_p \equiv (\eta - \eta_s)/\eta_s C_p \tag{111}$$

where η and η_s are the solution and solvent viscosities, respectively. The polymer concentration C_p may be defined in a number of ways, including grams per milliliter, weight percent, or (monomeric) equivalents per liter. For polyelectrolytes the latter method has the advantage of being independent of the degree of neutralization or the type of counterion and thus more aptly specifies the macromolecular concentration for comparative purposes.

The concentration dependence of η_{sp} for a nonionic polymer can generally be described by the Huggins equation [58]

$$\eta_{sp}/C_p = [\eta] + k_H[\eta]^2 C_p \tag{112}$$

where $[\eta]$ is the intrinsic viscosity. The Huggins constant k_H is a rheological measure of the interaction between macromolecules in a given polymer–solvent system, assuming that polymer configuration remains constant with dilution. Values of k_H in most nonionic systems are typically of the order of 0.5 [55, 57].

For polyelectrolytes, the variation in η_{sp}/C_p with polymer concentration depends a great deal on the solvent environment, and the above approach does not always hold. In the subsequent discussion, three types of solvent environment will be considered: (a) salt-free aqueous solutions, (b) aqueous solutions containing added salt, and (c) aqueous/nonaqueous mixtures, with or without salt.

(*a*) *Salt-Free Solutions*

Figure 5.9 shows the typical concentration dependence of a linear flexible polyelectrolyte, in pure water and with increasing concentration of added salt [59]. To describe the increase in η_{sp}/C_p with dilution in salt-free solution, Fuoss [60] proposed the following relationship:

$$\eta_{sp}/C_p = A/(1 + BC_p^{1/2}) + D \tag{113}$$

where A, B, and D are empirical constants, with D in most cases being negligible. The constant A is identified with the value of $[\eta]$ in salt-free solution. The values of $[\eta]$ obtained for salt-free polyelectrolyte solutions are found to correlate best with the values predicted for polymer models based on rigid rods or hydrodynamically equivalent prolate spheroids [55].

It was pointed out by Eirich [61] that the monotonous increase of η_{sp}/C_p with $1/C_p^{1/2}$ should terminate at some small value of C_p after maximum extension of the polyelectrolyte is attained, since with further dilution, the reduced viscosity would fall. Indeed, maxima in η_{sp}/C_p versus C_p curves have been seen for a number of salt-free polyelectrolyte systems, but in many of these cases, the observed maxima may have been due to other factors, such as impurities [55, 62] or the shear dependence of viscosity, since the shear rate in standard capillary viscometry increases upon dilution with solvent [55, 57]. It has also been suggested [63] that the Eirich limit of maximum extension may coincide with infinite dilution, making Eq. (113) the correct description of the dilute-solution viscosity behavior of pure polyelectrolytes.

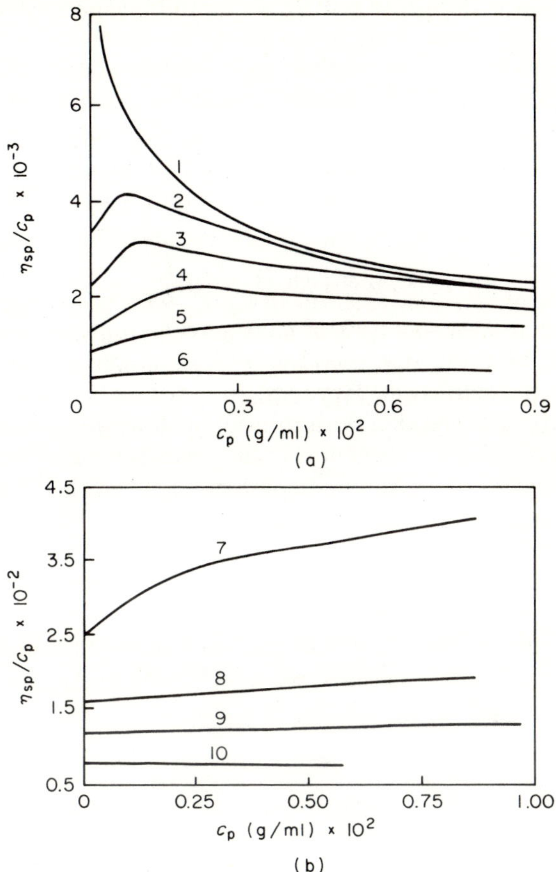

FIG. 5.9. Reduced viscosity versus polymer concentration for PVSA–K at different values of added electrolyte (HCl) content (a) at low C_s and (b) at high C_s; (1) water, (2) 3.297×10^{-4} *N*, (3) 7.615×10^{-4} *N*, (4) 1.995×10^{-3} *N*, (5) 4.998×10^{-3} *N*, (6) 4.916×10^{-2} *N*, (7) 9.970×10^{-2} *N*, (8) 5.394×10^{-1} *N*, (9) 1.000 *N*, (10) 3.55 *N* [59].

(*b*) *Solutions Containing Added Salt*

Figure 5.9 shows that η_{sp}/C_p decreases with added salt concentration C_s. The position of the maximum in η_{sp}/C_p shifts to higher values of C_p and eventually disappears, the plots becoming linear at moderate concentrations of added salt. At higher salt concentrations yet, the dependence of η_{sp}/C_p on C_p resembles the behavior of uncharged linear polymer solutions and can be described by the Huggins equation (112).

The nature of the concentration dependence of reduced viscosity depends on the method of dilution. Four methods of dilution can be considered: (a) constant added electrolyte content, as discussed above and illustrated in Fig. 5.9, (b) constant relative electrolyte content, (c) constant total ionic strength (isoionic dilution), (d) constant activity of added electrolyte.

Dilution at constant relative electrolyte content (i.e., with pure solvent, leaving the ratio $y = C_s/C_p$ constant) can be described by a modified form of the Fuoss equation [64], i.e.,

$$(\eta_{sp}/C_p)_y = [\eta]_y/(1 + BC_p^{1/2}) \tag{114}$$

This relationship predicts a series of η_{sp}/C_p versus C_p curves for different values of y, with the curve at $y = 0$ [Eq. (113)] a special case of the general formula. This dependence was expressed in a somewhat different way by Oosawa [12], who showed that with dilution at different values of y, plots of $\log \eta_{sp}/C_p$ versus $\log C_p$ were linear and parallel down to very small values of C_p, i.e., with only the intercept dependent on the value of y.

Isoionic dilution is usually done in the modified form introduced by Pals and Hermans [65], where the sum $z = C_s + rC_p$ is kept constant during dilution. The factor r is determined empirically so as to make the initial dependence of η_{sp}/C_p on C_p linear, i.e., to fit the data to a form of the Huggins equation (112). r can be identified with the free fraction of counterions [66], and thus z represents the total effective ionic strength. An example of this procedure is illustrated in Fig. 5.10. In this case, rather than forcing a linear

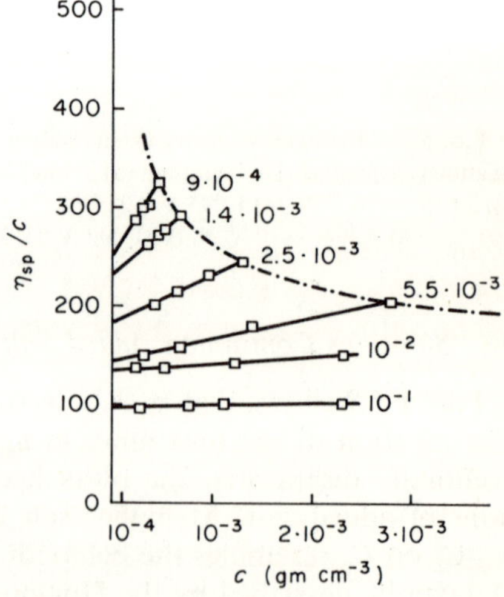

FIG. 5.10. η_{sp}/C_p versus C_p for CMC–Na at different values of total effective ionic strength (given in moles per liter on each curve) [67].

fit, the value of r was set equal to the osmotic coefficient ϕ, representing the effective degree of ionization of the polymer [67]. The linearity of all the curves indicates the validity of this dilution procedure over a range of ionic strengths from 10^{-4} to 10^{-1} M. At very high ionic strengths, the curves become identical with those obtained by dilution at constant C_s. At lower values of z, the use of the same value of ϕ results in nonlinear η_{sp}/C_p versus C_p plots, i.e., r can no longer be identified with ϕ, which probably indicates a change in configuration. The envelope of points for which $C_s = 0$ gives the dilution curve for salt-free solution, and by application of Eq. (113) allows the calculation of $[\eta]_0$.

Dilution at constant activity of added electrolyte provides a means of determining polymer–solvent interaction parameters in a three-component system (polymer, salt, solvent) without knowledge of specific polymer–electrolyte interactions. Instead of dilution at constant concentration of added salt, solutions are dialyzed at each dilution against a constant activity of small electrolyte [17]. This procedure also leads to linear plots of η_{sp}/C_p versus C_p but with different slopes, and thus different values of k_H, than obtained by simple dilution at constant salt content, which does not account for variations in polymer–electrolyte interaction with polymer concentration.

Huggins constants $(k_H)_z$ obtained from isoionic dilution curves are also different from those obtained by dilution at constant electrolyte content, but are related by the following [65]

$$(k_H)_z = (k_H)_{C_s} - \frac{r}{[\eta]} \frac{d \ln[\eta]}{dC_s} \tag{115}$$

Since $d \ln[\eta]/dC_s$ is normally negative, $(k_H)_z > (k_H)_{C_s}$.

At large values of small electrolyte activity, the values of k_H obtained for polyelectrolytes (by any dilution method) are of comparable magnitude to those found for uncharged polymers. With decreasing salt concentration, k_H is normally observed to increase; for example, Moan and Wolff found that in the NaCMC system [67], k_H increased linearly with κ^{-1}, the radius of the Debye–Hückel atmosphere [see Eq. (15)], which is representative of the zone of polymer–polymer interaction.

In some polyelectrolyte systems, extremely large values of k_H (>100) at very low ionic strengths have been reported [55]. Values of this magnitude cannot be accounted for by hydrodynamic interaction alone [3] and thus probably include a contribution due to a changing polymer configuration, in which case k_H would no longer directly represent a polymer–polymer interaction coefficient.

(c) *Aqueous/Nonaqueous Mixtures*

Variations in dilute-solution viscosity behavior have also been investigated in aqueous/nonaqueous mixtures. The configuration of poly(methacrylic acid) in ethanol–0.002 *N* HCl mixtures was studied by Priel and Silberberg [68]. They found that $[\eta]$ characteristically passed through two minima and two maxima with increasing ethanol content. At the same time, in spite of the presence of HCl to suppress the polyelectrolyte effect, great variations in k_H were observed from negative values to a high of ~ 70. The unusual variations in k_H were attributed to the influence of alcohol on intermolecular bonding in PMA.

The dilute-solution behavior of (ethyl acrylate)–(acrylic acid) copolymer in various aqueous and nonaqueous solvents was studied by Tan and Gasper [69] and by Rauscher *et al.* [70]. The former work examined the unperturbed dimensions of a 3:1 EA/AA copolymer in two θ-solvents—an organic solvent (methyl butyrate at 75.0°C), and an aqueous system (1.2 *N* NaCl at 34.5°C, pH 7). The unperturbed dimensions of the 3:1 copolymer in the organic medium, where it is unionized, were found to be 1.3 to 1.4 times the value obtained for the fully ionized copolymer in the aqueous medium. This contrasts with the influence of ionization on the unperturbed dimensions of the homopolymer, poly(acrylic acid), as described in Section Al (see Table I). Rauscher *et al.* found that in a 55:45 EA/AA copolymer, unperturbed dimensions in organic and aqueous media were approximately equal.

It should be recalled that (Section B1) the rubber elasticity of an ionizable gel was predicted to vary inversely with the change in polymer dimensions on ionization. In this case, for a gel formed from the 3:1 copolymer, one would predict an increase in Young's modulus on ionization, as opposed to the decrease that occurs on ionization of a PAA gel.

2. Dependence of $[\eta]$ on Molecular Weight

Intrinsic viscosity/molecular weight relationships are commonly described by one of the following two relationships. The first is the Mark–Houwink equation [1]

$$[\eta] = KM^{\nu} \tag{116}$$

where K and ν are empirical constants. In a θ solvent, $\nu = 0.5$. The second is the Stockmayer–Fixman equation [71]

$$[\eta]/M^{1/2} = K_0 + 0.51\Phi BM^{1/2} \tag{117}$$

where $\Phi = 2.6 \times 10^{21}$ and B is related to the second virial coefficient, the θ condition being realized when $B = 0$. The quantity $K_0M^{1/2}(=[\eta]_\theta)$ determines the unperturbed dimensions, and therefore the term on the left-hand side of Eq. (117) is proportional to $\alpha_\eta{}^3$, as discussed in Section A2. Examples of the use of this relation for partially and fully neutralized PAA were given in Figs. 5.3 and 5.4, showing in each case that the extrapolation procedure leads to a common intercept, and thus indicating the validity of employing this procedure for estimating the unperturbed dimensions over a wide range of added electrolyte.

Similar results have been obtained with the PVSA [17] and the PSSA [18, 72] systems. For all of these polyelectrolyte systems and, in general, with increasing concentration or activity of added salt at constant temperature, the value of the exponent ν in the Mark–Houwink equation decreases, reaching the value 0.5, corresponding to the θ condition, at some finite salt content. At the same time, the value of B in the Stockmayer–Fixman equation approaches 0, as seen in Figs. 5.3 and 5.4, for example. Recently, however, a sulfonate-containing polyelectrolyte system, poly(3-methacryloyloxypropane-1-sulfonic acid), PMOSA, was described in which the θ condition could not be reached at any realizable concentration of added 1:1

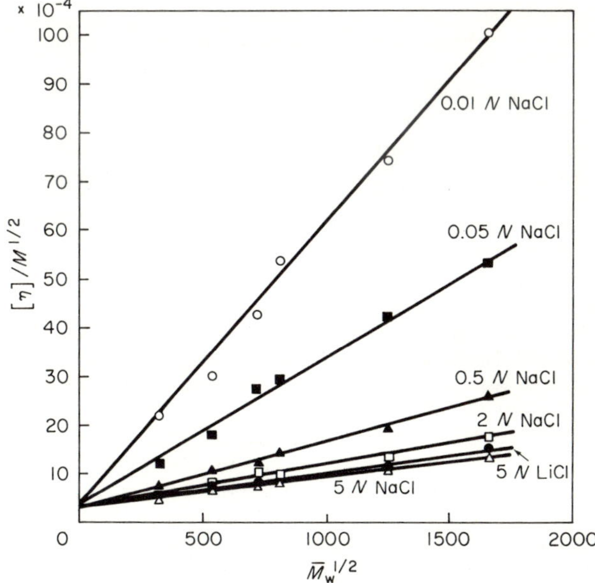

FIG. 5.11. $[\eta]/M_w^{1/2}$ versus $M_w^{1/2}$ for PMOSA–Na in salt solutions [73].

electrolyte [73]. This is illustrated in the Stockmayer–Fixman plots shown in Fig. 5.11. Expansion coefficients α_η calculated from these plots suggested that the concentration of NaCl required to reach the θ condition would be in excess of 50 N, indicating a remarkable toleration of this polyelectrolyte to added salt.

Other aspects of the dilution-solution viscosity of polyelectrolytes, including dependence on molecular weight, electrolyte concentration, and shear rate, have been discussed in Chapter III, Section C, and from a theoretical point of view, in Section A of this chapter.

REFERENCES

1. P. J. Flory, "Principles of Polymer Chemistry." Cornell Univ. Press, Ithaca, New York, 1953.
2. M. V. Volkenstein, "Configurational Statistics of Polymer Chains." Wiley, New York, 1963.
3. C. Tanford, "Physical Chemistry of Macromolecules." Wiley, New York, 1962.
4. M. Kurata and W. H. Stockmayer, *Adv. Polym. Sci.* **3,** 196 (1963).
5. H. Yamakawa, "Modern Theory of Polymer Solutions." Harper, New York, 1971.
6. H. Morawetz, "Macromolecules in Solution" (2nd ed.). Wiley, New York, 1975.
7. T. M. Birshtein and O. B. Ptitsyn, "Conformations of Macromolecules." Wiley (Interscience), New York, 1967.
8. P. J. Flory, "Statistical Mechanics of Chain Molecules." Wiley (Interscience), New York, 1969.
9. S. A. Rice and M. Nagasawa, "Polyelectrolyte Solutions." Academic Press, New York, 1961.

10a. A. Katchalsky, Z. Alexandrowicz, and O. Kedem, *in* "Chemical Physics of Ionic Solutions" (B. E. Conway and R. G. Barradas, eds.) p. 295. Wiley, New York, 1966.

10b. A. Katchalsky, *Pure Appl. Chem.* **26,** 327 (1971).

11. V. Crescenzi, *Adv. Polym. Sci.* **5,** 358 (1968).
12. F. Oosawa, "Polyelectrolytes." Dekker, New York, 1971.
13. G. Manning, *Ann. Rev. Phys. Chem.* **23,** 117 (1972).
14. J. A. Semlyen and P. J. Flory, *Trans. Faraday Soc.* **62,** 2622 (1976).
15. S. Newman, W. R. Krigbaum, C. Laugier, and P. J. Flory, *J. Polym. Sci.* **14,** 451 (1954).
16. A. Takahashi and M. Nagasawa, *J. Am. Chem. Soc.* **86,** 544 (1964).
17. H. Eisenberg and D. Woodside, *J. Chem. Phys.* **36,** 1844 (1962).
18. A. Takahashi, T. Kato, and M. Nagasawa, *J. Phys. Chem.* **71,** 2001 (1967).
19. U. P. Strauss and P. Ander, *J. Phys. Chem.* **66,** 2235 (1962).
20. U. P. Strauss and P. Wineman, *J. Am. Chem. Soc.* **80,** 2366 (1958).
21. G. Allegra, S. Brückner, and V. Crescenzi, *Eur. Polym. J.* **8,** 1255 (1972).
22. W. Kuhn, O. Künzle, and A. Katchalsky, *Helv. Chim. Acta* **31,** 1994 (1948).
23. A. Katchalsky and S. Lifson, *J. Polym. Sci.* **11,** 409 (1956).
24. H. Morawetz, *in* "Polyelectrolyte Solutions" (S. A. Rice and M. Nagasawa, eds.) Chapter V. Academic Press, New York, 1961.
25. S. A. Rice and F. E. Harris, *J. Phys. Chem.* **58,** 733 (1954).
26. I. Noda, T. Tsuge, and M. Nagasawa, *J. Phys. Chem.* **74,** 710 (1970).

27. E. Sélégny (ed.), "Polyelectrolytes." Reidel, Boston, 1974.
28. J. J. Hermans and J. T. G. Overbeek, *Rec. Trav. Chim.* **67,** 761 (1948).
29. S. Lifson, *J. Polym. Sci.* **23,** 431 (1957).
30. P. J. Flory, *J. Chem. Phys.* **21,** 162 (1953).
31. O. B. Ptitsyn, *Vysokomol. Soedin.* **A3,** 1673 (1961); transl. in *Polym. Sci. USSR* **3,** 1061 (1962).
32. M. Fixman, *J. Chem. Phys.* **36,** 3123 (1962).
33. M. Kurata, W. H. Stockmayer, and A. Roig, *J. Chem. Phys.* **33,** 151 (1960).
34. Z. Alexandrowicz, *J. Chem. Phys.* **46,** 3789 (1967).
35. M. Fixman, *J. Chem. Phys.* **41,** 3772 (1964).
36. M. Kurata, *J. Polym. Sci. Part C,* **15,** 347 (1966).
37a. Z. Alexandrowicz, *J. Chem. Phys.* **47,** 4377 (1967).
37b. Z. Alexandrowicz, *J. Phys. Chem.* **75,** 442 (1971).
38. M. Kurata, *J. Polym. Sci. Part A-2,* **6,** 1607 (1968).
39. Z. Alexandrowicz, *J. Chem. Phys.* **49,** 1599 (1969).
40. H. Fujita, K. Okita, and T. Norisuye, *J. Chem. Phys.* **47,** 2723 (1967).
41. A. Takahashi and M. Nagasawa, *J. Am. Chem. Soc.* **86,** 543 (1964).
42. Z. Alexandrowicz and Y. Accad, *J. Chem. Phys.* **54,** 5338 (1971).
43. W. Kuhn, B. Hargitay, A. Katchalsky, and H. Eisenberg, *Nature* **165,** 515 (1950).
44. A. Eisenberg and M. King, *Macromolecules* **4,** 204 (1971).
45. L. R. G. Treloar, "The Physics of Rubber Elasticity." Oxford Univ. Press, London, 1958.
46. M. King, Ph.D. thesis, McGill University, 1973.
47a. P. J. Flory, *J. Chem. Phys.* **17,** 303 (1949).
47b. Symposium on Physical Structure of the Amorphous State, *Polym. Preprints* **15**(2), (1974).
48. W. Kuhn, A. Ramel, D. H. Walters, G. Ebner, and H. J. Kuhn, *Adv. Polym. Sci.* **1,** 540 (1960); W. Kuhn, A. Ramel, and D. H. Walters *in* "Size and Shape Changes of Contractile Polymers" (A. Wasserman, ed.) Chapter II. Pergamon, New York, 1960.
49. A. Katchalsky, S. Lifson, I. Michaeli, and M. Zwick, *in* "Size and Shape Changes of Contractile Polymers" (A. Wasserman, ed.) Chapter I. Pergamon, New York, 1960.
50. W. Kuhn, *Helv. Chim. Acta* **44,** 927, 1017 (1961).
51. W. Kuhn, S. Müller, H. J. Kuhn, and A. Eisenberg, *Makromol. Chem.* **62,** 40 (1963).
52. K. J. Smith, Jr., *J. Polym. Sci., Polym. Phys.* **12,** 7 (1974).
53. J. Hasa, M. Ilavský, and K. Dusek, *J. Polym. Sci., Polym. Phys.* **13,** 253 (1975).
54. J. Hasa and M. Ilavský, *J. Polym. Sci., Polym. Phys.* **13,** 263 (1975).
55. T. Kurucsev, *Rev. Pure Appl. Chem.* **14,** 147 (1964).
56. M. Nagasawa, *J. Polym. Sci., Polym. Symp.* **49,** 1 (1975).
57. H. Eisenberg, "Biological Macromolecules and Polyelectrolytes in Solution." Oxford Univ. Press (Clarendon), London, 1976.
58. M. L. Huggins, *J. Am. Chem. Soc.* **64,** 2716 (1942).
59. M. Nagasawa and I. Kagawa, *J. Polym. Sci.* **25,** 61 (1957).
60a. R. M. Fuoss, *J. Polym. Sci.* **3,** 603 (1948).
60b. R. M. Fuoss, *J. Polym. Sci.* **4,** 96 (1949).
61. F. Eirich, *Disc. Faraday Soc.* **11,** 153 (1951).
62. R. L. Darskus, D. O. Jordan, T. Kurucsev, and M. L. Martin, *J. Polym. Sci., Part A* **3,** 1941 (1965).
63. R. A. Mock and C. A. Marshall, *J. Polym. Sci.* **13,** 263 (1954).
64. H. Terayama, *J. Polym. Sci.* **15,** 575 (1955).
65. D. T. F. Pals and J. J. Hermans, *Rec. Trav. Chim.* **71,** 458 (1952).
66. H. Terayama and F. T. Wall, *J. Polym. Sci.* **16,** 357 (1956).

67. M. Moan and C. Wolff, *Makromol. Chem.* **175,** 2881 (1974).
68. Z. Priel and A. Silberberg, *J. Polym. Sci., Part A-2* **8,** 689 (1970).
69. J. S. Tan and S. P. Gasper, *J. Polym. Sci., Polym. Phys.* **12,** 1785 (1974).
70. R. J. Rauscher, L. J. Garfield, and R. W. Connelly, *J. Polym. Sci., Polym. Symp.* **45,** 165 (1974).
71. W. H. Stockmayer and M. Fixman, *J. Polym. Sci., Part C* **1,** 137 (1963).
72. A. Raziel and H. Eisenberg, *Israel J. Chem.* **11,** 183 (1973).
73. J. S. Tan and S. P. Gasper, *J. Polym. Sci., Polym. Phys.* **13,** 1705 (1975).

Author Index

Numbers in parentheses are reference numbers and indicate that an author's work is referred to although his name is not cited in the text. Numbers in italics show the page on which the complete reference is listed.

L

M

S

T

U

V

W

Y

Z

Subject Index

E

F

G

R

S

T

U

V

X

A
B 7
C 8
D 9
E 0
F 1
G 2
H 3
I 4
J 5